Benchmark Papers in Geology Series

Editor: Rhodes W. Fairbridge—Columbia University

A selection from the published volumes in this series

MAGNETIC STRATIGRAPHY OF SEDIMENTS / *James P. Kennett*
MEGACYCLES: Long-Term Episodicity in Earth and Planetary History / *G. E. Williams*
TERRESTRIAL TRACE-FOSSILS / *William A. S. Sarjeant*
NANNOFOSSIL BIOSTRATIGRAPHY / *Bilal U. Haq*
STRATIGRAPHY: Foundations and Concepts / *Barbara M. Conkin and James E. Conkin*

Related Titles

CONCEPTS AND METHODS OF BIOSTRATIGRAPHY / *E. G. Kauffman and J. E. Hazel*
BIOSTRATIGRAPHY OF FOSSIL PLANTS / *D. A. Dilcher and T. N. Taylor*
THE ENCYCLOPEDIA OF SEDIMENTOLOGY / *R. W. Fairbridge and J. Bourgeois*

STRATIGRAPHY
Foundations and Concepts

Edited by

BARBARA M. CONKIN
University of Louisville

and

JAMES E. CONKIN
University of Louisville

A Hutchinson Ross Benchmark® Book

VNR VAN NOSTRAND REINHOLD COMPANY

Benchmark Papers in Geology, Volume 82
Library of Congress Catalog Card Number: 84-2215
ISBN: 0-442-21747-1

Manufactured in the United States of America.

Published by Van Nostrand Reinhold Company Inc.
135 West 50th Street
New York, New York 10020

Van Nostrand Reinhold Company Limited
Molly Millars Lane
Wokingham, Berkshire RG11 2PY, England

Van Nostrand Reinhold
480 Latrobe Street
Melbourne, Victoria 3000, Australia

Macmillan of Canada
Division of Gage Publishing Limited
164 Commander Boulevard
Agincourt, Ontario MIS 3C7, Canada

15 14 13 12 11 10 9 8 7 6 5 4 3 2 1

Library of Congress Cataloging in Publication Data
Main entry under title:
Stratigraphy: foundations and concepts.
"A Hutchinson Ross Benchmark book."
Includes index.
1. Geology, Stratigraphic—Addresses, essays, lectures.
I. Conkin, Barbara M. II. Conkin, James Elvin.
QE651.S78 1984 551.7 84-2215
ISBN 0-442-21747-1

CONTENTS

Series Editor's Foreword ix
Preface xi
Contents by Author xii

Introduction 1

PART I: STENO (1669) to SMITH (1796)

Editors' Comments on Papers 1, 2, and 3 8

1 **WOODWARD, J.:** *An Essay toward a Natural History of the Earth: And Terrestrial Bodies, Especially Minerals: As also of the Sea, Rivers, and Springs. With an Account of the Universal Deluge: And of the Effects That It Had upon the Earth.* 16
R. Wilkin, London, 1695, pp. 6–12

2 **MICHELL, J.:** The Course and Phenomena of Earthquakes 24
Royal Soc. London Philos. Trans. **51**(2):582–587 (1760)

3 **WHITEHURST, J.:** Of the Strata in Derbyshire and Other Parts of England 31
An Inquiry into the Original State and Formation of the Earth; Deduced from Facts and the Laws of Nature, W. Bent, London, 2nd ed., 1786, pp. 178–181

PART II: CUVIER AND BRONGNIART (1808) TO CONYBEARE AND PHILLIPS (1822)

Editors' Comments on Papers 4 Through 7 36

4A **CUVIER, G., and A. BRONGNIART:** Essay on the Mineralogic Geography in the Neighborhood of Paris 42
Translated from *Mus. Natl. Hist. Nat. Ann.* **11:**307–310, 312 (1808)

4B **CUVIER, G., and A. BRONGNIART:** Essay on the Mineralogic Geography in the Neighborhood of Paris 44
Translated from *Jour. Mines* **23:**426–427, 434–435 (1808)

5 **PARKINSON, J.:** Observations on some of the Strata in the Neighbourhood of London, and on the Fossil Remains contained in them 46
Geol. Soc. London Trans. **1:**324–326 (1811)

6 **CUVIER, G.:** *Essay on the Theory of the Earth* 49
Translated from the French by Robert Kerr; William Blackwood, John Murray, Robert Baldwin, Edinburgh, 1813, pp. 12–15, 49–60, 180–181

7 **GREENOUGH, G. B.:** Essay I. On Stratification — **67**
A Critical Examination of the First Principles of Geology; in a Series of Essays
Longman, Hurst, Rees, Orme, and Brown, London, 1819, pp. 1, 9, 25, 88-90

PART III: PHILLIPS (1829) TO D'ORBIGNY (1852)

Editors' Comments on Papers 8 Through 19 — **74**

8 **PHILLIPS, J.:** Organic Remains of the Eastern Part of Yorkshire — **85**
Illustrations of the Geology of Yorkshire; or, a Description of the Strata and Organic Remains of the Yorkshire Coast: Accompanied by a Geological Map, Sections, and Plates of the Fossil Plants and Animals J. Phillips, York, 1829, pp. 111-118

9A **LYELL, C.:** Different Circumstances under Which the Secondary and Tertiary Formations May Have Originated — **94**
Principles of Geology, Being an Attempt to Explain the Former Changes of the Earth's Surface, by Reference to Causes Now in Operation, vol. 3, John Murray, London, 1833, pp. 30-34

9B **LYELL, C.:** Classification of Tertiary Formations in Chronological Order — **99**
Principles of Geology, Being an Attempt to Explain the Former Changes of the Earth's Surface, by Reference to Causes Now in Operation, vol. 3, John Murray, London, 1833, pp. 45-61

10 **MURCHISON, R. I.:** On the Silurian System of Rocks — **116**
London Edinburgh Philos. Mag., ser. 3, **7**(37):46-50 (1835)

11 **HUNTON, L.:** Remarks on a Section of the Upper Lias and Marlstone of Yorkshire, Showing the Limited Vertical Range of the Species of Ammonites, and Other Testacea, with Their Value as Geological Tests. — **121**
Geol. Soc. London Trans., ser. 2, **5**:215-218, 221 (1836)

12 **FITTON, W. H.:** Observations on Some of the Strata between the Chalk and the Oxford Oolite, in the South-east of England. — **126**
Geol. Soc. London Trans., ser. 2, 4(Pt. II):103-105, 328-334, Plate Xb (1836)

13A **GRESSLY, A.:** Geologic Observations on the Jura of Soleure — **137**
Translated from *Soc. Helv. Sci. Nat. Mem.* **2**:10-12, 20-22 (1838)

13B **GRESSLY, A.:** Geologic Observations on the Jura of Soleure — **140**
Translated from *Soc. Helv. Sci. Nat. Mem.* **4**:165 (1840)

13C **GRESSLY, A.:** Geologic Observations on the Jura of Soleure — **141**
Translated from *Soc. Helv. Nat. Mem.* **5**:174-175 (1841)

14 **LYELL, C.:** On the Different Ages of the Aqueous Rocks — **142**
Elements of Geology, John Murray, London, 1838, frontispiece, pp. 271-279

15A **MURCHISON, R. I.:** Description of the Map — **153**
The Silurian System, Founded on Geological Researches in the Counties of Salop, Hereford, Radnor, Montgomery, Caermarthen, Brecon, Pembroke, Monmouth, Gloucester, Worcester, and Stafford; with Description of the Coal-Fields and Overlying Formations, Part I, John Murray, London, 1839, pp. xxvii-xxviii

15B **MURCHISON, R. I.:** Introduction 155
The Silurian System, Founded on Geological Researches in the Counties of Salop, Hereford, Radnor, Montgomery, Caermarthen, Brecon, Pembroke, Monmouth, Gloucester, Worcester, and Stafford; with Description of the Coal-Fields and Overlying Formations, Part I, John Murray, London, 1839, pp. 3–12

15C **MURCHISON, R. I.:** Introductory View of the Distribution of Organic Remains in the Older Formations 165
The Silurian System, Founded on Geological Researches in the Counties of Salop, Hereford, Radnor, Montgomery, Caermarthen, Brecon, Pembroke, Monmouth, Gloucester, Worcester, and Stafford; with Description of the Coal-Fields and Overlying Formations, Part II, John Murray London, 1839, pp. 579–586

16 **SEDGWICK, A., and R. I. MURCHISON:** On the Physical Structure of Devonshire, and on the Subdivisions and Geological Relations of Its Older Stratified Deposits, &c. 173
Geol. Soc. London Trans. **5:**633–634, 688–702 (1840)

17 **D'ORBIGNY, A.:** Conclusions Relative to the Separation of Geologic Stages, and to the Special Faunas Which They Enclose 190
Translated from *Cours Elémentaire de Paléontologie et de Géologie Stratigraphiques,* vol. I, Masson, Paris, 1849, pp. 149–156

18 **D'ORBIGNY, A.:** Geologic Summary on the Cephalopodes of the Jurassic Terrains 194
Translated from *Terrains Jurassiques, vol. I. Cephalopods,* Paléontologie Française, Paris, 1850, pp. 600–611

19 **D'ORBIGNY, A.:** General Geologic Deductions Drawn from Fossil Species 198
Translated from *Cours Elémentaire de Paléontologie et de Géologie Stratigraphiques,* vol. II, bk. 1, Masson, Paris, 1852, pp. 256–257

PART IV: OPPEL (1856) TO BUCKMAN (1902)

Editors' Comments on Papers 20 Through 28 202

20 **OPPEL, A.:** The Jurassic System of England, France, and Southwestern Germany, Divided and Compared According to Its Individual Members 209
Translated from *Wüerttemb. Naturwiss. Jahresh.,* 1856, pp. 123–124, 135

21 **QUENSTEDT, F. A.:** Historical Notes 211
Translated from *Der Jura,* Laupp, Tubingen, 1858, pp. 17–24

22 **BRONN, H. G.:** On the Laws of Evolution of the Organic World during the Formation of the Crust of the Earth 215
Ann. Mag. Nat. Hist., ser. 3, **4:**(20)89–90 (1859)

23 **PHILLIPS, J.:** Anniversary Address of the President 216
Geol. Soc. London Quart. Jour. **16:**xxxii–xxxiii, xxxv–xli, xliii–xlv (1860)

24 **LAPWORTH, C.:** On the Tripartite Classification of the Lower Palaeozoic Rocks 226
Geol. Mag., n.s., **6**(1):1–15 (1879)

25 **DELAPPARENT, A. A. C.:** Paleontologic Method: Divisions of the Sedimentary Series **241**
Translated from *Traité de Géologie*, Savy, Paris, 1885, pp. 704–712

26 **BUCKMAN, S. S.:** On the Bajocian of the Sherbourne District: Its Relation to Subjacent and Superjacent Strata **249**
Geol. Soc. London Quart. Jour. **49:**480–483, 518 (1893)

27 **BUCKMAN, S. S.:** On the Grouping of Some Divisions of So-called 'Jurassic' Time **253**
Geol. Soc. London Quart. Jour. **54:**442–444 (1898)

28 **BUCKMAN, S. S.:** The Term 'Hemera' **256**
Geol. Mag. **9:**554–557 (1902)

PART V: CHAMBERLIN (1898) TO GRABAU (1936)

Editors' Comments on Papers 29 Through 32

29 **CHAMBERLIN, T. C.:** The Ulterior Basis of Time Divisions and the Classification of Geologic History **269**
Jour. Geology **6**(5):449–462 (1898)

30 **CHAMBERLIN, T. C.:** Diastrophism as the Ultimate Basis of Correlation **283**
Jour. Geology **17**(8):685–693 (1909)

31 **ULRICH, E. O.:** Revision of the Paleozoic Systems **292**
Geol. Soc. Am. Bull. **22:**289–292, 394–411, 448–459, 461–464, 532–535, 574–577, 579–585, 590–593, 595–597 (1911)

32 **ARKELL, W. J.:** The Classification of the English Jurassic Rocks, and the Partition of Jurassic Time **351**
The Jurassic System in Great Britain, Clarendon Press, Oxford, 1933, pp. 29–36

Author Citation Index **357**
Subject Index **359**
About the Editors **365**

SERIES EDITOR'S FOREWORD

The philosophy behind the Benchmark Papers in Geology is one of collection, sifting, and rediffusion. Scientific literature today is so vast, so dispersed, and, in the case of old papers, so inaccessible for readers not in the immediate neighborhood of major libraries that much valuable information has been ignored by default. It has become just so difficult, or so time consuming, to search out the key papers in any basic area of research that one can hardly blame a busy person for skimping on some of his or her "homework."

This series of volumes has been devised, therefore, as a practical solution to this critical problem. The geologist, perhaps even more than any other scientist, often suffers from twin difficulties—isolation from central library resources and immensely diffused sources of material. New colleges and industrial libraries simply cannot afford to purchase complete runs of all the world's earth science literature. Specialists simply cannot locate reprints or copies of all their principal reference materials. So it is that we are now making a concerted effort to gather into single volumes the critical materials needed to reconstruct the background of any and every major topic of our discipline.

We are interpreting "geology" in its broadest sense: the fundamental science of the planet Earth, its materials, its history, and its dynamics. Because of training in "earthy" materials, we also take in astrogeology, the corresponding aspect of the planetary sciences. Besides the classical core disciplines such as mineralogy, petrology, structure, geomorphology, paleontology, and stratigraphy, we embrace the newer fields of geophysics and geochemistry, applied also to oceanography, geochronology, and paleoecology. We recognize the work of the mining geologists, the petroleum geologists, the hydrologists, and the engineering and environmental geologists. Each specialist needs a working library. We are endeavoring to make the task of compiling such a library a little easier.

Each volume in the series contains an introduction prepared by a specialist (the volume editor)—a "state of the art" opening or a summary of the object and content of the volume. The articles, usually some twenty to fifty reproduced either in their entirety or in significant extracts, are selected in an attempt to cover the field, from the key papers of the last century to fairly recent work. Where the original works are in foreign languages, we have endeavored to locate or commission translations. Geologists, because of their global subject, are often acutely aware of the oneness of our world.

The selections cannot therefore be restricted to any one country, and whenever possible an attempt is made to scan the world literature.

To each article, or group of kindred articles, some sort of "highlight commentary" is usually supplied by the volume editor. This commentary should serve to bring that article into historical perspective and to emphasize its particular role in the growth of the field. References, or citations, wherever possible, will be reproduced in their entirety—for by this means the observant reader can assess the background material available to that particular author, or, if desired, he or she too can double check the earlier sources.

A "benchmark," in surveyor's terminology, is an established point on the ground that is recorded on our maps. It is usually anything that is a vantage point, from a modest hill to a mountain peak. From the historical viewpoint, these benchmarks are the bricks of our scientific edifice.

RHODES W. FAIRBRIDGE

PREFACE

This book gathers works dating from the late seventeenth through early twentieth centuries that exemplify the development of stratigraphy as the science and the art of organizing the world's strata into a chronological succession; a science and art that remain to be perfected. We concentrate in this volume more on the discovery and development of principles than on the details of the stratigraphic succession itself. Innumerable papers and books have been written on various aspects of stratigraphy from the time it was first realized that meaning could be derived from strata and fossils, to the present time with the *International Stratigraphic Guide* (1976) and its attendant discussion and repercussion. Distillation of this immense stratigraphic literature to a book this size meant eliminating many significant and interesting papers.

We decided against including more recent twentieth century works for the foundation and framework of classic stratigraphy were largely nineteenth century productions. With stratigraphy in its present somewhat unsettled state, it would be premature and presumptuous on our part to attempt to select certain modern works as definitive.

ACKNOWLEDGMENTS

We have made use of several libraries during the preparation of this book, including those at Indiana University, Southern Illinois University, University of Cincinnati, Smith College, and Princeton University. We thank in particular the librarian of the Geology Library at Ohio State University, Margery Tibbetts, and the Interlibrary Loan Librarian at the University of Louisville, Ruth C. Culpepper, for their special assistance, as well as the following at the Geology Library at Princeton who were so helpful: David C. Stager, Patricia Gaspari-Bridges, Emily Ruben, and Steven H. Jaffe.

Translations of German papers, for which we are most grateful, were made by Ursula Barber.

We also acknowledge suggestions made by Hollis D. Hedberg of Princeton University for works to be included, even though, as he rightly noted, the task of selecting papers would be rather "frightening" and "overwhelming." Helpful suggestions of some lesser-known works were offered by Kennard B. Bork of Denison University.

We were fortunate to enjoy the hospitality of Winnie and Al Fischer while at Princeton, which made the research time doubly pleasurable. We also appreciate the information, advice, and encouragement so kindly given by Al.

Finally, we thank Rhodes W. Fairbridge for offering us the opportunity of working on this volume in 1976, and we thank the publishers for their patience over these seven years.

CONTENTS BY AUTHOR

Arkell, W. J., 351
Brongniart, A., 42, 44
Bronn, H. G., 215
Buckman, S. S., 249, 253, 256
Chamberlin, T. C., 269, 283
Cuvier, G., 42, 44, 49
deLapparent, A. A. C., 241
d'Orbigny, A., 190, 194, 198
Fitton, W. H., 126
Greenough, G. B., 67
Gressly, A., 137, 140, 141
Hunton, L., 121
Lapworth, C., 226
Lyell, C., 94, 99, 142
Michell, J., 24
Murchison, R. I., 116, 153, 155, 165, 173
Oppel, A., 209
Parkinson, J., 46
Phillips, J., 85, 216
Quenstedt, F. A., 211
Sedgwick, A., 173
Ulrich, E. O., 292
Whitehurst, J., 31
Woodward, J., 16

STRATIGRAPHY

INTRODUCTION

Stratigraphy as a science came into being in the early 1800s, essentially simultaneously in England and France where fossiliferous strata lie superposed and exposed over thousands of square miles, but its roots reach back into the late 1600s in England and Italy. In fact, except for mineralogy, the science of geology in its youth was stratigraphy.

Table 1 presents important works, arranged in chronological sequence by topic, that we have selected from among hundreds written between 1669 and 1936. They all are referred to in this volume and mark what we believe to be the most significant progressive steps in the conceptual development of stratigraphy. The first mentioned name under a topic is the earliest writer we have found from this period; subsequent listings present significant development of that concept. Many of the writers discussed various other topics in these works besides those under which they are listed.

We note first the recognition of the existence and significance of strata by Steno (1669) and Woodward (Paper 1) and the contemporaneous discovery of the importance of fossils by Lister (1671), Woodward (Paper 1), and Hooke (1705). In this early stage of development of stratigraphy, we find the roots of the concept of facies fossils in Woodward (Paper 1); the facies concept developed extensively in the late 1700s and early 1800s, prior to Gressly's first use of the term *facies* in 1838 (Paper 13A).

In the last half of the 1700s, several basic descriptions of fossiliferous strata appeared in England, France, Germany, and Italy. These descriptions were accompanied by considerable premature theorizing concerning earth's history as revealed by these strata. With respect to these theories, Zittel (1901, 41, 42) quoted Buffon's sarcastic remark that "geologists must feel like the ancient Roman augurs who could not meet each other without laughing." The more sophisticated or cautious geologists, especially those who founded the Geological Society of London in 1807, called for a moratorium on theorizing while data sufficient to this end could be gathered, which it was with

Table 1. Chronological development of important stratigraphic concepts from 1669 to 1936.

STRATA	FOSSILS	FACIES	CHRONOLOGY	FORMATION	SYSTEM	UNCONFORMITY	DISCONFORMITY	STAGE	ZONE
Steno, 1669	Lister, 1671								
Woodward, 1695	Woodward, 1695	Woodward, 1695	Hooke, 1705						
	Strachey, 1719								
Lehmann, 1756; Michell, 1760	Michell, 1760; Guettard, 1765	Rouelle, 1763-1764; Guettard, 1765	Arduino, 1759	Fuchsell, 1762					
	Soulavie, 1780	Soulavie, 1780							
Whitehurst, 1786	Whitehurst, 1786	Lavoisier, 1789		Werner, 1785					
	Smith, 1791-1796					Hutton, 1795			
Geol. Soc. London, 1807; Parkinson, 1811; Richardson, 1811; Cuvier, 1812; Greenough, 1819; d'Halloy, 1822	Cuvier and Brongniart, 1808; Parkinson, 1811; Cuvier, 1812; Schlotheim, 1813; Brocchi, 1814; Smith, 1815; 1817	Cuvier and Brongniart, 1808; Parkinson, 1811	Schlotheim, 1813	Cuvier and Brongniart, 1808			Cuvier and Brongniart, 1808	Cuvier and Brongniart, 1808	
		d'Halloy, 1822		Smith, 1815					
	Phillips, 1829; Lyell, 1833; Hunton, 1838; Lyell, 1838; Sedgwick and Murchison, 1839	Fitton, 1836; Lyell, 1838; Gressly, 1838; Sedgwick and Murchison, 1839; d'Orbigny, 1849-1852	Lyell, 1833	Conybeare and Phillips, 1822	Murchison, 1835; 1839; Sedgwick and Murchison, 1839	Sedgwick, 1830; d'Halloy 1831	Lyell, 1833		Phillips, 1829
				Fitton, 1836					Lyell, 1833
				Murchison, 1835; 1839					Hunton, 1836
								d'Orbigny, 1841-1852	d'Orbigny, 1841-1852
	Quenstedt, 1858; Phillips, 1860		Bronn, 1858; Phillips, 1860				Bronn, 1858	Oppel, 1856-1858	Oppel, 1856-1858
									Phillips, 1860
			Lapworth, 1879		Lapworth, 1879	Lapworth, 1879			
	deLapparent, 1885		deLapparent, 1885		deLapparent, 1885	deLapparent, 1885	deLapparent, 1885	deLapparent, 1885	
Chamberlin, 1898			Buckman, 1893-1902		Chamberlin, 1898		Chamberlin, 1898		Buckman, 1893-1902
							Grabau, 1905		
Ulrich, 1911	Ulrich, 1911		Ulrich, 1911		Chamberlin, 1909		Ulrich, 1911; Grabau, 1913		
	Arkell, 1933		Arkell, 1933				Grabau, 1936	Arkell, 1933	Arkell, 1933

great gusto in the first half of the 1800s by wealthy travelers, country parsons and doctors, professors, farmers and landed gentry, surveyors, engineers, and others. A number of geological societies were formed at this time for the purpose of studying the strata and fossils for their own sake as well as for practical application of this knowledge.

Subsequent to Strachey's (1719, 1725) figuring of an angular unconformity in another context, and Hutton's famous conceptualization and discovery in 1787 of same in support of his theory of the earth, the most important step with respect to understanding unconformities was the recognition by Cuvier and Brongniart in 1808 (Paper 4B) of the type of unconformity eventually termed *disconformity* by Grabau (1913).

During the early 1800s the concepts of "formation" and its fraternal twin "system" were developed in Germany and England respectively, along with attempts to develop a chronology to compartmentalize earth's history into various *eras, periods, epochs,* and *stages,* terms used quite interchangeably for several decades. At this time, earth history was often envisioned, based on the gross features of the stratigraphic record, as a series of relatively quiet episodes of deposition separated by shorter times of natural or more mysterious disturbance. Working out the details of this historical stage play by intensive study of the strata and fossils occupied many geologists and paleontologists during the middle portion of the nineteenth century. By this time, the major outlines of the geological time scale had been drawn, but it was usually compressed into an awkwardly short time span that resulted in considerable distortion of thought when one attempted to blend strict uniformitarian concepts with the evidence seen in the stratigraphic record; an example is found in d'Orbigny's (Paper 18) stages that unsuccessfully (of course) partook of both uniformitarianism (with his concept of stage as an epoch like the present) and catastrophism (with the events that supposedly caused their universal separation).

The last basic stratigraphic concept to evolve in the 1800s was that of the zone, which originated with Phillips (Paper 8) and Hunton (Paper 11), though d'Orbigny (Paper 18) and Oppel (Paper 20) are most often credited with its conception; certainly Oppel did bring it to fruition. Henceforward, more and more refined correlations were attempted, while at the same time many stratigraphers attempted to locate worldwide boundaries based on the fossil record and prominent large-scale unconformities.

By the early 1900s, it had been well established that earth history was long, that there had been many marine transgressions and regressions on every continent, and that, in spite of the immensely long documented record, there were numerous breaks or gaps. The complexity of the stratigraphic record seemed to obscure the rhythm of change, and the recognition of innumerable lithofacies and biofacies seemed to preclude broad correlation. The sedimentary crust seemed less like a neat stack of blankets and more like randomly arranged piles of patchwork crazy quilts. Evidence of some degree of tectonism at one place or another almost throughout geologic time made the older idea of long periods of earthly quiet punctuated by short times of revolution seem much too simple and naive. Diachronism was seen everywhere—in formations, boundaries, and faunas—so that correlation on a fine scale seemed impossible; some geologists despaired of ever correlating precisely with fossils and vainly turned to radiometric dating in the hope that this method could fix time points within the stratigraphic record. However, others such as Chamberlin (Paper 29; Paper 30), Ulrich (Paper 31), and Grabau (1936) believed that reasonably precise worldwide chronostratigraphic boundaries could be drawn because even though oceanic crust had yet to be explored, these workers (based on facts at hand in the early 1900s) were able to unite internal physical change, external geographic change, and biological change in order to marshal evidence for a rhythm in the stratigraphic record. Chamberlin proclaimed diastrophism as the ultimate basis of correlation inasmuch as it is changes in the relative levels of land and sea that control deposition and ultimate preservation of sediment, creation of unconformities of greater and lesser magnitude, geographic distribution of life forms, and promotion by natural selection of evolutionary change—collectively, the essentials of stratigraphy.

Worldwide pulsations (eustatic transgressions and regressions) are surely a reality; even though they can be modified or obscured by coincident tectonism, the nature of the sediments would be changed, thus aiding in recognition of boundaries. While geologists like Chamberlin and Grabau did not know (and could not then know) precisely what caused the changes in the stratigraphic record, they could, by analyzing the evidence preserved in the stratigraphic sequences, infer something of the source and nature of the changes. Certainly we can now believe that the activities of sea-floor spreading (activation and deactivation of ridges and translation of horizontal crustal movement) will affect large parts of the world to measureable extents and that the effects of these activities become embodied in the stratigraphic record.

For boundary determination, the practical tool is the small-scale (with respect to time) but widespread (geographically) disconformity that we (Conkin and Conkin, 1973) have named *paracontinuity.* A paracontinuity displays itself as a slight but significant physical discontinuum coincident with a slight but significant biologic discontinuum that thus allows for boundary determination on both physical and biological bases. In those areas on the craton, classically considered to display gradational or transition sequences in which the most complete section is presented, a paracontinuity separates the very youngest portion of the older unit (ranging from eonothem to bed) from the very oldest portion of the younger unit (ranging from eonothem to bed). In composite stratotype sections in which the upper and lower limits of biozones are most fully ascertainable, a paracontinuity constitutes essentially a chronostratigraphic boundary. The precise position of paracontinuities between chronostratigraphic packages of rocks is determined, and subsequent correlation of these units is effected, on the basis of classic paleontology and stratigraphy, involving variously, superposition, faunal and/or floral succession, and evidence of physical happenings (such as magnetic reversals, pyroclastics, and evidence of slight but significant surfaces of discontinuity).

The generalized worldwide chronostratigraphic column exists, of course, only in the human mind. The broad framework of this column is in place, and it is a colossal monument to the devoted labors of generations of stratigraphers and paleontologists. However, innumerable gaps remain in this gargantuan three-dimensional jigsaw puzzle. Geochronologic lines are not substantive but conceptual only, and in practice, they are not completely coincident with recognizable chronostratigraphic boundaries; only paracontinuities are close approximations of time lines and bracket the most precisely determinable chronostratigraphic units. Paracontinuities vary in their representation of the magnitude of time loss from place to place (locally to intercontinentally). However, the deviations are on the whole insignificant in magnitude because the second hand of the absolute time clock measures rather large units, and the lost time represented by a paracontinuity can be accommodated well within the individual seconds of the chronologic scale.

A minutely detailed study of the stratigraphic-paleontologic context of local stratotypes has been clearly seen for nearly 200 years to be the foundation for the construction of the chronostratigraphic column. Studies other than those based on minute scrutiny of this context are not presently applicable to the task of greater refinement of this column.

REFERENCES

Conkin, J. E., and B. M. Conkin, 1973, The Paracontinuity and the Determination of the Devonian-Mississippian Boundary in the Type Lower Mississippian Area of North America, *Univ. Louisville Stud. Paleont. Stratigr.*, no. 1, 36p.

Grabau, A. W., 1913, *Principles of Stratigraphy*, A. G. Seiler and Co., New York.

Grabau, A. W., 1936, Oscillation or Pulsation, *16th Internat. Geol. Congr. Washington,* Rept. 1, pp. 539–553.

Hooke, R., 1705, *Posthumous Works of Robert Hooke, V. Discourses of Earthquakes, Their Causes and Effects, and Histories of Several,* Waller, London.

Lister, M., 1671, A letter of Mr. Martin Lister . . . on that of Mr. Steno Concerning Petrified Shells, *Royal Soc. Philos. Trans.* **5:**645, 646.

Steno, N., 1669, *De solido intra solidum naturaliter contento dissertationis prodromus,* The Star, Florence.

Strachey, J., 1719, Observation on the Strata in the Coal-Mines of Mendip, in Somersetshire, *Royal Soc. Philos. Trans.* **30:**968–973, plate XI.

Strachey, J., 1725, An Account of the Strata in Coal-Mines, etc., *Royal Soc. Philos. Trans.* **33:**395–398, plate II.

Zittel, K. von, 1901, *History of Geology and Paleontology,* Scribners, New York.

Part I

STENO (1669) TO SMITH (1796)

Editors' Comments on Papers 1, 2, and 3

1 **WOODWARD**
Excerpt from *An Essay toward a Natural History of the Earth: And Terrestrial Bodies, Especially Minerals: As also of the Sea, Rivers, and Springs. With an Account of the Universal Deluge: And of the Effects That It Had upon the Earth*

2 **MICHELL**
Excerpt from *The Course and Phenomena of Earthquakes*

3 **WHITEHURST**
Excerpt from *Of the Strata in Derbyshire and other parts of England*

Between 1669 and 1796, several basic steps were taken that would lead to the development, in the 1800s, of the science of stratigraphy. These steps were the recognition and acceptance of the concepts of strata, fossils (as preexisting life and as indexes to the strata), unconformities, and formations and the possibility of constructing an earth chronology.

The first discussion of strata from this period is found in Steno's (1669) *Prodromus*. Although Zittel (1901, 26) said this work was nearly unknown before 1832 when translated into French by Elie de Beaumont, the *Prodromus* had been put into English by Oldenburg in 1671 and read by Lister. In *A letter of Mr. Martin Lister . . . on that of Mr. Steno, concerning Petrified Shells,* Lister (1671, 646) disagreed with Steno's belief in the natural origin of fossils, believing them to be *lapides sui generis,* in support of which he noted that "quarries of different stone yield us quite different sorts or species of shells." Some (e.g., Zittel, 1901, 17) have cited this as the earliest reference to layers being characterized by different fossils, but Lister meant only to disprove Steno by showing that inasmuch as the different mineral types of fossils were like the rocks enclosing them, they must have been formed in these rocks.

The first work that should rightly be called a benchmark in stratigraphy is John Woodward's *Essay* (Paper 1), said by Clark and Hughes (1890) to be a forerunner of a larger work never published.

Woodward sent a "List of Queries" to some 500 people around the world (Porter, 1977, 226) asking for information on fossils and minerals and their position in the strata and then based his conclusions on these data and his own careful study and observations. His paleontological data were published later in his *Fossils of All Kinds Digested into a Method* (1728) and *A Catalogue of the English Fossils in the Collections of J. Woodward, M. D.* in two volumes (1728, 1729). This latter work was studied by William Smith (Phillips, 1844, 24). Clark and Hughes (1890), biographers of Smith, found that "Woodward was far in advance of his age: and, had not his mind been predisposed to theory, he might have anticipated, by a century, the discoveries of William Smith" (p. 173). Woodward's will funded the Woodwardian Professorship at Cambridge, and his collections, though not arranged stratigraphically (Sedgwick, 1831, 275), formed the nucleus of the Woodwardian Museum, now incorporated in the Sedgwick Museum. Certainly Woodward influenced Smith.

In his *Essay* (Paper 1), Woodward (pp. 6–12) describes the strata and the "order and manner of position" of fossils therein. He observed in the preface:

> There are vast Multitudes of Shells, and other Marine Bodies, found at this day incorporated with and lodged in all sorts of Stone . . . . that these are found . . . . from near the Surface. . . . downwards to the greatest Depth we ever dig or lay it open, and this in all Parts of it quite round the Globe: that the said terrestrial Matter is disposed into Strata or Layers, placed one upon another, in like manner as any earthy Sediment, settling down from a Fluid in great quantity, will naturally be. [p. iii]

The concept of facies fossils was grasped by Woodward (Paper 1, 26) when he noted that shellfish of the deep were called *Pelagiae* and that those residing nearer the shores were called *Littorales* by naturalists. We later find Guettard (1765), Lavoisier (1789), and Desmarest (1794) reintroducing this terminology and making use of the concept, as did Lamarck (1802).

Robert Hooke's *Posthumous Works* (1705, written in 1688) introduced (pp. 334, 335) the idea of using fossils as "Media or Criteria of Chronology" and referred to them as "curious Medals of former Ages" (p. 372). Hooke was considerably more prophetic than he knew when he stated that "it is very difficult to read them, and to raise a *Chronology* out of them . . . yet 'tis not impossible, but that, by the help of those joined to *other means and assistances of Information* [last italics ours], much may be done" (p. 411). Three centuries later, raising a chronology is still difficult but not impossible.

The first application of fossils to the solution of a practical stratigraphic problem is found in the cross sections published by Strachey (1719, 1725), who used fossils to correlate coal "veins" offset by a "ridge" (fault). The cross sections also present the earliest illustrations of an unconformity, although Eyles (1969) believed its significance escaped Strachey. Nearly sixty years would pass before Hutton in 1787 sought out and interpreted an angular unconformity in the "junction of primary and secondary strata" at the Isle of Arran (Hutton, 1795, 432). Another unconformity was seen at Jedburgh in the same year and a third, the following year, at Siccar Point (Hutton, 1795, 458).

The earliest publication to describe and figure a stratigraphic succession is that of Lehmann (1756) wherein thirty layers of Permian rocks in the Harz Mountains were detailed and illustrated with a cross section. This work demonstrated that strata succeed each other in a definite order and set the stage in Germany for the wide application of descriptive rock divisions. Nearly at the same time, Füchsel (1762) recognized that the rock strata might be grouped into a formation that represented an epoch in the history of the earth.

Concurrently, Arduino (1759) divided the rocks of northern Italy into the Primitive ("Monti minerali o primitivi, Monti primarj"), Secondary ("Monti secundarj"), Tertiary ("Monti terziarj"), and Volcanic divisions and described their composition, surface outcrop, and order of superposition. Ronconi (1865, cited in Adams, 1938) said Arduino was the first to see the importance of the "application of paleontology and chemistry to the chronology of the formations" (p. 373).

The previous reference to Guettard's work (1765) was to his *On the Accidents That Have Befallen Fossil Shells Compared with Those Which Are Found to Happen to Shells Now Living in the Sea,* wherein he showed that beds of fossil shells on land are analogous to present sea floors in every respect. Geikie (1905) called Guettard's work one of the most illustrious in geologic literature, one which opened a new field of investigation of earth history and provided the methodology for doing so. Geikie believed others used Guettard's work without giving him sufficient credit.

In a contemporary work, Michell (Paper 2) traced and established the order of English strata and made tables, about 1788 or 1789, one of which was published in 1810 by Farey. Sedgwick (1831) noted that Michell was intimately acquainted with the Woodward collection inasmuch as he occupied the Woodwardian chair from 1762 to 1767. Michell gave a clear theoretical definition of stratification (Paper 2, 582–587, sec. 38–49), noting that "without a general knowledge of the *fossil bodies* of a large tract of country, it is hardly possible" to trace the strata. Sedgwick (1831, 275), in his address upon awarding the Wollaston Medal to William Smith, believed Michell did not mean organic remains by his use of "fossil" inasmuch as in Michell's day the word *fossil* usually meant mineral. Past life was referred to as "extraneous fossils" or "organic fossils." However, since Michell used the expression *fossil bodies* and was familiar with organic fossils, we believe he was indeed referring to these and, thus, was among the first to propose the use of fossils for correlation. Michell's paper was published later in an *Abridgement of the Philosophical Transactions of 1760* (pp. 448–456), without mention of fossils, under the title *Conjectures Concerning the Cause, and Observations on the Phenomena of Earthquakes . . . ;* the existence of these two versions of this work evidently has not always been recognized.

In France, Soulavie (1780–1784), adopting Hooke's (1705) ideas, divided the calcareous rocks of the Vivarais in southern France into five epochs or ages, each with a distinct assemblage of fossils, although some of the oldest fossils were found in the next age along with some similar to modern types, but then the oldest ones were entirely absent in the third age. At the same time, he noted that certain fossilliferous shales of the fourth age contained fossils similar to those of the three calcareous ages and therefore presumed these were more or less equivalent in age, another early glimmering of a facies concept. Soulavie stated (p. 350) that "the progress of time and, above all, of increased observation will augment the number of epochs which I have given, and fill up the blanks, but they will not change the relative places which I have assigned to these epochs" (cited in Geikie, 1905, 340).

We mention Werner (1786) at this point largely because of his fame, but we do not quote him. Werner's direct efforts have been said to have retarded geologic progress (Geikie, 1905, 203), but indirectly, he served to advance the science by inspiring others to trace strata over wide areas and by his unreasonable attitude in the basalt controversy that spawned productive argument (Geikie, 1905; Schneer, 1969). Geikie (1905, 24), nevertheless, cited Amie Boué's belief that Werner's death contributed greatly to the progress of geology in Germany. Ospovat (Schneer, 1969, 242–256) reflected light onto parts of Werner's ideas that he believed have often been misunderstood and oversimplified. Ospovat (Schneer, 1969) believed Werner's *Kurze Klassification* (1786) established the principle of geologic succession as the basis for geological research and gave the term *formation* petrographical rather than time significance inasmuch as different kinds of rocks (formations) could be formed at different times and were grouped into suites, each of which represented a period in the history of sedimentation.

Hedberg (1969) pointed out (as had Zittel, 1901, 56) that Bergman's *Physisk beskrifning* (1766) was the basis in large part for Werner's (1786) classification and was even relied on by Hutton. Thus, it is difficult to determine just where most credit is due for stratigraphic progress during this period. In any event, various English and French geologists' field studies of local conditions of deposition and their emphasis on the use of fossils for interpretation of the strata soon eclipsed Werner's concepts of formations.

Lavoisier (1789) and Desmarest (1794) both used ideas concerning pelagic and littoral sediments and fossils presented by Rouelle in lectures from 1762 to 1769 (Carrozi, 1965, 72). Lavoisier believed that slow oscillations of sea level produced the different strata in basins such as the Seine, and he thought that the number of oscillations was revealed by the number of littoral and pelagic strata. According to Geikie (1905, 344), Lavoisier's distinction between pelagic and littoral organisms was afterward adopted by Lamarck (1802), but Carrozi (1965) said Lavoisier's work went unnoticed. In any event, Woodward (Paper 1) had already published the idea of "pelagiae" and "littorales." We do notice in these papers some of the groundwork being laid for Cuvier and Brongniart's (Paper 4A, Paper 4B, 1811) work as well as for d'Orbigny's later (Papers 17, 18, 19) concepts of stages and ideas of periods of quiet deposition punctuated by times of upheaval.

It may seem negligent merely to mention a geologist as famous as Hutton, who is usually credited with being the first to teach clearly that the past history of the earth is to be explained by the present (1788, 233). However, we find Whitehurst (Paper 3) and even Woodward (Paper 1) expressing similar ideas. Hutton was evidently the first to realize that the relationships to be seen in an unconformity would give evidence of the cycles in earth history (1788, 1795), although Jameson (1805) first applied the term *unconformity* to Hutton's "junction of primary and secondary strata" (Tomkeieff, 1962). The term *unconformity* was long used only in the sense of an angular unconformity and was not even in general use until taken up by Lyell in the 1830s and De La Beche in the 1840s, by which time the existence of disconformities was recognized. Of course, stratigraphic boundaries were often originally identified in part by the presence of angular unconformities, but these breaks are not significant in precise boundary determination.

Not until 1808 (Paper 4B, 434, 435) did Cuvier and Brongniart describe a disconformity as Lyell later did (Paper 9A, 32, 33), while Grabau used the expression *disconformable relations* in 1905 and coined the term *disconformity* in 1913.

Putting aside the well-publicized works of Werner (1786) and Hutton (1788, 1795), we would select Whitehurst's recently reprinted

An Inquiry into the Original State and Formation of the Earth, etc. (1778; 2nd ed., 1786) as being of prime importance (Paper 3). Challinor (1971, 64, 65) called it "one of the main landmarks in the progress of knowledge in British geology, being packed with significant facts and just inferences." Whitehurst was involved in the Lunar Society of Birmingham, which was visited by Hutton (Porter, 1977, 97). This society established the importance of geology for the development of industry, agriculture, and transportation. Whitehurst intended to "derive the nature of things from Causes truly existent" (p. ii). He stressed the regularity, gradualness, and continuity of nature's actions in the progressive formation of strata and found evidence in his Derbyshire studies that the strata had been formed horizontally but had then undergone disruption. Whitehurst stated that fossils "were entombed in the bowels of the earth, in successive periods of time" (p. 59). He pointed out (pp. 178-181) that in the regular succession of strata, if one knows the "incumbent stratum" one knows the "inferior ones" and that records of strata are needed to "render the subject of subterraneous geography of general use." These statements antedate those of Hutton and Smith whose works were promoted and so well publicized in the 1800s by Playfair (1802), Phillips (1844), Sedgwick (1831), Geikie (1897), and others. Whitehurst (Paper 3, p. 131) intended to deposit and arrange specimens in the British Museum in their natural order, which Sheppard (1920, 98) said was subsequently done by Smith.

Smith unquestionably made use of the principles of superposition, original lateral extension, and faunal succession, but he did not discover them as his nephew, John Phillips, believed (Paper 8, 4). However, Smith was the first to apply them so widely, as in his maps, but Porter (1977, 168) notes evidence that Smith made scant use of the principle of faunal succession. In the years when Smith was completing his major work and struggling to find support for its publication, some investigators like Farey (1807) made use of his ideas and work, while others including Parkinson (Paper 5) and Townsend (1813) gave him full credit. Smith's earliest notes and writings (1791-1796), on strata and correlation by fossils are known to us now (Phillips, 1844; Sheppard, 1920; Cox, 1942) but were not made public at the time they were written. Smith wrote in 1796 of the "wonderful order and regularity with which Nature has disposed of these singular productions [fossils], and assigned to each class its particular stratum" (Phillips, 1844, 17). In 1815 Smith referred to fossils as "indices" (Davis, 1943, 141, 144) and appears to be the first (1817) to have used the term *stratigraphical,* a precursor of the noun *stratigraphy.* Taken altogether, Smith's various works formed a practical foundation for future descriptive stratigraphy.

REFERENCES

Adams, F. A., 1938, *Birth and Development of the Geological Sciences,* Dover, New York (republished in 1954).

Arduino, G., 1759, Letters to Antonio Vallisnieri, in *Nuovo raccolta di opuscoli scientifici e filologici del padre abate Angiolo Calogiera,* vol. 6, pp. 99–180.

Bergman, T., 1766, *Physisk beskrifning öfver Jord-klotet pa cosmographiska Sällskapets vägnar forfattad,* Uppsala.

Carrozi, A. V., 1965, Lavoisier's Fundamental Contribution to Stratigraphy, *Ohio Jour. Sci.* **65**(2):71–85.

Challinor, J., 1971, *The History of British Geology, a Bibliographical Study,* David and Charles, Newton-Abbot.

Clark, J. W., and T. M'K. Hughes, 1890, *The Life and Letters of The Reverend Adam Sedgwick,* vol. I, Cambridge University Press, Cambridge.

Cox, L. R., 1942, New Light on William Smith and His Work, *Yorkshire Geol. Soc. Proc.* **25**(1):1-99.

Cuvier, G., and A. Brongniart, 1811, *Essai sur la géographie minéralogique des environs de Paris,* Baudouin, Paris.

Davis, A. G., 1943, The Triumvirate: A Chapter in the Heroic Age of Geology, *Croydon Nat. Hist. Sci. Soc. Proc.* **2:**122–146.

Desmarest, 1794, Geographie Physique, *Encyclopédie Méthodique* **1**(64):409–431.

Eyles, V. A., 1969, The Extent of Geological Knowledge in the Eighteenth Century and the Methods by Which It was Diffused, in *Toward a History of Geology,* C. J. Schneer, ed., M.I.T. Press, Cambridge, Mass., pp. 159–183.

Farey, J., 1807, On the Dislocations of the Strata of the Earth, *Philos. Mag.* **28:**120.

Farey, J., 1810, Copy of a List of the Principal British Strata, by the Late Rev. John Michell, *Philos. Mag.* **36:**102–104.

Füchsel, G. C., 1762, Historia terrae et maris, ex historia Thuringiae per montium descriptionem erecta, *Acta Acad. Elect. Moguntinae Scientiarum* (Erfurt) **2:**44–209.

Geikie, Sir A., 1897, *The Founders of Geology,* Macmillan, London.

Geikie, Sir A., 1905, *The Founders of Geology,* 2nd ed. Reprint, Dover, New York, 1962.

Grabau, A. W., 1905, Physical Characters and History of Some New York Formations, *Science* (n.s.) **22**(565):528–535.

Grabau, A. W., 1913, *Principles of Stratigraphy,* A. G. Seiler and Co., New York.

Guettard, J., 1765, On the Accidents That Have Befallen Fossil Shells Compared With Those Which Are Found to Happen to Shells Now Living in the Sea, *Acad. Sci. Trans.,* pp. 189, 329, 399.

Hedberg, H. D., 1969, The Influence of Torbern Bergman (1735–1784) on Stratigraphy, *Stockholm Contr. Geology* **20**(2):19–47.

Hooke, R., 1705, *Posthumous Works of Robert Hooke, V. Discourses of Earthquakes, Their Causes and Effects, and Histories of Several,* Waller, London.

Hutton, J., 1788, Theory of the Earth, or an Investigation of the Laws Observable in the Composition, Dissolution, and Restoration of Land Upon the Globe, *Royal Soc. Edinburgh Trans.* **1:**109-304

Hutton, J., 1795, *Theory of the Earth With Proofs and Illustrations,* vol. I, Bell, Bradfute, Blackwood, Edinburgh.

Jameson, R., 1805, *A Mineralogical Description of the County of Dumfries,* Bell, Bradfute, Blackwood, Edinburgh.

Lamarck, J. B., 1802, *Hydrogéologie,* Agasse, Paris.

Lavoisier, A., 1789, Observations générales sur les couches horizontales, qui ont été deposées par la mer, et sur les consequences, qu'on peut tirer de leurs dispositions, relativement à l'ancienneté du globe terrestre, *Acad. Sci. Mem.,* pp. 351-371.

Lehmann, J. G., 1756, *Versuch einer Geschichte von Flötz-Gebürgen, etc.,* Lange, Berlin.

Lister, M., 1671, A Letter of Mr. Martin Lister . . . on that of Mr. Steno Concerning Petrified Shells, *Royal Soc. Philos. Trans.* **5:**645, 646.

Phillips, J., 1844, *Memoirs of William Smith, LL.D., Author of the "Map of the Strata of England and Wales",* Murray, London.

Playfair, J., 1802, *Illustrations of the Huttonian Theory of the Earth,* Creech, Edinburgh.

Porter, R., 1977, *The Making of Geology—Earth Science in Britain 1660–1815,* Cambridge University Press, 278p.

Schneer, C. J., 1969, *Toward a History of Geology,* M.I.T. Press, Cambridge.

Sedgwick, A., 1831, Presidential Address, *Geol. Soc. London Proc.* **20:**270-316.

Sheppard, R., 1920, *William Smith, His Maps and Memoirs,* A. Brown and Sons, Hull.

Smith, W., 1817, *Stratigraphical System of Organized Fossils with Reference to the Specimens of the Original Collection in the British Museum Explaining Their State of Preservation and Their Use in Identifying the British Strata,* Williams, London.

Steno, N. 1669, *De solido intra solidum naturaliter contento dissertationis prodromus,* The Star, Florence.

Strachey, J., 1719, Observation on the Strata in the Coal-Mines of Mendip, in Somersetshire, *Royal Soc. Philos. Trans.* **30:**968-973, plate XI.

Strachey, J., 1725, An Account of the Strata in Coal-Mines, etc., *Royal Soc. Philos. Trans.* **33:**395-398, plate II.

Tomkeieff, S. I., 1962, Unconformity—An Historical Study, *Geol. Assoc. Proc.* **73:**383-417.

Townsend, J., 1813, *The Character of Moses Established for Veracity as an Historian, etc.,* Longman, Hurst, Rees, Orme, and Brown, Bath and London.

Werner, A. G., 1786, *Kurze Klassification und Beschreibungder verschiedenen Gebirgsarten,* Abhandlungen der Böhmischen Gesellschaft der Wissenschaften, 28p.

Woodward, J., 1728, *Fossils of All Kinds Digested into a Method,* London.

Woodward, J., 1728-1729, An Attempt Towards a Natural History of the Fossils of England, etc., in *A Catalogue of the English Fossils in the Collections of J. Woodward, M.D.,* 2 vols. London.

Zittel, K. von, 1901, *History of Geology and Paleontology,* Scribners, New York.

1

Reprinted from pages 6-12 of *An Essay Toward a Natural History of the Earth: And Terrestrial Bodies, Especially Minerals: As also of the Sea, Rivers, and Springs. With an Account of the Universal Deluge: And of the Effects That It Had upon the Earth,* R. Wilkin, London, 1695, 277p.

AN ESSAY TOWARD A NATURAL HISTORY OF THE EARTH: AND TERRESTRIAL BODIES, ESPECIALLY MINERALS: AS ALSO OF THE SEA, RIVERS, AND SPRINGS. WITH AN ACCOUNT OF THE UNIVERSAL DELUGE: AND OF THE EFFECTS THAT IT HAD UPON THE EARTH

J. Woodward

[*Editors' Note:* In the original, material precedes and follows this excerpt.]

The Result of this was, that in time I was abundantly assured, that the Circumstances of *these Things* in remoter Countries were much the same with those of ours *here:* that the Stone, and other terrestrial Matter, in *France, Flanders, Holland, Spain, Italy, Germany, Denmark, Norway*, and *Sweden*, was distinguished into *Strata*, or Layers, as it is in *England:* that those *Strata* were divided by parallel Fissures: that there were enclosed in the Stone, and all the other denser kinds of terrestrial Matter, great numbers of Shells, and other Productions of the Sea, in the same manner as in that of this Island. To be short, by the same means I got sufficient intelligence that *these Things* were found in like manner in *Barbary*, in *Egypt*, in *Guiney*, and other parts of *Africa*; in *Arabia, Syria, Persia, Malabar, China*, and other *Asiatick* Provinces; in *Jamaica, Barbadoes, Virginia, New-England, Brasil, Peru*, and other Parts of *America*. But I reserve the more *particular Relation* hereof to its proper place.

So that though my own Observations were confined to *England*, yet by this means I was made acquainted with the

the ſtate of *theſe Bodies* in other Countries; even in almoſt all parts of the World wherewith the *Engliſh* maintain any Commerce or Correſpondence: and learn'd from all hands, that the ſtate of them *there* was conformable to that of ours *here*, in the main, and as far as I ſhall lay any ſtreſs upon it in my Concluſions; which indeed are not built upon any Niceties, or ſolitary and uncommon Appearances, but on the moſt imple and obvious Circumſtances of theſe terreſtrial Bodies.

As to the Certainty and Accurateneſs of my Obſervations, thus much may modeſtly and very truly be ſaid, that I do not offer any one before I had firſt thoroughly and clearly informed my ſelf in all material Circumſtances of it, and had opportunity of obſerving it in more places than one, that I might be ſatisfied there was nothing caſual or contingent in any of thoſe Circumſtances. This will not be thought an over-great exactneſs, or any thing more than was needful, by *thoſe* who have noted how much Philoſophy hath ſuffered by the neglect and overſight of ſome Naturaliſts in this

this respect. A transient and perfunctory Examination of things, frequently leads Men into considerable Mistakes, which a more correct and rigorous Scrutiny would have detected and avoided. The truth is, I have been the more scrupulous and wary in regard the Inferences drawn from these Observations are of some importance: 'twas but necessary that the Foundation should be firm, when a Superstructure of Bulk and Weight was to be raised upon it. And therefore I advance nothing from any Observation that was not made with this Caution, and that any Man may not, as well as my self, without any great pains, inform himself of the truth of: And as long as the next Cole-pit, or Mine, the next Quarry, or Chalk-pit, will give abundant Attestation to what I write, *these* are so ready and obvious in almost all places, that I need not be any where far to seek for a Compurgator; and to these I may very safely appeal.

Concerning the Observations themselves therefore, there cannot well arise any doubt but what may be easily satisfied: And what I propose in this Essay being founded upon these Observations,

ons, every Reader will be judge of the truth and probability of it, and whether that which I do ſo propoſe naturally follows from them or not.

I ſhall diſtribute *them* into two general Claſſes or Sections, whereof the *former* will comprehend my Obſervations upon all the Terreſtrial Matter that is naturally *diſpoſed into Layers*, or *Strata*; ſuch as our common Sandſtone, Marble, Cole, Chalk, all ſorts of Earth, Marle, Clay, Sand, Gravel, with ſome others.

Of this various Matter, thus formed into *Strata*, the far greateſt part of the Terreſtrial Globe conſiſts, from its Surface downwards to the greateſt depth we ever dig or mine. And it is upon my Obſervations on *this* that I have grounded all my general Concluſions concerning the Earth; all that relate to its Form; all that relate to the *Univerſal* and *other* Deluges; in a word, all that relate to the ſeveral Viciſſitudes and Alterations that it hath yet undergone. Nay, upon the ſame Obſervations I have alſo founded ſeveral Concluſions touching Metalls, Spar, and other Minerals, which are found lodged either in theſe *Strata*, amongſt he the Sand, Chalk, Earth, and the reſt; or in the perpendicular Intervals of the *Strata* of Stone, Marble, or other ſolid Matter.

For upon the *particular* Obſervations on the ſaid Metallick and Mineral Bodies, (which are the Subjects of the *ſecond* Section,) I have not founded any thing but what purely and immediately concerns the Natural Hiſtory of thoſe Bodies.

To proceed therefore to the Account of my Obſervations upon Sand-ſtone. And in *theſe*, though I do not neglect to note the ſeveral Kinds or Varieties of it: Free-ſtone, Ragg-ſtone, Lime-ſtone, and the reſt: the different Hardneſs, or Solidity, of each: as alſo its Colour, Texture, and the peculiar matter which conſtitutes it: yet I confine my ſelf more ſtrictly to conſider the manner how 'tis diſpoſed in the Earth: the *Strata*, into which, by means of *horizontal* || and parallel Fiſſures, it is divided: the Order and Number of theſe *Strata*: their ſituation in reſpect of the Horizon: the Thick-

|| *I call thoſe Fiſſures, which diſtinguiſh the Stone into Strata,* Horizontal *ones; and thoſe which interſect theſe* Perpendicular: *not ſo much with reſpect to the* preſent *ſite of the Strata, which (as I ſhall ſhew) is altered in many places, as to its* original *ſituation; concerning which, ſee* Part 2. Conſect. 5.

Thickneſs, Depth, and other Circumſtances of each: the Interruptions of the *Strata*, I mean the *perpendicular* || Fiſſures, which interſect the horizontal ones: the different Capacity or Largeneſs of theſe perpendicular Intervals: their Diſtances from each other: and the Spar, and other Mineral and Metallick Matter, uſually contained in them.

But, becauſe I ſaw that Deductions of conſiderable import and conſequence might be drawn from them, I have with great Care and Intention obſerved the Condition of ſuch heterogeneous Bodies, which I found immerſed and included in the maſs of this Sandſtone; particularly the Shells of Oyſters, Muſcles, Scallopes, Cockles, Periwinkles, and very many other marine Productions. I have, I ſay, very diligently noted all Circumſtances of theſe Shells: the vaſt Numbers of them; the ſeveral kinds that are thus lodged in the ſubſtance of the Stone: the Order and Manner of their poſition in it: the ſeveral Depths at which they are found: the Matter which they contain in them, and wherewith their Cavities are uſually filled.

Theſe

These Obſervations about Stone are ſucceeded by others, of like nature, concerning Marble, Cole, and Chalk: their Fiſſures: the Situation of their *Strata*: the Shells, and other heterogeneous Bodies lodged therein.

In the next place, thoſe which concern Marle, Clay, the ſeveral kinds of Earth, Sand, Gravel, and ſome other Foſſils: the Shells and other like Bodies, lodged in their *Strata*: the Poſition of thoſe *Strata*: their Order; their Diſtinctions from each other, by the difference of the Matter of each, and by its different Conſiſtence and Colour; the *Strata* of theſe laxer kinds of Matter being not ordinarily divided from each other by interpoſition of horizontal Fiſſures; as thoſe of Stone, and ſuch other ſolid Matter, conſtantly are.

2

Reprinted from *Royal Soc. London Philos. Trans.* **51**(2):582–587 (1760)

THE COURSE AND PHENOMENA OF EARTHQUAKES

J. Michell

[*Editors' Note:* In the original, material precedes and follows this excerpt.]

38. The earth then (as far as one can judge from the appearances), is not compofed of heaps of matter cafually thrown together, but of regular and uniform ftrata. Thefe ftrata, though they frequently do not exceed a few feet, or perhaps a few inches, in thicknefs, yet often extend in length and breadth for many miles, and this without varying their thicknefs confiderably. The fame ftratum alfo preferves a uniform character throughout, though the ftrata immediately next to each other are very often totally different. Thus, for inftance, we fhall have, perhaps, a ftratum of potters clay; above that, a ftratum of coal; then another ftratum of fome other kind of clay; next, a fharp grit fand ftone; then clay again; next, perhaps, fand ftone again; and coal again above that; and it frequently happens, that none of thefe exceed a few yards in thicknefs. There are, however, many inftances, in which the fame kind of matter is extended to the depth of fome hundreds

hundreds of yards; but in all theſe, a very few only excepted, the whole of each is not one continued maſs, but is again ſubdivided into a great number of thin laminæ, that ſeldom are more than one, two, or three feet thick, and frequently not ſo much.

39. Beſide the horizontal diviſion of the earth into ſtrata, theſe ſtrata are again divided and ſhattered by many perpendicular fiſſures, which are in ſome places few and narrow, but oftentimes many, and of conſiderable width. There are alſo many inſtances, where a particular ſtratum ſhall have almoſt no fiſſures at all, though the ſtrata both above and below it are conſiderably broken: this happens frequently in clay, probably on account of the ſoftneſs of it, which may have made it yield to the preſſure of the ſuperincumbent matter, and fill up thoſe fiſſures which it originally had; for we ſometimes meet with inſtances in mines, where the correſpondent fiſſures in an upper and lower ſtratum are interrupted in an intermediate ſtratum compoſed of clay, or ſome ſuch ſoft matter.

40. Though theſe fiſſures do ſometimes correſpond to one another in the upper and lower ſtrata, yet this is not generally the caſe, at leaſt not to any great diſtance: thoſe clefts, however, in which the larger veins of the ores of metals are found, are an exception to this obſervation; for they ſometimes paſs through many ſtrata, and thoſe of different kinds, to unknown depths.

41. From this conſtitution of the earth, *viz.* the want of correſpondence in the fiſſures of the upper and lower ſtrata, as well as on account of thoſe ſtrata which are little or not at all ſhattered, it will come to paſs,

pafs, that the earth cannot eafily be feparated in a direction * perpendicular to the horizon, if we take any confiderable portion of it together; but in the horizontal direction, as there is little or no adhefion between one ftratum and another, it may be feparated without difficulty.

42. Thofe fiffures which are at fome depth below the furface of the earth, are generally found full of water; but all thofe that are below the level of the fea, muft always be fo, either from the oozing of the fea, or rather of the land waters between the ftrata.

43. The ftrata of the earth are frequently very much bent, being raifed in fome places, and depreffed in others, and this fometimes with a very quick afcent or defcent; but as thefe afcents and defcents, in a great meafure, compenfate one another, if we take a large extent of country together, we may look upon the whole fet of ftrata, as lying nearly horizontally. What is very remarkable, however, in their fituation, is, that from moft, if not all, large tracts of high and mountainous countries, the ftrata lie in a fituation more inclined to the horizon, than the country itfelf, the † mountainous countries being generally,

* What I faid before of thofe deep clefts, in which metals are found, will not affect this conclufion; for they are confiderably different from either perpendicular or plane fections of earth; they are frequently interrupted by ftrata of clay, or other foft matter; and they are, in moft parts, either filled up with rubbifh, or with ores and fpars, that adhere as firmly to the rocks on both fides, as if they compofed one continued ftratum with them.

† It feems very probable, from many appearances, not only that the mountainous countries are formed out of the lower ftrata of the earth, but that fometimes the higheft hills in them are formed out

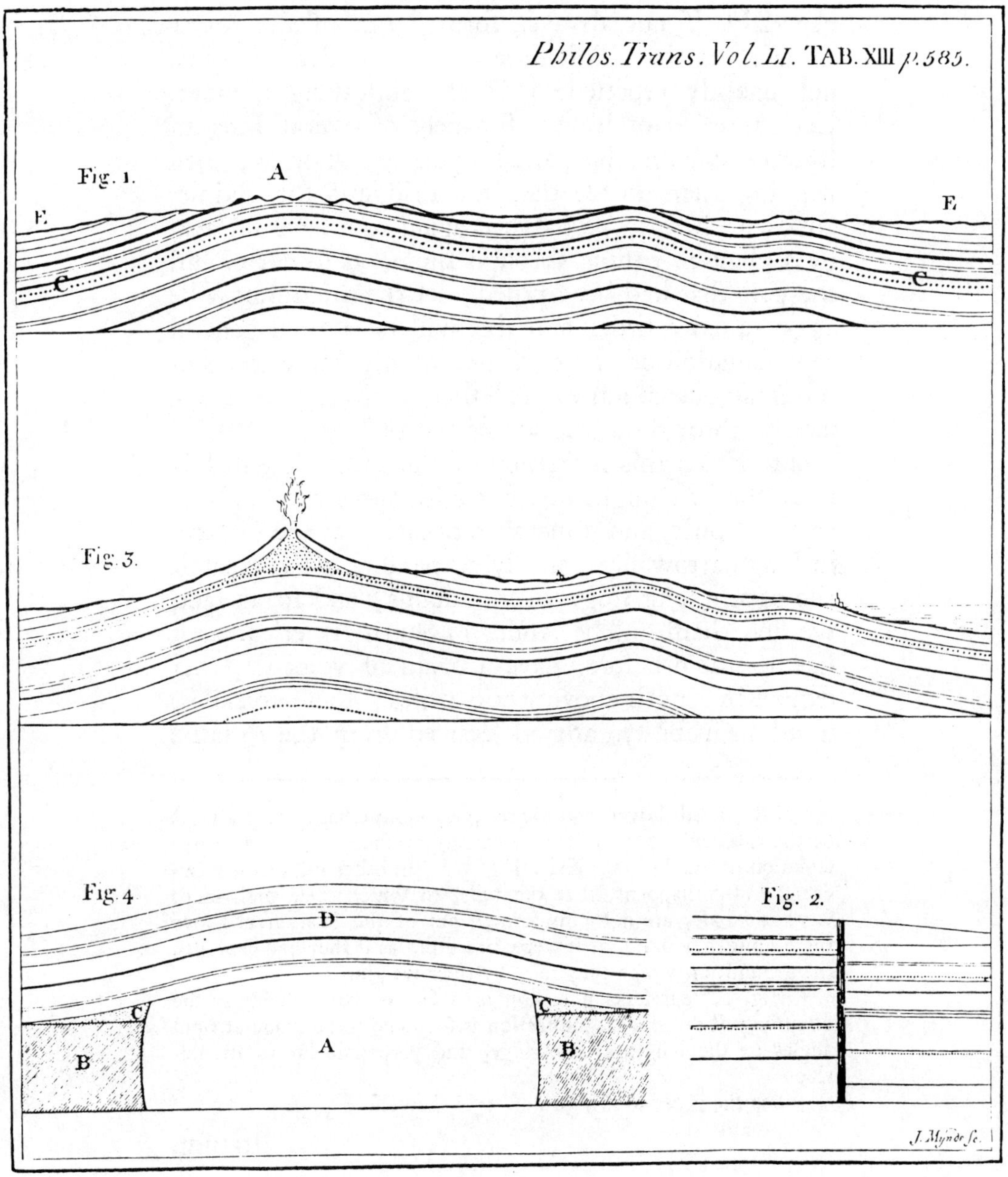
Philos. Trans. Vol. LI. TAB. XIII p. 585.
Fig. 1.
A
E
E
C
C
Fig. 3.
Fig. 4
D
C
C
A
B
B
Fig. 2.
J. Mynde sc.

generally, if not always, formed out of the lower ſtrata of earth. This ſituation of the ſtrata may be not unaptly repreſented in the following manner. Let a number of leaves of paper, of ſeveral different ſorts or colours, be paſted upon one another; then bending them up together into a ridge in the middle, conceive them to be reduced again to a level ſurface, by a plane ſo paſſing through them, as to cut off all the part that had been raiſed; let the middle now be again raiſed a little, and this will be a good general repreſentation of moſt, if not of all, large tracts of mountainous countries, together with the parts adjacent, throughout the whole world *.

44. From this formation of the earth, it will follow, that we ought to meet with the ſame kinds of earths, ſtones, and minerals, appearing at the ſurface, in long narrow ſlips, and lying parallel to the greateſt riſe of any long ridges of mountains; and ſo, in fact, we find them. The Andes in South America, as it has been ſaid before, have a chain of volcanos, that extend in length above 5000 miles: theſe volcanos, in all probability, are all derived from the † ſame ſtratum.

---

out of ſtrata ſtill lower than the reſt, which, perhaps, may always be the caſe, where they have volcanos in them. [See a repreſentation of this in Tab. XIII. Fig. 3.] In other inſtances, however, it often happens, that the hills, to which theſe high lands ſerve as a baſe, are not only formed out of the ſtrata next above them, but they ſtand, as it were, in a diſh, as if they had depreſſed the ground, on which they reſt, by their weight.

* Fig. 1. repreſents a ſection of a ſett of ſtrata, lying in the ſituation juſt deſcribed: the ſection is ſuppoſed to be made at right angles to the length of the ridge, and perpendicular to the horizon.

† See the notes to art. 36 and 53. See alſo Fig. 3.

ſtratum. Parallel to the Andes, is the Sierra, another long ridge of mountains, that run between the Andes and the ſea; and "theſe two ridges of mountains run "within ſight of one another, and almoſt equally, "for above a thouſand leagues together *," being each, at a medium, about twenty leagues wide. The gold and ſilver mines wrought by the Spaniards, are found in a tract of country parallel to the direction of theſe, and extending through a great part of the length of them.

45. The ſame thing is found to obtain in North America alſo. The great lakes, which give riſe to the river St. Laurence, are kept up by a long ridge of mountains, that run nearly parallel to the eaſtern coaſt. In deſcending from theſe towards the ſea, the ſame ſets † of ſtrata, and in the ſame order, are generally met with throughout the greateſt part of their length.

46. In Great Britain, we have another inſtance to the ſame purpoſe, where the direction of the ridge varies about a point from due north and ſouth, lying nearly from ‡ N. by E. to S. by W. There are many more inſtances of this to be met with in the world, if we may judge from circumſtances, which make it highly probable, that it obtains in a great number of places, and in ſeveral they ſeem to put it almoſt out of doubt.

47. The reader is not to ſuppoſe, however, that, in any inſtances, the higheſt riſe of the ridge, and

* See Acoſta's Natural Hiſtory of the Indies.

† See Lewis Evans's Map and Account of North America.

‡ Of this I could give many undoubted proofs, if it would not too far exceed the limits of my preſent deſign, and which, for that reaſon, I am obliged to omit.

the

the inclination of the ſtrata from thence to the countries on each ſide, is perfectly uniform; for they have frequently very conſiderable inequalities, and theſe inequalities are ſometimes ſo great, that the ſtrata are bent for ſome ſmall diſtance, even the contrary way from the general inclination of them. This often makes it difficult to trace the appearance I have been relating, which, without a general knowlege of the foſſil bodies of a large tract of country, it is hardly poſſible to do.

48. At conſiderable diſtances from large ridges of mountains, the ſtrata, for the moſt part, aſſume a ſituation nearly level; and as the mountainous countries are generally formed out of the lower ſtrata, ſo the more level countries are generally formed out of the upper ſtrata of the earth.

49. Hence it comes to paſs, that, in countries of this kind, the ſame ſtrata are found to extend themſelves a great way, as well in breadth as in length: we have an inſtance of this in the chalky and flinty countries of England and France, which (excepting the interruption of the Channel, and the clays, ſands, *&c.* of a few counties) compoſe a tract of about three hundred miles each way.

3

Reprinted from pages 178–181 of *An Inquiry into the Original State and Formation of the Earth; Deduced from Facts and the Laws of Nature,* 2nd ed., W. Bent, London, 1786.

Of the Strata in Derbyſhire and other parts of England.

J. Whitehurst

THE following obſervations on the *ſtrata* of the earth, are not only intended as an illuſtration of the preceding inquiries, into the original ſtate and formation of the earth, and the great changes it has undergone; but likewiſe to repreſent the advantages ariſing from a Syſtem of Subterraneous Geography—A ſcience not merely ſpeculative, but ſubſervient to the purpoſes of human life; by leading mankind to the diſcovery of many things of great utility which lied concealed in the lower regions of the earth.

The arrangement of the *ſtrata* in general is ſuch, that they invariably follow each other, as it were, in alphabetical order, or as a ſeries of numbers; whatever may be their different denominations.

I do not mean to inſinuate that the *ſtrata* are alike in all the different regions of the earth, either with reſpect to thickneſs or quality; for experience ſhews the contrary. But that the order of the *ſtrata* in each particular part, how much ſoever they may differ, as to quality,

quality, yet they follow each other in a regular ſucceſ-ſion, both as to thickneſs and quality: inſomuch that by knowing the incumbent *ſtratum*, together with the arrangement thereof in any particular part of the earth, we come to a perfect knowledge of all the inferior beds, ſo far as they have been previouſly diſcovered in the adjacent country; of which the following pages contain many inſtances.

With reſpect to the *ſtrata* accompanying coal, I have obſerved ſome inſtances which apparently contradict that general rule, though they do not in reality.

The mountainous part of Derbyſhire affords innumerable inſtances of the uniform arrangement of the *ſtrata*, and alſo of their various productions of animal, vegetable, or mineral ſubſtances, or rather of the figures or impreſſions of the two former. And theſe diſcoveries have aroſe from the numerous perforations through the *ſtrata* in ſearch of minerals, and alſo from the violent convulſions which they have ſuffered, anterior to hiſtory or tradition; whereby the *ſtrata*, which lay in the lower regions of the earth, were thrown up to the ſurface.

I ſay by means of ſuch operations of nature and art, the thickneſs, quality, and poſition of the *ſtrata*, have been truly aſcertained, as repreſented in the ſections given at the end of the work.

And

And in order to render the different denomination of the *ſtrata* obvious by inſpection, I have attempted to aſſign a particular character to each of them, by hatched lines, as colours are repreſented in coats of arms; conſidering that a faithful repreſentation of the *ſtrata,* to be a matter of much importance, not only to the maintainers of mines, but alſo to practical miners; preſuming they may be enabled thereby, to proſecute their mineral reſearches with more propriety and advantage to themſelves and to the public.

It is not only in mineral operations that faithful records of the *ſtrata* are requiſite, but alſo in attempting to diſcover either ſalt or freſh-water ſprings, coals, &c. therefore in order to render the ſubject of ſubterraneous geography of general uſe; whether pits are dug for water or minerals, ſuch records will be found beneficial: for the more general the obſervations, the more concluſive the inferences thence deduced. I am very doubtful whether the mode I have choſen is the beſt poſſible for that purpoſe, and therefore, ſhould gladly receive any hints tending towards the improvement of ſo valuable a branch of Science.

It was my intention to have depoſited ſpecimens of each *ſtratum,* with its productions, in the Britiſh Muſeum, arranged in the ſame order upon each other, as they are in the earth; being perſuaded that ſuch a plan would

would convey a more perfect idea of Subterraneous Geography, and of the various bodies contained in the earth, than words or lines can poffibly exprefs: and though I have not hitherto been able to complete this defign, yet I am ftill in hopes of doing it fome future day.

It is not my intention to enter into a minute defcription of the various ftones, minerals, petrefactions, &c. but rather to reprefent the general ftate and condition of the *ftrata*, and the changes they have undergone from various caufes. Neither do I affume to myfelf the fole honour of the following obfervations, having principally obtained them from feveral experienced miners; and particularly from Mr. George Tiffington, late of Winfter.

I have, indeed, very affiduoufly endeavoured to afcertain the truth of their reports by fubterraneous vifits, &c. and have alfo made fome difcoveries unnoticed by any other perfon before me: therefore, I am not confcious of any mifreprefentations, arifing either from neglect or a theoretical influence; neither am I infenfible that much more remains to be done, and that in works of this nature, errors are unavoidable.

[*Editors' Note:* In the original, material follows this excerpt.]

Part II

CUVIER AND BRONGNIART (1808) TO CONYBEARE AND PHILLIPS (1822)

Editors' Comments on Papers 4 Through 7

4A CUVIER and BRONGNIART
Excerpts from *Essay on the Mineralogic Geography in the Neighborhood of Paris*

4B CUVIER and BRONGNIART
Excerpts from *Essay on the Mineralogic Geography in the Neighborhood of Paris*

5 PARKINSON
Excerpt from *Observations on some of the Strata in the Neighborhood of London, and on the Fossil Remains contained in them*

6 CUVIER
Excerpts from *Essay on the Theory of the Earth*

7 GREENOUGH
Excerpts from *Essay I. On Stratification*

The first two most significant events for the development of stratigraphy in the period from 1807 through 1822 are the founding of the Geological Society of London in 1807 and the publication (in two journals) of Cuvier and Brongniart's *Essai* (Papers 4A and 4B). The aims of the Geological Society were to collect facts about the strata and fossils (rather than to theorize about earth's origin and history); to promote, integrate, and publish diverse researches; and to develop collections of specimens. The founders wished not only to communicate the results of their research but also to attempt to maintain opinions that conformed with facts presented by nature. Two decades later, in his presidential address, Murchison (1832) referred to the society as "a brotherhood united for purposes so great, and knit together by . . . lofty and enduring sympathies" (p. 386). Porter (1977, 129, 130) remarked, this "emphasis upon the sequence of strata and life, a long time-scale, naturalistic explanation, the progressive development of the Earth, were continuous with—not reactions against—interpretations which had been steadily crystallizing during the last third of the eighteenth century," which were rooted in turn in earlier eighteenth and seventeenth century works, some of which we have mentioned in Part I.

Cuvier and Brongniart's famous *Essai* (Paper 4A, Paper 4B) (said by Cuvier, Paper 6, 175, to be almost entirely the work of Brongniart), contains much evidence of stratigraphic sophistication. The work describes the rocks of the Paris Basin from the chalk up through the Tertiary. Their formations differ (Paper 4B, 435) "by chemical nature, by the kind of stratification, and above all by the fossils found there." Within a formation, they found (Paper 4A, 307, 308) a remarkable "constancy in the order of superposition of the thinnest beds," even if some beds present in one place were lacking or thinner elsewhere. Cuvier and Brongniart (Paper 4A, 308) further noted the gradual disappearance of fossils characteristic of one bed in overlying beds and their gradual replacement by new fossils. This certainly does not sound like Catastrophism. In addition, they noted (Paper 4A, 309) that shells in the lower beds are "more separated from species actually living than those of the upper beds," a fact later put to use by others, including Lyell (Paper 9B).

These authors employed the term *stage* (Paper 4B, 426, 427) for different deposits, one of which was divided into two distinct lateral equivalents—in effect, facies (though they did not use that term). Their stages could be subdivided mineralogically into several distinct kinds of beds.

The first description and interpretation of a disconformity were given by Cuvier and Brongniart (Paper 4B, 434), and a naturalistic cause was given for this break between the two formations, a break representing "perhaps a long space of time." However, these authors did have a problem with their concept of geologic time because they could not relate the evidence of relatively great change seen in these rocks (Paper 4A, 312) to the apparent lack of change during their lifetime, and thus, they concluded that "things happened . . . very different from those in our present seas."

The first volume of the *Transactions of the Geological Society of London,* published in 1811, contains Parkinson's *Observations* (Paper 5) in which he used (and acknowledged) Smith's method of identifying fossils with particular strata and quoted Cuvier and Brongniart (Paper 4A; 1811) and thus succeeded them and, to some extent, emulated them. Challinor (1971, 90) remarked that Parkinson's work presented the first detailed application in Britain of the principle that strata are characterized by their own peculiar fossils. Parkinson (pp. 325, 326) described the chalk and overlying beds and referred, as had Cuvier and Brongniart, to gradual vertical change in genera and species. He made several astute observations concerning fossils, including the possibility of reworked fossils, when he noted that two imperfect specimens never found before in a particular stratum might have been "washed out of some of the more ancient strata, and lodged by accident in the bed where they were found" (p. 339). He further cautioned against another source of error "whilst appropriating fossils to their respective strata . . . where, at the line of junction of two strata, the animals of the one are found within the borders of the other stratum; a circumstance by no means difficult to be conceived or explained" (p. 339).

Parkinson (Paper 5, 326) predicted that "geology may derive considerable assistance, from an examination of fossils, made in connexion with that of the strata to which they belong," and (p. 334) advised comparison of fossil and recent shells by an experienced conchologist, a task undertaken by Brocchi (1814), Deshayes (1830–1831), and Lyell (Paper 9B).

We find it interesting that Parkinson (Paper 5) of the Geological Society of London rather than Cuvier, the supposed catastrophist, alluded to the "abruption" (erosion) of strata (p. 354) "by that catastrophe, instances of the astonishing force of which have been already noticed," for example, animal bones in gravel (p. 341), and "The whole of this island displays evident marks of its stratification having . . . suffered considerable disturbance, from some prodigious and mysterious power" (p. 326).

Another contemporary opinion as to the importance of stratigraphy was given by Richardson (1811), who wrote the following in *On the Strata of Mountains:* "Enumeration of the strata of which any country is composed, with an account of the changes they have undergone, and the operations that have been performed upon them, forms the true geological history of that country, and an accumulation of many of these detached histories affords the best materials for a general history of the physical world" (p. 368).

We have referred to three of the basic stratigraphic works of Cuvier and Brongniart (Papers 4A, 4B; 1811). Cuvier followed these in 1812 with his *Essay on the Theory of the Earth* (Paper 6), translated into English by Kerr in 1813. In this, we find Cuvier using the terms *catastrophe* and *revolution* but on the whole sounding not at all like the arch-catastrophist, a role in which he has often been cast (e.g., Berry, 1968, 43, 44). In fact, Cuvier was more a uniformitarian where he referred (p. 13) to the gradual disappearance of the more ancient species and, with respect to sea level changes (p. 14), wrote that different parts of continents have risen by degrees and that lands laid dry can sink or the sea can rise. He stated that the changes in strata were not due to "a gradual and general retreat of the waters [an allusion to the Neptunist view], but to successive irruptions [transgressions] and retreats" (p. 15). Cuvier (pp. 50–56) advocated a uniformitarian approach to the study of fossils and, along with the members of the Geological Society of London, called for an investigation of the general laws affecting the position of fossils within their strata (p. 51). Moreover, he wrote (pp. 58–60) not of universal catastrophes, with subsequent recreation of life, but of local annihilations due to natural causes, several of which he suggested, and of repopulation by migration and subsequent expansion of faunas that had continued to live elsewhere. Again, Cuvier stated his belief in the value of "a consecutive history of such singular deposits" (p. 180) over conjecture. As did Parkinson, Cuvier noted (pp. 177, 178) the similarity of Tertiary shells to those in recent strata and the sea and urged their comparison.

Concurrently, in Germany, Schlotheim (1813, 5, 6) wrote of the stratigraphic significance of fossils as evidence of the time of origin of the strata, relatively and perhaps in years; as evidence of whether revolutions were general or local; and as evidence of the development of life. Schlotheim (p. 9) strongly stressed the need for accurate determination of species that appeared to belong exclusively to certain formations.

In Italy, Brocchi's *Conchiologia Fossile Subapennina* (1814) exemplified Schlotheim's desideratum with its specific descriptions and beautiful illustrations of Italian Tertiary mollusks. Brocchi noted the great similarity of Tertiary and extant species of the Mediterranean and Adriatic, as well as the differences between Italian fossils and those described from the Paris Basin. Brocchi's work was, of course, known to Lyell, and it served as a foundation for Lyell's later (Paper 9) subdivision of the Tertiary "epoch" into "periods", stratigraphic terminology not standardized at that time. Like Cuvier (Paper 6), Brocchi explained the differences between Secondary and Tertiary faunas as due to gradual extinction but went further by laying down principles of evolution along continuous lines of descent according to natural laws of growth and decay.

Strata had by this time been studied more and less diligently for over one hundred years, but this branch of geology had not been given its name. This was largely because, except for mineralogy, geology was stratigraphy. Although Smith used the adjectival form in his *Stratigraphical System of Organized Fossils* in 1817, the earliest use of the word, as *stratigraphie,* of which we know is that by d'Orbigny in 1852 (Paper 19, 257).

Even though stratigraphy had developed as a science in its own right by this time, Greenough (Paper 7) noted that, in the whole range of geological investigation, no subject was more pregnant with controversy "than whether a given mass be, or be not stratified" (p. 1). This was followed (pp. 88–90) by Greenough's series of unanswered questions regarding the subject and a call for definition.

In France, d'Halloy (1822, 356; written in 1813) believed, with regard to the construction of his geologic map of France, that he should group the rocks according to their epoch of formation (determined by fossils) inasmuch as there could be lateral change in the mineral composition of formations (systems), or change in "aspect." He further noted the tendency of one system (formation) to develop "at the expense of another." D'Halloy believed this approach to mapping to be "much more advantageous for the progress of science" (p. 356) inasmuch as the fossils are less changeable than the sediments—again, the recognition of facies changes as well as reiteration (pp. 363, 364) of correlating strata in distant countries by means of their fossils. We believe that the modern mapping procedure with emphasis on lithofacies stratigraphic units suffers by comparison with d'Halloy's use of biostratigraphic units for mappable units; the preoccupation with facies in modern mapping discourages the search for the natural breaks that separate genetic packages of strata.

Besides the term *stratum,* another term used in a variety of ways at this time was *formation.* Conybeare and Phillips undertook to define it in 1822 in their *Outlines of the Geology of England and Wales,* which work was said by Woodward (1907, 82) to have done "more than any other independent work at that date to advance the study of geology" in England. Conybeare and Phillips wrote:

> When, however, the observer commences his attempt to trace more in the detail the succession of these mineral beds and masses, he will at first find himself perplexed by their almost infinite numbers; but he will soon discover that these individual strata are arranged together, in such a manner as to afford natural and easy grounds for classing them in a limited number of series, each series comprehending numerous individual strata naturally allied and associated together . . . . each series being characterised by the repetition of its own peculiar members. Such series are called FORMATIONS: . . . and this division must appear perfectly natural, inasmuch as the beds composing each formation, being identical in character, must have resulted from the same order of causes. [1822, iv, v]

With this definition, a formation is seen as a genetic unit and, as such, would be set off, we believe, from other formations by discontinuities of the magnitude of paracontinuities (Conkin and Conkin, 1973).

REFERENCES

Berry, W. B. N., 1968, *Growth of a Prehistoric Time Scale Based on Organic Evolution,* W. H. Freeman and Co., San Francisco.

Brocchi, G., 1814, Conchiologia Fossile Subapennina, Stamperia Reale, Milan.

Challinor, J., 1971, *The History of British Geology, a Bibliographic Study,* David and Charles, Newton-Abbot.

Conkin, J. E., and B. M. Conkin, 1973, The Paracontinuity and the Determination of the Devonian-Mississippian Boundary in the Type Lower Mississippian Area of North America, *Univ. Louisville Stud. Paleont. Stratigr.,* no. 1, 36p.

Conybeare, W. D., and W. Phillips, 1822, Outlines of the Geology of England and Wales, part i, Phillips, London.

Cuvier, G., and A. Brongniart, 1811, *Essai sur la géographie minéralogique des environs de Paris,* Baudouin, Paris.

Deshayes, G. P., 1830–1831, Tableau comparatif des espèces de coquilles vivantes, etc., *Geol. Soc. France Bull.* **1**(8):185–187.

d'Halloy, J. B. d'O., 1822, Observations sur un essai de carte géologique de la France, des Pays-Bas et des contrées voisines, *Ann. Mines* **7:**353–376.

Murchison, R. I., 1832, Presidential Address, *Geol. Soc. London Proc.,* pp. 362–386.

Porter, R., 1977, *The Making of Geology—Earth Science in Britain 1660–1815,* Cambridge University Press.

Richardson, W., 1811, On the Strata of Mountains, *Philos. Mag.* **37:**367–369.

Schlotheim, E. F., 1813, Taschenbuch für die gesammte Mineralogie (Leonhard), vol. III, pp. 2–134.

Smith, W., 1817, *Stratigraphical System of Organized Fossils With Reference to the Specimens of the Original Collection in the British Museum Explaining Their State of Preservation and Their Use in Identifying the British Strata,* Williams, London.

Woodward, H. B., 1907, *The History of the Geological Society of London,* London.

4A

ESSAY ON THE MINERALOGIC GEOGRAPHY IN THE NEIGHBORHOOD OF PARIS

G. Cuvier and A. Brongniart

These excerpts were translated expressly for this Benchmark volume by Barbara Conkin, University of Louisville, from Mus. Natl. Hist. Nat. Ann. ***11**:307–310, 312 (1808).*

The coarse limestone does not always lie immediately above the clay; it is often separated from it by a more or less thick bed of sand. We cannot say if this sand belongs to the limestone formation or to that of the clay. We have not found shells therein in the few places where we have observed it, which would connect it to the clay formation; but the lowest limestone bed being ordinarily sandy and always filled with shells, we do not know yet if the sand is different from the first or if it is the same deposit. What would make us suspect that it is different is that the sand of the clays that we have seen is generally rather pure, whether colored red or bluish gray. It is refractory and often very coarse grained.

The limestone formation aside from this sand is composed of alternating beds of coarse, more or less hard limestone, or argillaceous marl and even of clay foliated in very thin beds, and of calcareous marl. However, it is not necessary to believe that these diverse beds were placed at random and without rules; they always follow in the same order of superposition in the considerable extent of terrain we have examined. There are in it sometimes several that are lacking or that are very thin; but a bed that is lower in one district never becomes upper in another.

This constancy in the order of superposition of the thinnest beds, and over an extent of 12 myriameters (120,000 meters) at least, is, it seems to us, one of the most remarkable facts that we have established during our research. It must result for arts and for geology in consequences all the more interesting as they are more certain.

The means we have employed to recognize, in the midst of so great a number of limestone beds, a bed previously observed in a distant district is taken from the nature of the fossils enclosed in each bed. These fossils are always generally the same in corresponding beds and present notable differences of species from one system of beds to another. This sign of recognition, up to the present, has not fooled us.

It is not necessary to believe nevertheless that the difference of one bed from another is as sharp as that of the chalk and the limestone. If it was thus, one would have that many particular formations, but the fossils characteristic of one bed become less numerous in the upper bed and disappear altogether in others, or are replaced little by little by new fossils that had not previously appeared.

We shall indicate, following this, the principal systems of beds that one can observe in the coarse limestone. One will find the complete description detailed in the Memoire, bed by bed, from numerous quarries that we have observed in order to get the results we present today.

The lowest beds of the limestone formation are the more characterized: they are very sandy and often even more sandy that calcareous. . . .

The shelly limestone composing it, and even the sand that sometimes replaces it, nearly always encloses some chlorite as powder or as grains. . . . It owes its color to iron; it is found only in the lower layers; one sees it neither in the chalk or in the clay nor in the middle or upper limestone beds, and one can regard its presence as a sure index to the nearness of the plastic clay and by consequence of the chalk. But what characterizes this system of beds even more particularly is the prodigious quantity of fossil shells that fill it. In order to give an idea of the number of species these beds contain, it will suffice to say that M. Defrance has found there more than 600 species, which have all been described by M. de Lamarck. We will point out that most of these shells are much more separated from species actually living than those of the upper beds. [*Editors' Note:* Material has been omitted at this point.]

Another particular characteristic of the shells of this bed is that they are for the most part whole and well preserved, that they detach easily from their rock, and finally that many among them have their nacreous state preserved. It is in all the preceding places and in others less remarkable that we have recognized that the sandy calcareous beds that enclose these shells immediately follow the plastic clay covering the chalk; and it is from these multiplied observations that we have verified the generality of the rule we have just established. [*Editors' Note:* Material has been omitted at this point.]

The results of the observations we have just reported are (1) that the fossils of the coarse limestone have been deposited slowly and in a tranquil sea, since these fossils are deposited there in regular and distinct beds; that they are not indistinctly mixed, and that most there are in a state of perfect preservation, however delicate these shells; that even the points of spiny shells are very often entire; (2) that these fossils are entirely different from those of the chalk; (3) that in proportion as the beds of this formation were deposited, the number of species of shells always diminishes until the moment where no more are found in it. The waters that formed these beds either no longer enclosed them or have lost the property of preserving them.

Certainly things happened in these seas very differently than they happen in our present seas in these respects: layers no longer are formed; the species of shells are always the same there, in the same localities. One does not see, for example, since the time when one fishes for oysters on the coast of Cancale, that these shells have disappeared in order to be replaced by other species.

4B

ESSAY ON THE MINEROLOGIC GEOGRAPHY IN THE NEIGHBORHOOD OF PARIS

G. Cuvier and A. Brongniart

These excerpts were translated expressly for this Benchmark volume by Barbara Conkin, University of Louisville, from Jour. Mines ***23:****426–427, 434–435 (1808).*

Thus, one can portray the materials that compose the Paris basin, in the sense in which we limit it, as having been deposited in a vast hollow space, in a kind of vast gulf, the shores of which were of chalk.

Perhaps this gulf made an entire circle, a kind of great lake; but we cannot know this inasmuch as the southwest borders of the shore have been covered, as well as the materials they contain, by a great sandy layer of which we have spoken before.

Moreover, this great sandy layer is not the only thing that has covered the chalk. [*Editors' Note:* Material has been omitted at this point.]

First we shall describe the *chalk,* the more ancient of the materials we have in our region. We shall end with the sandy bed, the newest of our geologic products. Between these two extremes, we shall discuss some less extensive, but more varied materials, that filled this great cavity of chalk before the layer of sand was deposited over all.

These materials can be divided into two stages. The first, which covers the chalk everywhere where it was not sufficiently elevated, and which has filled all the bottom of the gulf, subdivides itself into two parts equal in level and placed not one on the other but end to end; to wit: the layers of siliceous limestone without shells and the layer of coarse shelly limestone.

We know the limits of this stage of the chalk coast rather well because it was not covered at all; but these same limits are masked in several places by the second stage and by the great sandy layer that forms the third, which covers a great part of the other two. [*Editors' Note:* Material has been omitted at this point.]

Such are the great masses of which our district is composed and that form the different stages. In subdividing each stage, however, one can attain yet more precision, and one obtains more rigorous mineralogic determinations, which give up to ten distinct kinds of beds, of which we are first going to present a rapid enumeration.

[*Editors' Note:* Material has been omitted at this point.]

If we compare the descriptions we have just given of the beds of chalk and of plastic clay, we shall notice (1) not only that one finds no fossils in the clay that one encounters in the chalk but also that one does not even find any fossils; and (2) that there is not an insensible passage between the chalk and clay, since the parts of the clay bed nearest the chalk do not enclose more chalk than the other parts.

It seems to us that one can conclude from these observations, first, that the liquid that deposited the plastic clay bed was very different than the liquid that deposited the chalk, since it obviously does not contain carbonated lime and since no animals were living there that lived in the waters depositing the chalk.

Second, there has necessarily been a break and perhaps even a long space of time between the deposition of the chalk and that of the clay, since there is no transition between these two kinds of terrain. The kind of breccia with fragments of chalk and groundmass of clay that we have noticed at Meudon even seems to prove that the chalk was already solid when the clay was deposited. This earth is insinuated between the fragments of chalk produced on the surface of the chalky terrain by the movement of water or by another cause.

The two sorts of terrain we have just described have thus been produced in very different and even very separate circumstances. They are the result of the most distinct and the most characteristic formations that one can find in geognosy, since they differ by chemical nature, by the kind of stratification, and above all, by the fossils found therein.

5

Reprinted from *Geol. Soc. London Trans.* **1**:324-326 (1811)

Observations on some of the Strata in the Neighbourhood of London, and on the Fossil Remains contained in them.

By James Parkinson, Esq.

Member of the Geological Society.

THE study of fossil organized remains has hitherto been directed too exclusively to the consideration of the specimens themselves; and hence has been considered rather as an appendix to botany and zoology, than as (what it really is) a very important branch of geological inquiry.

From a comparison of fossil remains with those living or extant beings to which they bear the closest analogy, great resemblances and striking differences are at the same time perceivable. In some instances the generic characters materially differ, but in most they very closely correspond; whilst the specific characters are very rarely found to agree, except when the fossil appears to have existed at, comparatively, a late period. Of man, who constitutes a genus by himself, not a single decided remain has been found in a fossil state.

Chemical analysis has been called in to the aid of the naturalist, in order to account for the perfect state of preservation observable in remains organized with the most exquisite delicacy, and which there is every reason for supposing to have been readily decomposable in their recent state. From this investigation we learn the manner in which these memorials of the old world, so interesting and so frail have been preserved. Some have been impregnated with calcareous matter, others with siliceous, and others with iron or copper pyrites.

But these facts, however important and interesting, cannot, when considered by themselves, add much to our knowledge respecting the formation and structure of the earth. To derive any information of consequence from them, on these subjects, it is necessary that their examination should be connected with that of the several strata, in which they are found.*

Already have these examinations, thus carried on, taught us the following highly instructive facts. That exactly similar fossils are found in distant parts of the same stratum, not only where it traverses this island, but where it appears again on the opposite coast: that, in strata of considerable comparative depth, fossils are found, which are not discovered in any of the superincumbent beds: that some fossils, which abound in the lower are found in diminishing numbers through several of the superincumbent, and are entirely wanting in

* This mode of conducting our inquiries was long since recommended by Mr. W. Smith, who first noticed that *certain fossils are peculiar to, and are only found lodged in, particular strata;* and who first ascertained the constancy in *the order of superposition*, and *the continuity of the strata of this island.* It will appear from the following quotation, that these observations have lately also occurred to Messrs. Cuvier and Brongniart whilst examining into the nature of the strata of the neighbourhood of Paris. "Cette constance dans l'ordre de superposition des couches les plus minces, et "sur une étendue de 12 myriamètres au moins, est, selon nous, un des faits les plus "remarquables que nous ayons constatés dans la suite de nos recherches. Il doit en "résulter pour les arts et pour la géologie des conséquences d'autant plus intéres- "santes, qu'elles sont plus sûres.

"Le moyen que nous avons employé pour reconnoitre au milieu d'un si grand nombre "de lits calcaires, un lit dejà observé dans un canton très-éloign , est pris de la nature "des fossiles renfermés dans chaque couche, ces fossiles sont toujours généralement "les mêmes dans les couches correspondantes, et présentent des differences d'espèces "assez notables d'un système des couches à un autre système. C'est un signe de recon- "noissance qui jusqu'à présent ne nous a pas trompés." Annales du Muséum d'histoire naturelle, tome XI. p. 307.

the uppermost strata: that some fossils, occurring in considerable numbers in one stratum, become very rare in the adjacent portion of the next superincumbent stratum, and afterwards are lost: that fossils of one particular genus, which exist abundantly in the lower strata, and occur in several of the superincumbent ones, are not found in the three highest strata; whilst one species of that genus, but which has not been found in a fossil state, exists in our present seas: and lastly, that most of the remains which are abundant in the superior strata, are not at all found in the lower. These general facts lead us to hope, that geology may derive considerable assistance, from an examination of fossils, made in connexion with that of the strata to which they belong.

The following is an attempt to investigate on this plan some of the upper strata in the vicinity of the metropolis with their contained fossils; and, although by no means complete, it will, it is hoped, induce others, who possess superior abilities and opportunities, not only to re-examine more correctly these strata, but to extend their researches to the subjacent strata.

The whole of this island displays evident marks of its stratification having, since its completion, suffered considerable disturbance, from some prodigious and mysterious power. By this power all the known strata, to the greatest depths that have been explored, have been more or less broken and displaced; and in some parts have been so lifted, that some of the lowest of these have been raised to the surface; whilst portions of others, to a very considerable depth and extent, have been entirely carried away.*

[*Editors' Note:* In the original, material follows this excerpt.]

* See several essays on this subject in the Philosophical Magazine, by Mr. Farey, and the Report on Derbyshire. vol. I. p. 105.

Also a Letter on the alterations which have taken place in the structure of rocks, on the surface of the basaltic country in the counties of Derry and Antrim, by William Richardson, D.D. Phil. Trans. 1808.

6

ESSAY

ON THE

Theory of the Earth.

TRANSLATED FROM THE FRENCH OF

M. CUVIER,

PERPETUAL SECRETARY OF THE FRENCH INSTITUTE, PROFESSOR AND ADMINISTRATOR OF THE MUSEUM OF NATURAL HISTORY, &c. &c.

BY

ROBERT KERR, F.R.S. & F.A.S. EDIN.

WITH

MINERALOGICAL NOTES,

AND

AN ACCOUNT OF CUVIER'S GEOLOGICAL DISCOVERIES,

BY PROFESSOR JAMESON

EDINBURGH:
PRINTED FOR WILLIAM BLACKWOOD, SOUTH BRIDGE-STREET, EDINBURGH; AND JOHN MURRAY, ALBEMARLE-STREET, AND ROBERT BALDWIN, PATERNOSTER-ROW, LONDON.

1813.

Reprinted from pages 12-15, 49-60, and 180-181 of *Essay on the Theory of the Earth,* Robert Kerr, trans., William Blackwood, John Murray, Robert Baldwin, Edinburgh, 1813

[*Editors' Note:* In the original, material precedes this excerpt.]

§. 5. *Proofs that such Revolutions have been numerous.*

If we institute a more detailed comparison between the various strata and those remains of animals which they contain, we shall soon discover still more numerous differences among them, indicating a proportional number of changes in their condition. The sea has not always deposited stony substances of the same kind. It has observed a regular succession as to the nature of its deposits; the more ancient the strata are, so much the more uniform and extensive are they; and the more recent they are, the more limited are they, and the more variation is observed in them at small distances. Thus the great catastrophes which have produced revolutions in the basin of the sea, were preceded, accompanied, and followed by changes in the nature of the fluid and of the substances which it held in solution; and when the surface of the seas came to be divided by islands and projecting ridges, different changes took place in every separate basin.

Amidst these changes of the general fluid, it

must have been almost impossible for the same kind of animals to continue to live:—nor did they do so in fact. Their species, and even their genera, change with the strata; and although the same species occasionally recur at small distances, it is generally the case that the shells of the ancient strata have forms peculiar to themselves; that they gradually disappear, till they are not to be seen at all in the recent strata, still less in the existing seas, in which, indeed, we never discover their corresponding species, and where several even of their genera are not be found; that, on the contrary, the shells of the recent strata resemble, as it respects the genus, those which still exist in the sea; and that in the last-formed and loosest of these strata there are some species which the eye of the most expert naturalist cannot distinguish from those which at present inhabit the ocean.

In animal nature, therefore, there has been a succession of changes corresponding to those which have taken place in the chemical nature of the fluid; and when the sea last receded from our continent, its inhabitants were not very different from those which it still continues to support.

Finally, if we examine with greater care these remains of organized bodies, we shall discover, in the midst even of the most ancient secondary strata, other strata that are crowded with animal or vegetable productions, which belong to the land and to fresh water; and amongst the more recent strata, that is, the strata which are nearest the surface, there are some of them in which land animals are buried under heaps of marine productions. Thus the various catastrophes of our planet have not only caused the different parts of our continents to rise by degrees from the basin of the sea, but it has also frequently happened, that lands which had been laid dry have been again covered by the water, in consequence either of these lands sinking down below the level of the sea, or of the sea being raised above the level of the lands. The particular portions of the earth also which the sea has abandoned by its last retreat, had been laid dry once before, and had at that time produced quadrupeds, birds, plants, and all kinds of terrestrial productions; it had then been inundated by the sea, which has since retired from it, and left it to be occupied by its own proper inhabitants.

The changes which have taken place in the

productions of the shelly strata, have not, therefore, been entirely owing to a gradual and general retreat of the waters, but to successive irruptions and retreats, the final result of which, however, has been an universal depression of the level of the sea.

[*Editors' Note:* Material has been omitted at this point.]

§ 21. *Statement of the Nature and Conditions of the Problem to be solved.*

To quit the language of mathematics, it may be asserted, that almost all the authors of these systems, confining their attention to certain difficulties by which they were struck more forcibly than by others, have endeavoured to solve these in a way more or less probable, and have allowed others to remain unnoticed, equally numerous and equally important. For example, the only difficulty with one consisted in explaining the change which had taken place on the level of the seas; with another it consisted in accounting for the solution of all terrestrial substances in the same fluid; and with a third, it consisted in shewing how animals that were natives of the torrid, could live under the frigid zone. Exhausting the whole of their ingenuity on these questions, they conceived that they had done every thing that was necessary, when they had contrived some method of answering them; and yet, while they neglected all the other phenomena, they did not always think of determining with precision the measure and extent of those which they attempted to explain. This is peculiarly the case in regard to the secondary stratifications, which constitute, however, the most difficult and most important portion of the problem. It has hardly ever been attempted carefully to ascertain the superpositions of their strata, or the connections of these strata with the species of animals and of plants whose remains they inclose.

Are there certain animals and plants peculiar to certain strata, and not found in others? What are the species that appear first in order, and those which succeed? Do these two kinds of species ever accompany one another? Are there alternations in their appearances; or, in other words, does the first species appear a second time, and does the second species then disappear? Have these animals and plants lived in the places where their exuviæ are found, or have they been brought there from other places? Do all these animals and plants still continue to live in some part of the earth, or have they been totally or partially destroyed? Is there any constant connection between the antiquity of the strata, and the resemblance or non-resemblance of the extraneous fossils, to the animals and plants that still exist? Is there any connection in regard to climate, between the extraneous fossils and the still living organized bodies which most nearly resemble them? May it be concluded, that the transportation of these living organized bodies, if such a thing ever happened, has taken place from north to south, or from east to west; or was it effected by means that irregularly scattered and mingled them together? And, finally, is it still possible to distinguish the epochs of these transportations, by attentively examining the strata which inclose the remains, or are imprinted by their forms?

If, from the want of sufficient evidence, these questions cannot be satisfactorily answered, how shall we be able to explain the causes of the presently existing state of our globe? It is certain, that so far from any of these points being as yet completely established, naturalists seem to have scarcely any idea of the propriety of investigating facts before they construct their systems. The cause of this strange procedure may be discovered, by considering that all geologists hitherto have either been mere cabinet naturalists, who had themselves hardly paid any attention to the structure of mountains, or mere mineralogists, who had not studied in sufficient detail the innumerable diversity of animals, and the almost infinite complication of their various parts and organs. The former of these have only constructed systems; while the latter have made excellent collections of observations, and have laid the foundations of true geological science, but have been unable to raise and complete the edifice.

§ 22. *Of the Progress of Mineral Geology.*

The purely mineralogical portion of the great problem of the Theory of the Earth has been investigated with admirable care by Saussure, and has been since explained in an astonishing degree by Werner, and by the numerous enlightened pupils of his school.

The former of these celebrated philosophers, by a laborious investigation of the most inaccessible mountain districts during twenty years of continual research, in which he examined the Alps on all sides, and penetrated through all their defiles, has laid open to our view the entire disorder of the primitive formations, and has clearly traced the boundaries by which they are distinguishable from the secondary formations. The other equally celebrated geologist, taking advantage of the numerous excavations in the most ancient mining district in the world, has fixed the laws which regulate the succession of strata, pointing out their respective antiquity in regard to each other, and tracing each of them through all its changes and metamorphoses. From him alone we date the commencement of real geology, so far as respects the mineral natures of the strata: But neither he nor Saussure has defined the species of organized extraneous fossils in each description of the strata with that accuracy which has become necessary, now that the number of animals already known has become so great.

Other naturalists, it is true, have studied the fossil remains of organized bodies; they have collected and represented them by thousands, and their works certainly will serve as a valuable storehouse of materials. But, considering these fossil plants and animals merely in themselves, instead of viewing them in their connection with the theory of the earth; or regarding their petrifactions and extraneous fossils as mere curiosities, rather than as historical documents; or confining themselves to partial explanations of the particular bearings of each individual specimen; they have almost always neglected to investigate the general laws affecting their position, or the relation of the extraneous fossils with the strata in which they are found.

§ 23. *Of the Importance of Extraneous Fossils, or Petrifactions, in Geology.*

The importance of investigating the relations of extraneous fossils with the strata in which they are contained is quite obvious. It is to them alone that we owe the commencement even of a Theory of the Earth; as, but for them, we could never have even suspected that there had existed any successive epochs in the formation of our earth, and a series of different and consecutive operations in reducing it to its present state. By them alone we are enabled to ascertain, with the utmost certainty, that our earth has not always been covered over by the same external crust because we are thoroughly assured that the organized bodies to which these fossil remains belong must have lived upon the surface, before they came to be buried, as they now are, at a great depth. It is only by means of analogy, that we have been enabled to extend to the primitive formations, the same conclusions which are furnished directly for the secondary formations by the extraneous fossils; and if there had only existed formations or strata in which there were no extraneous fossils, it could never have been asserted that these several formations had not been simultaneous.

It is also owing to these extraneous fossils, slight as is the knowledge we have hitherto acquired respecting them, that we have been enabled to discover the little that we yet know concerning the revolutions of our globe. From them we have learned, that the strata, or at least those which contain their remains, have been quietly deposited in a fluid; that the variations of the several strata must have corresponded with the variations in the nature of the fluid; that they have been left bare by the transportation of this fluid to some other place; and that this fact must have happened more than once. Nothing of all this could have been known with certainty, without the aid of extraneous fossils.

The study of the mineralogical part of geology, though not less necessary, and even a great deal more useful to the practical arts, is yet much less instructive so far as respects the objects of our present enquiry. We remain in utter ignorance respecting the causes which have given rise to the variety in the mineral substances of which the strata are composed. We are ignorant even of the agents which may have held some of these substances in a state of solution; and it is still disputed respecting several of them, whether they have owed their origin to the agency of water or of fire. After all, philosophers are only agreed on one point, which is, that the sea has changed its place; and this could never have been certainly known, but for the existence of extraneous fossils. These fossils, then, which have given rise to the theory of the earth, have at the same time furnished its principal illustrations—the only ones, indeed, that have as yet been generally received and acknowledged.

This is the consideration by which I have been encouraged to investigate the subject of extraneous fossils. But the field is extensive; and it is only a very inconsiderable portion of it that can be cultivated by the labour of a single individual. It was necessary, therefore, to select a particular department, and I very soon made my choice. That class of extraneous fossils, which forms the peculiar subject of this Essay, engaged my attention at the very outset, because it is evidently the most fertile in affording precise results, yet at the same time less known than others, and richer in new objects of research. *

§ 24. *High Importance of investigating the Fossil Remains of Quadrupeds.*

It is obvious that the fossil remains of the bones of quadrupeds must lead to more rigorous conclusions than any other remains of organized bodies, and that for several reasons.

In the first place, they indicate much more clearly the nature of the revolutions to which

* My work on this subject will clearly show how far this enquiry is yet new, notwithstanding the excellent labours of Camper, Pallas, Blumenbach, Merk, Sœmmerring, Rosenmuller, Fischer, Faujus, and other learned men, whose works I have most scrupulously cited in such of my chapters as their researches are connected with.

they have been subjected. The remains of shells certainly indicate that the sea has once existed in the places where these collections have been formed: But the changes which have taken place in their species, when rigorously enquired into, may possibly have been occasioned by slight changes in the nature of the fluid in which they were formed, or only in its temperature, and may even have arisen from other accidental causes. We can never be perfectly assured that certain species, and even genera, inhabiting the bottom of the sea, and occupying certain fixed spaces for a longer or shorter time, may not have been driven away from these by other species or genera.

In regard to quadrupeds, on the contrary, every thing is precise. The appearance of their bones in strata, and still more of their entire carcases, clearly establishes that the bed in which they are found must have been previously laid dry, or at least that dry land must have existed in its immediate neighbourhood. Their disappearance as certainly announces that this stratum must have been inundated, or that the dry land had ceased to exist in that state. It is from them, therefore, that we learn with perfect certainty the important fact of the repeated irruptions of the sea upon the land, which the extraneous fossils and other productions of marine origin could not of themselves have proved; and, by a careful investigation of them, we may hope to ascertain the number and the epochs of those irruptions of the sea.

Secondly, the nature of the revolutions which have changed the surface of our earth, must have exerted a more powerful action upon terrestrial quadrupeds than upon marine animals. As these revolutions have consisted chiefly in changes of the bed of the sea, and as the waters must have destroyed all the quadrupeds which they reached, if their irruption over the land was general, they must have destroyed the entire class, or, if confined only to certain continents at one time, they must have destroyed at least all the species inhabiting these continents, without having the same effect upon the marine animals. On the other hand, millions of aquatic animals may have been left quite dry, or buried in newly-formed strata, or thrown violently on the coasts, while their races may have been still preserved in more peaceful parts of the sea, whence they might again propagate and spread after the agitation of the water had ceased.

Thirdly, this more complete action is also more easily ascertained and demonstrated; because, as the number of terrestrial quadrupeds is limited, and as most of their species, at least the large ones, are well known, we can more easily determine whether fossil bones belong to a species which still exists, or to one that is now lost. As, on the other hand, we are still very far from being acquainted with all the testaceous animals and fishes belonging to the sea, and as we probably still remain ignorant of the greater part of those which live in the extensive deeps of the ocean, it is impossible to know, with any certainty, whether a species found in a fossil state may not still exist somewhere alive. Hence some naturalists persist in giving the name of oceanic or pelagic shells to *belemnites* and *cornua-ammonis*, and some other genera, which have not hitherto been found, except in the fossil state, in ancient strata; meaning by this, that although these have not as yet been found in a living or recent state, it is because they inhabit the bottom of the ocean, far beyond the reach of our nets.

[*Editors' Note:* Material has been omitted at this point.]

It appears to me, that a consecutive history of such singular deposits would be infinitely more valuable than so many contradictory conjectures respecting the first origin of the world and other planets, and respecting phenomena which have confessedly no resemblance whatever to those of the present physical state of the world; such conjectures finding, in these hypothetical facts, neither materials to build upon, nor any means of verification whatever. Several of our geologists resemble those historians who take no interest in the history of France, except as to what passed before the time of Julius Cæsar. Their imaginations, of course, must supply the place of authentic documents; and accordingly each composes his romance according to his own fancy. What would become of these historians, if they had not been assisted in their combinations by the knowledge of posterior facts? But our geologists neglect exactly those posterior geological facts, which might at least, in some measure, dispel the darkness of the preceding times.

It would certainly be exceedingly satisfactory to have the fossil organic productions arranged in chronological order, in the same manner as we now have the principal mineral substances. By this, the science of organization itself would be improved; the developements of animal life; the succession of its forms; the precise determinations of those which have been first called into existence; the simultaneous production of certain species, and their gradual extinction;— all these would perhaps instruct us fully as much in the essence of organization, as all the experiments that we shall ever be able to make upon living animals: And man, to whom only a short space of time is allotted upon the earth, would have the glory of restoring the history of thousands of ages which preceded the existence of the race, and of thousands of animals that never were contemporaneous with his species.

END OF THE ESSAY.

7

A

CRITICAL EXAMINATION

OF

THE FIRST PRINCIPLES

OF

GEOLOGY;

IN

A SERIES OF ESSAYS.

By G. B. GREENOUGH,

PRESIDENT OF THE GEOLOGICAL SOCIETY,

F.R.S. F.L.S.

LONDON:

Printed by Strahan and Spottiswoode, Printers-Street;

FOR LONGMAN, HURST, REES, ORME, AND BROWN,

PATERNOSTER-ROW.

1819.

Reprinted from pages 1, 9, 25, and 88–90 of *A Critical Examination of the First Principles of Geology; in a Series of Essays,* Longman, Hurst, Rees, Orme, and Brown, London, 1819.

ESSAY I.

ON STRATIFICATION.

Stratum is a word so familiar to our ears that it requires some degree of manliness to acknowledge ourselves ignorant of its meaning: the sense in which it is used is, however, very far from being precise. Easy as it may seem to determine whether a given mass be, or be not stratified, there is, perhaps, in the whole range of geological investigation, no subject more pregnant with controversy.

[*Editors' Note:* Material has been omitted at this point.]

Whence this contrariety of opinion? Are our senses at variance, or our judgments? The cause I think is obvious. Every one uses the word stratum, no one enquires its meaning; the remedy is as obvious — definition.

[*Editors' Note:* Material has been omitted at this point.]

6. The term Stratification is by no means unconnected with theory. To constitute a Stratum it is not enough, in the opinion of many geologists, that a substance divides in parallel planes, unless it does so in consequence of the manner in which its particles were at first deposited from a fluid menstruum; now it is obvious that as long as the propriety of using the term depends on the idea we entertain of the manner in which strata were formed, unity of opinion can alone preclude confusion of language. Let us enquire, therefore, how far geologists agree in their opinions upon this subject.

[*Editors' Note:* Material has been omitted at this point.]

To conclude, then, let me ask, Where a rock is *stratified*, is it necessarily bounded by parallel surfaces? if so, let us hear no more of mantle-shaped, saddle-shaped, shield-shaped, fan-shaped, basin-shaped, trough-shaped stratification.

Are its surfaces necessarily parallel to those of the adjoining rock? if so, let us hear no more of unconformable and overlying stratification.

Is it sufficient that parallelism shall be found in a portion of the rock? Let us never hear of substances being unstratified. Or must it extend through the entire mass? Let us hear no more of strata.

The laminæ of flagstone, the folia of slate, are these strata? Are laminæ, four hundred yards thick, strata? Is there any assignable limit to their thickness or tenuity?

When one set of parallel planes crosses another, are both sets to be called strata, or neither, or only one of them? If one only, by what rule are we to be guided in distinguishing the real from the counterfeit?

Must the beds be so arranged as to convey to the observer the idea of deposition alternately suspended and renewed? If this is not necessary, how is the parallelism derived from stratification, to be distinguished from parallelism resulting from other causes ? and of what use is it to know whether a substance is stratified or not ? If it is necessary, where two observers have imbibed contrary impressions, how shall we determine which of the two is right ?

Let him who can answer these questions rest assured that he has a distinct idea of stratification.

Part III

PHILLIPS (1829) TO D'ORBIGNY (1852)

Editors' Comments on Papers 8 Through 19

8 PHILLIPS
Excerpt from *Organic Remains of the Eastern Part of Yorkshire*

9A LYELL
Excerpt from *Different Circumstances under Which the Secondary and Tertiary Formations May Have Originated*

9B LYELL
Classification of Tertiary Formations in Chronological Order

10 MURCHISON
Excerpt from *On the Silurian System of Rocks*

11 HUNTON
Excerpts from *Remarks on a Section of the Upper Lias and Marlstone of Yorkshire, showing the limited vertical range of the Species of Ammonites, and other Testacea, with their value as Geological Tests*

12 FITTON
Excerpts from *Observations on some of the Strata between the Chalk and the Oxford Oolite, in the South-east of England*

13A GRESSLY
Excerpts from *Geologic Observations on the Jura of Soleure (1838)*

13B GRESSLY
Excerpt from *Geologic Observations on the Jura of Soleure (1840)*

13C GRESSLY
Excerpt from *Geologic Observations on the Jura of Soleure (1841)*

14 LYELL
Excerpt from *On the Different Ages of the Aqueous Rocks*

15A MURCHISON
Description of the Map

15B MURCHISON
Introduction

15C MURCHISON
Introductory View of the Distribution of Organic Remains in the Older Formations

16 SEDGWICK and MURCHISON
Excerpts from *On the Physical Structure of Devonshire, and on the Subdivisions of its older stratified Deposits, &c.*

17 D'ORBIGNY
Excerpts from *Conclusions Relative to the Separation of Geologic Stages, and to the Special Faunas Which They Enclose*

18 D'ORBIGNY
Excerpts from *Geologic Summary on the Cephalopods of the Jurassic Terrains*

19 D'ORBIGNY
Excerpt from *General Geologic Deductions Drawn from Fossil Species*

During the period 1829 through 1852, several classic stratigraphic works including those by Sedgwick (1830), Lyell (Papers 9A, 9B, 14), Murchison (Papers 10, 15), Sedgwick and Murchison (Paper 16), and d'Orbigny (Papers 17, 18, 19) were published. Three less well known but sophisticated and valuable English works from this period are those by Phillips (Paper 8), Fitton (Paper 12), and Hunton (Paper 11). During this time also Gressly (Paper 13A) introduced the term *facies* for a concept that had been developing for well over sixty years.

John Phillips's *Illustrations* (Paper 8) and Fitton's *Observations* (Paper 12) have been extolled by Arkell (Paper 32, 8) as "undisputedly the prototypes of all regional stratigraphic treatises." Phillips (Paper 8, 44) perceived formations in the same way as Conybeare and William Phillips (1822) and stated that correlations can be made by comparing the series of strata "in mode of arrangement, in mineralogical composition, and organic contents," three criteria later given also by Lyell (Paper 14, 271-279). Phillips further stated, "This should teach us not to undervalue the evidence of organic remains, for these are always useful and often necessary guides to determine the affinities of detached portions of the strata; and, when viewed in combination with the substance and arrangement of the rocks, the results to which they lead may be confidently adopted" (pp. 44, 45). Using this method, Phillips was able to identify and trace the variable, in substance and thickness, stratum known as the "cornbrash" in the oölitic series "by organic fossils and geologic position alone" (p. 87). Phillips (111-118), in discussing fossils and their ecological, stratigraphical, and chronological value, pointed to the fact that this must be "founded on a minute scrutiny of their specific characters, and a careful register of their localities in the strata" (p. 113). However, he went beyond merely identifying a stratum by "particular species of fossils which are found in no other stratum" (p. 145) and arrived at the concept of an assemblage zone by identifying a formation "by the *occurrence together* in it of some fossils which are repeated in rocks above, and several others which are found in those beneath" (p. 145). He also advocated caution in the application of using fossils to distinguish strata, caution "which is not always observed" (p. 146).

The existence of a disconformable relationship of strata had been recognized by Cuvier and Brongniart (Paper 4B). It was again mentioned by Sedgwick in 1829 in his *On the Geological Relations and Internal Structure of the Magnesian Limestone,* (cited in Clark and Hughes, 1890, 295, 296) where he pointed out that angular unconformities are not worldwide, that "formations which in one country are unconformable, may in another be parallel to each other, and so intimately connected as to appear the production of one epoch (cited in Clark and Hughes, 1890, 295, 296). Sedgwick further advised, in regard to correlating strata where differences are found in the strata of different countries (cited in Clark and Hughes, 1890, 297), "Each country ought to be described without any accommodating hypothesis, according to the type after which it has been moulded. But in comparing unconnected deposits of remote countries, we must act on an opposite principle; learning to suppress all local phenomena, and to seize on those only which are coextensive with the objects we attempt to classify." This statement is similar to later ideas of John Phillips (Paper 23).

In France, d'Halloy (1831, 75) made use of fossils to recognize six successively older systems and pointed out that since each system is characterized by particular fossils, the rocks from two superposed (but not truly consecutive) systems can be recognized even if there is no unconformity (angular unconformity) between them—again, recognition of a disconformable relationship.

Lyell, the uniformitarian, explained (Paper 9A, 32, 33) the existence of disconformities following his statement (p. 30) of belief in the gradual extinction of certain animals and plants and the successive introduction of new species, a process which he believed goes on continuously. This process is coupled with the fact that deposition, and hence fossilization, goes on randomly and irregularly from place to place. Lyell (p. 33) saw "no necessary connexion" between physical and biological change, inasmuch as he did not at that time, of course, understand the role of physical environmental change in natural selection; he did (p. 56), however, refer to climatic change and destroying causes for change in species. Darwin later used Lyell's evidence of geologic change as one important factor in bringing about biological evolution through natural selection.

Subsequently, Lyell (Paper 14) described and illustrated the disconformable situation that exists in a paraconformity (Dunbar and Rodgers, 1957) wherein there is "the entire absence of one or more formations of intervening periods between two groups" (p. 278). Here, Lyell was not using "period" in the presently understood sense. As for the distinct breaks seen at angular unconformities, Lyell (Paper 9A), in explaining the causes of the superposition of successive formations having distinct mineral and organic characters, denied that they are due to great catastrophes or sudden revolutions and stated that "it is only by carefully considering the combined action of all causes of change now in operation, whether in the animate or inanimate world, that we can hope to explain such complicated appearances as are exhibited in the general arrangement of mineral masses" (p. 26). Lyell (Paper 9A) continued:

> Although the causes which occasion the transference of the places of sedimentary deposition are continually in action in every region, yet they are most frequent where subterranean movements alter, from time to time, the levels of land, and they must be immense during the successive elevations and depressions which must be supposed to accompany the rise of a great continent from the deep. . . . Without entering into more detailed explanations, the reader will perceive that, according to the laws now governing the aqueous and igneous causes, distinct deposits must, at different periods, be thrown down on various parts of the earth's surface, and that, in the course of ages, the same area may become, again and again, the receptacle of such dissimilar sets of strata. During intervening periods, the space may either remain unaltered, or suffer what is termed *denudation,* in which case a superior set of strata are removed by the power of running water, and subjacent beds are laid bare, as happens wherever a sea encroaches upon a line of coast. By such means, it is obvious that the discordance in age of rocks in contact must often be greatly increased.

> The frequent unconformability [meaning angular unconformity] in the stratification of the inferior and overlying formation is another phenomenon in their arrangement, which may be considered as a natural consequence of those movements that accompany the gradual conversion of part of an ocean into land. [pp. 29, 30]

One of Lyell's greatest contributions to stratigraphy is, of course, his work arranging the separated Tertiary formations in chronological order according to the proportion of fossil and recent shells, a work he began in 1828 but subsequently based largely on tables drawn up by Deshayes (1830–1831). In this work, Lyell (Paper 9B) defined *recent* as the time "which has elapsed since the earth has been tenanted by man. . . . All sedimentary deposits, all volcanic rocks, in a word, every geological monument, whether belonging to the animate or inanimate world, which appertains to this epoch, may be termed *recent*" (p. 52). Here, although he variously referred to the recent (p. 52) as a period, era, and epoch, Lyell conceptualized what is now called a geochronologic unit (Hedberg, 1976, 10) in a manner similar to that in which d'Orbigny (Paper 19) later envisioned his stages.

Lyell noted (Paper 9B) that while the proportion of recent species in each of the older Tertiary formations is one biochronological characteristic, "certain *peculiar species* of testacea also distinguish its deposits from all other strata" (p. 54), a zonal concept. In addition, although Lyell divided the Tertiary into epochs, he pointed out that they were:

> invented for the convenience of systematic arrangement, always expecting to discover intermediate gradations between the boundary lines that we have first drawn. . . . we may select certain Geological formations as typical of particular epochs . . . we then arrange the groups referred to the same period in chronological order, according as they deviate in their organic contents from the *normal* groups, or according as they approximate to the type of an antecedent or subsequent epoch. [p. 57]

Lyell further suggested that comparison of fossil shells from any Tertiary formation, from any place, with recent shells could give the age and that correlations might even be made by these numerical proportions even when no species are in common. Lyell continued, "We may afterwards avail ourselves of the dates thus established, as eras to which the monuments of preceding periods may be referred" (p. 58).

The term *system* was first used in the modern sense by Murchison in his *On the Silurian System of Rocks* (Paper 10, p. 46). He recognized the need for the definition of "boundaries of groups naturally distinct from each other" (p. 48) and thus proposed his Silurian System. At this same time, the concept of type locality was established when he named the Silurian formations "each distinguished by separate and *peculiar organic remains,* . . . after those localities where each of them could be best studied, and their places in the series most clearly established" (p. 46).

The work by Hunton (Paper 11) on the Lias formation of the Yorkshire coast was said by Arkell (Paper 32, 14) to be "a truly remarkable document" and one of the "first definite attempts to formulate ideas of detailed correlation solely by means of fossils" (p. 8); Arkell called Hunton's collecting "zonal in the full modern sense of the word" (p. 8). In this work, Hunton chose to use only those species whose limits of distribution were clearly defined (a practice later followed by Oppel in Paper 20, 123, 124) and noted (Paper 11, p. 216) that "of all organic remains, the Ammonites afford the most beautiful illustration of the subdivision of strata." Hunton was cognizant of the concept of "first appearance" (p. 217) and of "greatest development" of species.

His Section (p. 221) shows the "limited Distribution of certain Species of Testacea, and the Order Cephalopoda in particular."

The other regional stratigraphical treatise praised by Arkell (Paper 32) is that by Fitton (Paper 12). In this work, Fitton proposed (p. 329) the existence of marine deposits coeval with the Wealden freshwater beds—in other words, facies. He also discussed (p. 333) the significance of the similarity of deposits of different epochs. Fitton presented one of the earliest composite generalized sections with his "Abstract Section" (Paper 12, Fig. 13).

Investigators had realized for many years that deposits of the same age might be of differing lithologies or aspect, that certain fossils may be found only in certain kinds of rock, and that different faunas do not necessarily mean difference in age (a point of disagreement between Arkell and Buckman that is cited in Part IV). However, the term *facies* was not introduced until 1838 when Gressly (Paper 13A, p. 11) elucidated the lithological and paleontological variations seen in adjacent localities in his studies of Jurassic rocks of the Jura Mountains. Gressly restricted the definition of facies to units of unique petrologic and faunal aspect exhibited in the horizontal dimension of terrains and believed (p. 12) that the paleontologic variations seen in this horizontal dimension were similar to those seen in the vertical dimension. The French *terrains* were more or less comparable to German *formations* and English *systems*; American formations are seemingly more comparable to Gressly's facies. It is interesting to recall that the modern usage of *formation* in the United States is still largely a reference to lithofacies. Gressly (Paper 13A, 22) found transition beds or passage beds between two distinctive facies in some cases. He believed (Paper 13C, 174, 175) that careful attention to the recognition of possible facies would prevent miscorrelation in areas of discontinuous outcrops and would one day allow even for description of ancient geographies.

Lyell (Paper 14, 271) reiterated the "three principal tests by which we determine the age of a given set of strata" that had been outlined by Phillips in 1829 (Paper 8)—superposition, mineral composition, and organic remains—and added included fragments. He also referred (pp. 274, 275) to horizontal change in lithology (mineral aspect and composition) in some formations, as well as to change in fossils, but further stated that fossils may remain the same where lithology changes and vice versa. In the vertical dimension, Lyell noted a more rapid change in fossils than in the horizontal, though some species endured longer than others (p. 276) yet none reappears after dying out. This fact, Lyell asserted, "confers on fossils their highest value as chronological tests" (p. 276).

Murchison's *The Silurian System* (Paper 15) has been declared by Challinor (1971, 108) to be "probably the greatest work ever to be published embodying the results of a single piece of original research by one man." Murchison (pp. xxvii, xxviii) divided England and Wales into eight sedimentary groups that he named "systems" based on their "dimensions, lithological characters, and zoological contents." He extended the use of superposition and characteristic fossils to the older, more deformed rocks and thus confirmed the validity of these principles. At the same time, noting that the Silurian system of rocks is not of uniform lithology everywhere, he repeated his faith in "*zoological* contents," coupled with "*order of superposition,*" as the "only safe criteria of their *age*" and that allow, "through every lithological change" (pp. 8, 9) separation of the system into upper and lower divisions, as well as identification of the system in distant lands, even though "sometimes under different aspects" (p. 11).

Murchison (Paper 15C, 579–586) generalized concerning the distribution of fossils in the older formations and cited (p. 579) the misunderstanding that can arise from comparison of faunal lists, or from erroneous identifications and descriptions—perennial problems.

Eschewing catastrophism, Murchison stated that while some species "vanished with the lapse of time [and] others unsuited to sudden changes were destroyed" (p. 572), some continued to live through long epochs. Thus, he believed that only the "large natural groups of strata" (the systems) are characterized by entirely peculiar fossils.

The older strata, Murchison professed (pp. 583, 584), have an abundance of fossils but fewer species than the secondary and Tertiary formations. He marveled at their wide geographical occurrences in the Silurian and how "as we approach the younger deposits . . . their geographical distribution harmonizes more and more with that of existing nature" (p. 584). This last should be a result of sea-floor spreading and continental drift, though Murchison used this evidence to support a belief in a more equable earthly temperature in the Silurian (p. 585).

Murchison discussed the phenomenon of the gradual change in faunas and the lithologic, geologic, and zoologic transitions that led him to his belief in *a geological succession* with no evidence of general destruction. These gradual transitions that Murchison saw are in contrast to the angular unconformities exhibited elsewhere. Subtle breaks in his succession are no doubt present, but Murchison exclaimed "how striking a succession do they present—so various yet so uniform—so vast yet so connected" (p. 574). Murchison believed this to be the result of *one system of operations* and referred to a "deep-seated and widely extended explanatory cause. . . . Each great igneous eruption, carrying with it fresh materials for additional deposits to be spread out on the bed of the ocean, would in fact be auxiliary to the repression of similar eruptions in future, by adding new folds to the pre-existing crust which enveloped the central and heated nucleus" (p. 575). (One might perhaps read too much into this statement taken out of context as it is, but we find it interesting to read early writings like this in light of present knowledge and concepts.) Murchison ascribed the appearance of greater heat and deformation in the earlier history of the earth, as compared to "the *gradual* elevation and depression of *modern* continents" (p. 576) to this diminishing central heat (and surely there was more radioactive heat earlier than now).

Inasmuch as we have seen some confusion as to the date of publication of Sedgwick and Murchison's proposal of the term *Devonian System* (variously cited as 1837, 1839, and 1840), we note here that their intention to propose it was announced in 1839 in the *Proceedings of the Geological Society of London,* [**3**(63):121-123, before April]. In their paper written March 25, 1839, and published in April (evidently after the twenty-fourth), 1839 in the *London and Edinburgh Philosophical Magazine* [ser. 3, **14**(89):241-260], they proposed the name "Devonian System" (p. 259), having rushed this into print for reasons explained therein (pp. 253-256). In proposing the name, they stated that "it may be applied, without any contradiction of terms, to rocks of every variety of mineral structure which contain the characteristic series of organic remains" (p. 259). Then, on April 24, 1839, they read a different paper, which subsequently was published in 1840 (Paper 16) with nearly the same title as that written on March 25, 1839, in the *Transactions of the Geologic Society of London* (5:688-702). Sedgwick and Murchison stated, "This is undoubtedly the greatest change which has ever been attempted at one time, in the classification of

British rocks" (p. 691) though this distinction would be rivaled later by Lapworth's erection of the Ordovician System (Paper 24). Sedgwick and Murchison prophesied that this recognition of Devonian rocks by their fossils would "have a most important bearing upon the classification of ancient rocks of foreign countries" (p. 253). However, they explained and justified some variations in fossil occurrences by admitting (1839) that "when sediments of a given epoch have been accumulated under peculiar conditions, we must expect to find considerable variations in the form of animal life" (p. 257). They expanded this idea on pages 699-701 of Paper 16.

In France, d'Orbigny's research (1841-1852) (Sherborn, 1899) led him to distinguish twenty-seven sequential faunas in the fossil record and thus to divide the strata into twenty-seven units that he called stages, based on "the annihilation of a series of beings replaced by another" and on "the composition of the faunas, or the extinction of genera or families" (1842, 9). His Paleozoic stages were essentially the previously recognized systems.

While the separateness of stages is certainly marked where there is an angular unconformity, d'Orbigny (Paper 17, 149-152) also gave other indicators of their distinctiveness such as erosion surfaces, bone breccias, ferruginous deposits, basal conglomerates, and thick unfossiliferous basal beds. As he noted (pp. 154, 155), however, only fossils can give us the relative mutual age of the stages.

We present a translation of d'Orbigny's discussion of his ten Jurassic stages (Paper 18, 600-604) and excerpts of their descriptions (pp. 604-611). D'Orbigny (p. 601) pointed to the necessity of tracing out paleontological horizons wherever found to determine if they are merely local facies before establishing a stage. While it is true that species have limited ranges, d'Orbigny's reason for believing that they were limited was a result of his unsophisticated catastrophist bent. Since he believed his stages represented worldwide epochs, he chose widespread species for his special species and warned against a local occurrence that might be regarded incorrectly as "un fait important de stratigraphie" (Paper 19, 257). This is the earliest use of the word *stratigraphy* that we have encountered, although Weller (1960, 2) said the word was not coined until 1865, without saying by whom. Where d'Orbigny found layers with mixtures of species that he believed belonged to two different stages, he (Paper 19, 256) rejected this as being merely a local exception and advised searching for the real limits of the faunas in places where stages are well limited, retrogressive stratigraphic practice, even in his day, to say the least.

Three who have discussed d'Orbigny's contribution to stratigraphy are Quenstedt (Paper 21), Arkell (Paper 32), and Monty (1968). Quenstedt's criticism is quoted in Part IV. Arkell believed d'Orbigny's work to be mainly premature (Paper 32), and he stated that "much of the confusion in international nomenclature and correlation has been caused by d'Orbigny's attempt to take a short cut" (p. 10). Arkell noted that while d'Orbigny recognized 10 stages in the Jurassic, at least 100 have subsequently been named, and that "by their ardour the disciples of d'Orbigny have entirely defeated his ends. They have 'improved' his system until nothing is left but a meaningless complex of overlapping stages of differing values, not in the smallest degree more efficient than the old nomenclature which it was designed to replace" (p. 12).

Monty (1968), who believed zonal stratigraphy truly began with d'Orbigny, stated that "d'Orbigny appears as a pioneer who consolidated the bases of modern stratigraphy. He introduced basic ideas and systems which were polished in the second half of the 19th century. The major imperfection of d'Orbigny's zone [stage] is its much-too-large vertical range" (p. 699).

We find that d'Orbigny's work parallels or mimics that of Murchison to some extent with respect to the principles employed, including the classic principles of superposition and faunal succession, the existence of unconformities (angular or no), the use of type localities as the basis for names of stratigraphic units, and the discounting of lithology in correlation. Many of d'Orbigny's stage names followed Brongniart's (1829) (Arkell, Paper 32, 9). Therefore, d'Orbigny does not seem to have been quite such an originator but more an adapter, and of course, he did not originate the term *stage* as it had been used by Cuvier and Brongniart in 1808. Much of Monty's praise (1968, 699) might well have been applied to Murchison and his system work rather than to d'Orbigny. It is of interest to contrast d'Orbigny's ten entirely separate Jurassic stages (Paper 18, 603), "divisions that nature has traced with grand strokes on the entire globe," with Sedgwick and Murchison's (Paper 16, 702) single great Paleozoic sequence of systems, "each passing into the system next in order by almost imperceptible shades of difference." We believe the true nature of the stratigraphic succession is to be found between these two views, that at least a slight physical break (paracontinuity) exists between every stratigraphic unit that exhibits a slight faunal change.

REFERENCES

Brongniart, A., 1829, *Tableau des terrains qui composent l'écorce du globe, un essai sur la structure de la partie connue de la terre,* Paris.

Challinor, J., 1971, *The History of British Geology, a Bibliographical Study,* David and Charles, Newton-Abbot.

Clark, J. W., and T. M'K. Hughes, 1890, *The Life and Letters of The Reverend Adam Sedgwick,* vol. I, Cambridge University Press.
Conybeare, W. D., and W. Phillips, 1822, *Outlines of the Geology of England and Wales,* part i, Phillips, London.
Deshayes, G. P., 1830-1831, Tableau comparatif des espèces de coquilles vivantes, etc., *Geol. Soc. France Bull.* **1**(8):185-187.
d'Halloy, J. B. d'O., 1831, *Eléments de Géologie,* Levrault, Paris.
d'Orbigny, A., 1842, *Terrains Jurassique, Paleontologie Francaise,* Masson, Paris, p. 9.
Dunbar, C. O., and J. Rodgers, 1957, *Principles of Stratigraphy,* Wiley, New York.
Hedberg, H. D., 1976, *International Stratigraphic Guide,* Wiley, New York.
Monty, C. L. V., 1968, d'Orbigny's Concepts of Stage and Zone, *Jour. Paleont.* **42**(3):689-701.
Sedgwick, A., 1829, On the Geological Relations and Internal Structure of the Magnesian Limestone, *Geol. Soc. London Trans.* (2nd ser.), p. iii.
Sedgwick, A., and Murchison, R. I., 1839, Classification of the Older Stratified Rocks of Devonshire and Cornwall, *London Edinburgh Philos. Mag.* ser. 3, 14, 89, pp. 241-260.
Sherborn, D. C., 1899, On the Dates of the "Paleontologie Française" of d'Orbigny, *Geol. Mag.* **6:**223-225.
Weller, J. M., 1960, *Stratigraphic Principles and Practice,* Harper, New York.

8

ILLUSTRATIONS

OF THE

GEOLOGY OF YORKSHIRE;

OR, A

DESCRIPTION OF THE STRATA AND ORGANIC REMAINS

OF THE

YORKSHIRE COAST:

ACCOMPANIED BY A

GEOLOGICAL MAP, SECTIONS, AND PLATES OF THE FOSSIL PLANTS AND ANIMALS.

By JOHN PHILLIPS, F. G. S.

KEEPER OF THE MUSEUM OF THE YORKSHIRE PHILOSOPHICAL SOCIETY; HONORARY MEMBER OF THE PHILOSOPHICAL SOCIETIES OF YORKSHIRE, LEEDS, AND HULL.

Je ne doute pas, que dans peu d'années peut-être, je ne sois reduit a dire, que l'ouvrage que je termine aujourd'hui, et auquel j'ai consacré tant de travail, ne sera qu'un leger aperçu, qu'un premier coup d'œil jeté sur ces immenses créations des anciens temps.

CUVIER, OSSEMENS FOSSILES, Tome V. page 487.

YORK:

PRINTED FOR THE AUTHOR,

BY THOMAS WILSON AND SONS, HIGH-OUSEGATE.

1829.

Reprinted from pages 111-118 of *Illustrations of the Geology of Yorkshire; or, a Description of the Strata and Organic Remains of the Yorkshire Coast: Accompanied by a Geological Map, Sections, and Plates of the Fossil Plants and Animals,* J. Phillips, York, 1829, 192p.

ORGANIC REMAINS OF THE EASTERN PART OF YORKSHIRE

[*Editors' Note:* In the original, material precedes this excerpt.]

DISTRIBUTION OF FOSSILS.

MODERN naturalists have discovered in the earth the remains of several hundred different plants, and several thousand kinds of animals. The peculiarities of form and structure among fossils are as constant and defined as among the living productions of nature, and the species are often as well distinguished. Upon comparing them with existing races, it is discovered that they are generally quite distinct; so that the fossil tribes, in some degree, appear like a separate creation, and have been elegantly termed "organic remains of a former world." But though different in detail, the ancient and existing races of organic nature are alike in generalities, and analogous in essential points of structure; and forcibly urge us to conclude that they were destined for similar modes of life. In the present economy of nature, plants of particular structure are appointed to exist under particular circumstances; shells of certain forms are peculiar to water, and others live habitually on land; and, generally, so constant is the agreement in the structure and functions of organic beings, that from the one we may infer the other. Who, that views the striking general resemblance of fossil and recent bodies, and considers the similar accidents to which both have been exposed, can hesitate for a moment to admit that conclusions drawn from examination of the structure of fossils, are as valid as those which are inferred from recent examples. The principle of investigation is in both cases the same, viz. the inevitable accordance between the construction of the creature, and the uses for which it was created.

From examinations conducted on this principle, it is inferred that the secondary strata contains remains of marine, lacustrine, and terrestrial plants; of marine and fresh-water shells, crustacea, and fishes; and of aquatic and terrestrial reptiles, mammalia, and birds. This simple statement furnishes ground for most interesting deductions respecting the ancient condition of the globe. We cannot, indeed, determine what was the comparative extent of its seas, lakes, and dry land, but we may form very reasonable opinions concerning its temperature, and a tolerable history of its inhabitants at different periods. For as the order of successive position among the rocks is likewise that of their relative antiquity, the fossils collected from these rocks may be arranged in chronological order.

The fossils of Britain thus arranged, (according to the example of Mr. W. Smith,) present us with many curious and important results. The following instances are selected rather to shew the richness and beauty of the subject, than to include all that is known respecting it.

The organic reliquiæ of animals are more ancient than those of plants, for they lie in the slate rocks of Cornwall and North Wales, whilst no plants have yet been found in any rock older than the lower red sandstone. The most abundant fossil remains of plants belong to terrestrial tribes; but the animal reliquiæ are mostly of aquatic origin; and very few examples are known of any bones of terrestrial animals occurring in strata more ancient than those above the chalk.

The most ancient animal remains are those of bivalve shells, (spiriferæ,) such as are not known to exist at present. The most ancient fossil plants which appear in the lower carboniferous rocks, almost wholly belong to terrestrial genera of the natural monocotyledonous orders, filices, lycopodiaceæ, and equisetaceæ, and, by their analogy to existing tropical tribes, seem to demonstrate that the climate of these northern regions was then warmer than it is at present.

The fossil plants of the middle æra, which accompany the lias and oolitic rocks in Yorkshire and Sutherland, belong chiefly to the natural monocotyledonous orders, filices, lycopodiaceæ, equisetaceæ, and cycadeæ, but fragments of dicotyledonous plants also occur with them.

The least ancient group of fossil plants, which are enclosed in strata above the chalk, are a mingled suite of monocotyledonous and dicotyledonous tribes, both terrestrial and lacustrine, bearing considerable analogy to plants now in existence. The greater number of fossil shells are certainly marine, but those which lie in layers amidst the monocotyledonous plants of the carboniferous formation, belong almost wholly to fresh-water genera, now in existence. Other local aggregations of fresh-water shells occur in the upper part of the oolitic series of rocks; but a general deposit of this kind occurs among the most recent, and contains species very similar to those that now exist.

The greater portion of the most ancient fossil shells, &c. belong to genera now extinct, as the productæ, spiriferæ, pentameri, orthoceratites, trilobites, and many genera of crinoidea; and on the other hand, the least ancient of the fossils, though specifically distinct from existing races, are mostly included in the same genera.

But the most important results to geology, arising from the contemplation of organic remains, are founded on a minute scrutiny of their specific characters, and a careful register of their localities in the strata. It is not enough for the rigid accuracy of modern inquiry, to say that a given rock contains corals, shells, and bones of fishes; but we must know the particular species, and determine all the circumstances of their occurrence. The more exact and extended our researches on this subject become, the more clear will be our statements on the succession of created beings, the more certain our applications of zoological principles to determine the relative antiquity of rocks, and the more satisfactory our views of the formation of the strata. Works which, like the present, profess to describe the rocks and fossils of a particular district, lose a large portion of their utility if they are composed without reference to general principles. It is in such local catalogues that the man of enlarged views in geology ought to find the best evidence of important truths, and the means of correcting serious errors. For these important ends, it is necessary that every known locality in the strata should be recorded of every fossil. For want of this precaution, fossils have been often stated to be characteristic of a particular rock, when in truth they occur in several others; and thus a crowd of errors have been introduced, which have obscured the truths taught by Mr. Smith, and given occasion for denying that a comparison of their imbedded fossils is useful in identifying and discriminating the strata. Deeply impressed with the interest and importance of this subject, I have sought the means of placing it in a clear and correct light; and am not without hopes, that whether my views be received or rejected, my statements will be found unprejudiced, and, though incomplete, correct.

I shall now endeavour to investigate some of the general laws, respecting the relation of fossils to the strata, which are either already recognised and admitted among geologists, or unfolded in the following pages. The inquiry naturally divides itself into two parts, according as the strata are considered, with respect to their chemical and mineralogical composition, or their relative antiquity. Considering rocks as definite chemical compounds, (an assumption sufficiently exact in a limited district,) we may inquire if fossils of the same kind belong to strata of the same character.

A decisive answer in the affirmative will suggest itself to him who observes the agreement in this respect, between the transition limestone and the mountain limestone, in their bivalve shells and trilobites, between this latter rock and the oolites in their astreæ, turbinoliæ, and milleporæ, and between the oolites and the chalk, in some of their echini and terebratulæ. But this analogy vanishes altogether when we attempt to extend it to a considerable series of fossils; no other strata than the limestones exhibit it in a striking degree, and few tribes of organic remains can be quoted in illustration, except the radiaria. On the contrary, the shells of the mountain limestone, oolite, and chalk, are all entirely distinct from one another, and immediately suggest the second inquiry, to which we now proceed. What is the relation between the species of fossils, and the antiquity of their enveloping strata? That such a connexion between the age of a rock and its organic contents does certainly exist, and may plainly be recognised, will appear from a few facts which any one may verify by examining a good collection of Yorkshire fossils, or a sufficient suite of specimens from the same strata in other parts of England. The mountain limestone of the north-western dales of Yorkshire, abounds with crinoidea, productæ, spiriferæ, and bellerophontes, of which no single individual has ever been found in the strata of the eastern part of the county, which on the other hand, contain echini, trigoniæ, cucullææ, rostellariæ, and ammonites, to which there is nothing similar in the west. The partition between these groups of strata and their fossils is made by the red sandstone stratum, which, in Yorkshire at least, has never yielded one single organic fossil. The same observation has been made in other parts of England. Again, in the eastern part of Yorkshire itself, a complete partition of the same kind is made by the blue clays of the vale of Pickering, between the chalk on the south and the oolitic rocks on the north; both full of fossils, and those entirely different.

I am sure that these assertions will not be disputed by any person at all acquainted with geological phenomena, or accustomed to distinguish the characters of fossils. The consequence flowing from them is of the highest importance and interest; for since it thus appears, that a few shells brought from a quarry, are data sufficient to determine the geological relations of the rock, we are entitled to conclude, that in a given district the age and position of certain strata, or groups of strata, are infallibly indicated by their organic contents. These researches, commenced by Mr. Smith in England, have been extended with the same results over all parts of Europe, and a large portion of America, and therefore it is concluded that strata, or groups of strata, are to be discriminated in local regions, and identified in distant countries, by their imbedded organic remains.

Having thus obtained the general principle, let us endeavour to ascertain the extent of its applicability, and the precautions necessary to ensure accurate results.

So unequally are the different species of fossils distributed in the earth, that, whilst some are dispersed through several neighbouring strata, as clypeus clunicularis among the oolitic rocks, others are confined to one stratum, as ammonites calloviensis to the Kelloways rock, and some to a particular bed of stone, as the astreæ which characterise the coralline oolite.

It is, therefore, possible, by collecting numerous specimens procured from a limited district, to assign to each formation of strata, single stratum, or even characteristic bed of stone, all the fossils which have ever been discovered in it. Such catalogues being compared, formations, strata, and beds, may be found to differ from one another by the presence or absence of particular species. A given formation may possess species never found in any of those above, and it may be deficient in others which do occur above, and in like manner it may differ from those below. Hence it may be concluded,

1. That a formation or stratum may differ from all those above it, by the presence or absence of certain species, and from all those below it, by the presence or absence of other species :

2. That it may contain some particular species, unknown either above or below. We may add, that formations and strata may differ by the relative abundance or paucity of their imbedded fossils.

EXAMPLES.

1. The coralline oolite formation, as defined p. 32, appears to me to differ from all the formations above, by the presence of ammonites perarmatus, mya literata, and clypeus clunicularis, and by the absence of ostrea delta, hamites, and ananchytes ; and from all those below, by the

presence of spatangus ovalis,? and ammonites perarmatus, and the absence of productæ, axini, ammonites Walcottii, nerita costata, astarte minima, and terebratula digona.

2. Again, the Kelloways rock differs from all the strata above it and below it, by the presence of ammonites calloviensis, and gryphæa dilatata: no stratum in Yorkshire but the Kimmeridge clay contains ostrea delta; nor is gryphæa incurva found except in lias beds or lias boulders.

3. It is in the lower part of the coralline oolite that clypeus dimidiatus, and c. clunicularis abound, but melania striata belongs to the upper layers of that rock.

These are the principles of investigation which it is proposed to apply to the strata and fossils of the eastern part of Yorkshire, and to illustrate by the aid of the arranged catalogues which follow, and their accompanying plates.

I shall first present complete lists for each stratum, of all the fossils found in it, distinguishing those which are repeated in other strata, and referring to figures contained in this work; Mr. Sowerby's Mineral Conchology; Mr. Smith's Works; Mantell's Geology of Sussex; Young and Bird's Survey of the Yorkshire Coast; Parkinson's Organic Remains, &c.

After thus unfolding the evidence which I have collected on the subject, I shall venture to propose some inferences concerning the distribution of organic remains in the rocks of the country which I have undertaken to describe, and likewise endeavour to trace the extent of the agreement in this respect, between the strata of Yorkshire and those of Scotland, the southern counties of England, and several parts of the continent.

Lastly, I propose to combine the whole into one synoptic list of the organic remains mentioned in this work, arranged in the ascending order of their natural affinities; accompanied by references to their place in the scale of stratification. As a considerable portion of the fossils herein mentioned have not been previously described or figured, it has become necessary to assign names to them. In the execution of this duty, manuscript names, when attainable, have been generally adopted, and, instead of specific characters, a figure, of the natural size, or in the proportion expressed by the contiguous fractions, is given to illustrate every new species. New figures have sometimes been added of fossils but little known, highly important, or not before represented with proper fidelity; but this has been seldom judged necessary with respect to common species, for which a reference will be found to some generally known and accessible work on the subject.

[*Editors' Note:* Material has been omitted at this point.]

9A

Reprinted from pages 30-34 of *Principles of Geology, Being an Attempt to Explain the Former Changes of the Earth's Surface, by Reference to Causes Now in Operation,* vol. 3, John Murray, London, 1833, 398p.

DIFFERENT CIRCUMSTANCES UNDER WHICH THE SECONDARY AND TERTIARY FORMATIONS MAY HAVE ORIGINATED

C. Lyell

[*Editors' Note:* In the original, material precedes this excerpt.]

Change of Species everywhere in progress.—If, then, it be conceded, that the combined action of the volcanic and the aqueous forces would give rise to a succession of distinct formations, and that these would be sometimes unconformable, let us next inquire in what manner these groups might become characterized by different assemblages of fossil remains.

We endeavoured to show, in the last volume, that the hypothesis of the gradual extinction of certain animals and plants, and the successive introduction of new species, was quite consistent with all that is known of the existing economy of the animate world; and if it be found the only hypothesis which is reconcilable with geological phenomena, we shall have strong grounds for conceiving that such is the order of nature.

Fossilization of Plants and Animals partial.—We have seen that the causes which limit the duration of species are not confined, at any one time, to a particular part of the globe; and, for the same reason, if we suppose that their place is supplied, from time to time, by new species, we may suppose their introduction to be no less generally in progress. Hence, from all the foregoing premises, it would follow, that the change of species would be in simultaneous operation everywhere throughout the habitable surface of sea and land; whereas the fossilization of plants and animals must always be confined to those areas where new strata are produced. These areas, as we have proved, are always shifting their position, so that the fossilizing process, whereby the commemoration of the particular state of the organic world, at any given time, is effected, may be said to move about, visiting and revisiting different tracts in succession.

In order more distinctly to elucidate our idea of the working of this machinery, let us compare it to a somewhat analogous case that might easily be imagined to occur in the history of human affairs. Let the mortality of the population of a large country represent the successive extinction of species, and the births of new individuals the introduction of new species. While these fluctuations are gradually taking place everywhere, suppose commissioners to be appointed to visit each province of the country in succession, taking an exact account of the number, names, and individual peculiarities of all the inhabitants, and leaving in each district a register containing a record of this information. If, after the completion of one census, another is immediately made after the same plan, and then another, there will, at last, be a series of statistical documents in each province. When these are arranged in chronological order, the contents of those which stand next to each other will differ according to the length of the intervals of time between the taking of each census.

If, for example, all the registers are made in a single year, the proportion of deaths and births will be so small during the interval between the compiling of two consecutive documents, that the individuals described in each will be nearly identical; whereas, if there are sixty provinces, and the survey of each requires a year, there will be an almost entire discordance between the persons enumerated in two consecutive registers.

There are undoubtedly some other causes besides the mere quantity of time which may augment or diminish the amount of discrepancy. Thus, for example, at some periods a pestilential disease may lessen the average duration of human life, or a variety of circumstances may cause the births to be unusually numerous, and the population to multiply; or, a province may be suddenly colonized by persons migrating from surrounding districts.

We must also remind the reader, that we do not propose the above case as an exact parallel to those geological phenomena which we desire to illustrate; for the commissioners are supposed to visit the different provinces in rotation, whereas the commemorating processes by which organic remains become fossilized, although they are always shifting from one area to another, are yet very irregular in their movements. They may abandon and revisit many spaces again and again, before they once approach another district; and besides this source of irregularity, it may often happen, that while the depositing process is suspended, denudation may take place, which may be compared to the occasional destruction of some of the statistical documents before mentioned. It is evident, that where such accidents occur, the want of continuity in the series may become indefinitely great, and that the monuments which follow next in succession will by no means be equidistant from each other in point of time.

If this train of reasoning be admitted, the frequent distinctness of the fossil remains, in formations immediately in contact, would be a necessary consequence of the existing laws of sedimentary deposition, accompanied by the gradual birth and death of species.

We have already stated, that we should naturally look for a change in the mineral character in strata thrown down at distant intervals in the same place; and, in like manner, we must also expect, for the reason last set forth, to meet occasionally with sudden transitions from one set of organic remains to another. But the causes which have given rise to such differences in mineral characters, have no necessary connexion with those which have produced a change in the species of imbedded plants and animals.

When the lowest of two sets of strata are much dislocated over a wide area, the upper being undisturbed, there is usually a considerable discordance in the organic remains of the two groups; but this coincidence must not be ascribed to the agency of the disturbing forces, as if they had exterminated the living inhabitants of the surface. The immense *lapse of time* required for the development of so great a series of subterranean movements, has in these cases allowed the species also throughout the globe to vary, and hence the two phenomena are usually concomitant.

Although these inferences appear to us very obvious, we are aware that they are directly opposed to many popular theories respecting catastrophes; we shall, therefore, endeavour to place our views in a still clearer light before the reader. Suppose we had discovered two buried cities at the foot of Vesuvius, immediately superimposed upon each other, with a great mass of tuff and lava intervening, just as Portici and Resina, if now covered with ashes, would overlie Herculaneum. An antiquary might possibly be entitled to infer, from the inscriptions on public edifices, that the inhabitants of the inferior and older town were Greeks, and those of the modern, Italians. But he would reason very hastily, if he also concluded from these data, that there had been a sudden change from the Greek to the Italian language in Campania. Suppose he afterwards found *three* buried cities, one above the other, the intermediate one being Roman, while, as in the former example, the lowest was Greek, and the uppermost Italian, he would then perceive the fallacy of his former opinion, and would begin to suspect that the catastrophes, whereby the cities were inhumed, might have no relation whatever to the fluctuations in the language of the inhabitants; and that, as the Roman tongue had evidently intervened between the Greek and Italian, so many other dialects may have been spoken in succession, and the passage from the Greek to the Italian may have been very gradual, some terms growing obsolete, while others were introduced from time to time.

If this antiquary could have shown that the volcanic paroxysms of Vesuvius were so governed as that cities should be buried one above the other, just as often as any variation occurred in the language of the inhabitants, then, indeed, the abrupt passage from a Greek to a Roman, and from a Roman to an Italian city, would afford proof of fluctuations no less sudden in the language of the people.

So in Geology, if we could assume that it is part of the plan of nature to preserve, in every region of the globe, an unbroken series of monuments to commemorate the vicissitudes of the organic creation, we might infer the sudden extirpation of species, and the simultaneous introduction of others, as often as two formations in contact include dissimilar organic fossils. But we must shut our eyes to the whole economy of the existing causes, aqueous, igneous, and organic, if we fail to perceive *that such is not the plan of Nature.*

9B

Reprinted from pages 45-61 of *Principles of Geology, Being an Attempt to Explain the Former Changes of the Earth's Surface, by Reference to Causes Now in Operation,* vol. 3, John Murray, London, 1833, 398p.

CLASSIFICATION OF TERTIARY FORMATIONS IN CHRONOLOGICAL ORDER

C. Lyell

We explained in the last chapter the principles on which the relative ages of different formations may be ascertained, and we found the character to be chiefly derivable from superposition, mineral structure, and organic remains. It is by combining the evidence deducible from all these sources, that we determine the chronological succession of distinct formations, and this principle is well illustrated by the investigation of those European tertiary strata to the discovery of which we have already alluded.

It will be seen, that in proportion as we have extended our inquiries over a larger area, it has become necessary to intercalate new groups of an age intermediate between those first examined, and we have every reason to expect that, as the science advances, new links in the chain will be supplied, and that the passage from one period to another will become less abrupt. We may even hope, without travelling to distant regions,—without even transgressing the limits of western Europe, to render the series far more complete. The fossil shells, for example, of many of the Subalpine formations, on the northern limits of the plain of the Po, have not yet been carefully collected and compared with those of other countries, and we are almost entirely ignorant of many deposits known to exist in Spain and Portugal.

The theoretical views developed in the last chapter, respecting breaks in the sequence of geological monuments, will explain our reasons for anticipating the discovery of intermediate gradations as often as new regions of great extent are explored.

Comparative value of different classes of organic remains.

In the mean time, we must endeavour to make the most systematic arrangement in our power of those formations which are already known, and in attempting to classify these in chronological order, we have already stated that we must chiefly depend on the evidence afforded by their fossil organic contents. In the execution of this task, we have first to consider what class of remains are most useful, for although every kind of fossil animal and plant is interesting, and cannot fail to throw light on the former history of the globe at a certain period, yet those classes of remains which are of rare and casual occurrence, are absolutely of no use for the purposes of general classification. If we have nothing but plants in one assemblage of strata, and the bones of mammalia in another, we can obviously draw no conclusion respecting the number of species of organic beings common to two epochs ; or if we have a great variety, both of vertebrated animals and plants, in one series, and only shells in another, we can form no opinion respecting the remoteness or proximity of the two eras. We might, perhaps, draw some conclusions as to relative antiquity, if we could compare each of these monuments to a third ; as, for example, if the species of shells should be almost all identical with those now living, while the plants and vertebrated animals were all extinct ; for we might then infer that the shelly deposit was the most recent of the two. But in this case it will be seen that the information flows from a direct comparison of the species of corresponding orders of the animal and vegetable kingdoms,— of plants with plants, and shells with shells ; the only mode of making a systematic arrangement by reference to organic remains.

Although the bones of mammalia in the tertiary strata, and those of reptiles in the secondary, afford us instruction of the most interesting kind, yet the species are too few, and confined to too small a number of localities, to be of great importance in characterizing the minor subdivisions of geological formations. Skeleton of fish are by no means frequent in a good state of preservation, and the science of ichthyology must be farther advanced, before we can hope to determine their specific character with sufficient precision. The same may be said of fossil botany, notwithstanding the great progress that has recently been made in that department; and even in regard to zoophytes, which are so much more abundant in a fossil state than any of the classes above enumerated, we are still greatly impeded in our endeavour to classify strata by their aid, in consequence of the smallness of the number of recent species which have been examined in those tropical seas where they occur in the greatest profusion.

Fossil remains of testacea of chief importance. The testacea are by far the most important of all classes of organic beings which have left their spoils in the subaqueous deposits; they are the medals which nature has chiefly selected to record the history of the former changes of the globe. There is scarcely any great series of strata that does not contain some marine or freshwater shells, and these fossils are often found so entire, especially in the tertiary formations, that when disengaged from the matrix, they have all the appearance of having been just procured from the sea. Their colour, indeed, is usually wanting, but the parts whereon specific characters are founded remain unimpaired; and although the animals themselves are gone, yet their form and habits can generally be inferred from the shell which covered them.

The utility of the testacea, in geological classification, is greatly enhanced by the circumstance, that some forms are proper to the sea, others to the land, and others to freshwater,

Rivers scarcely ever fail to carry down into their deltas some land shells, together with species which are at once fluviatile and lacustrine. The Rhone, for example, receives annually, from the Durance, many shells which are drifted down in an entire state from the higher Alps of Dauphiny, and these species, such as *Bulimus montanus*, are carried down into the delta of the Rhone to a climate far different from that of their native habitation. The young hermit crabs may often be seen on the shores of the Mediterranean, near the mouth of the Rhone, inhabiting these univalves, brought down to them from so great a distance*. At the same time that some freshwater and land species are carried into the sea, other individuals of the same become fossil in inland lakes, and by this means we learn what species of freshwater and marine testacea coexisted at particular eras; and from this again we are able to make out the connexion between various plants and mammifers imbedded in those lacustrine deposits, and the testacea which lived in the ocean at the same time.

There are two other characters of the molluscous animals which render them extremely valuable in settling chronological questions in Geology. The first of these is a wide geographical range, and the second (probably a consequence of the former), is the superior duration of species in this class. It is evident that if the habitation of a species be very local, it cannot aid us greatly in establishing the contemporaneous origin of distant groups of strata, in the manner pointed out in the last chapter; and if a wide geographical range be useful in connecting formations far separated in space, the longevity of species is no less serviceable in establishing the relations of strata considerably distant from each other in point of time.

We shall revert in the sequel to the curious fact, that in tracing back these series of tertiary deposits, many of the existing species of testacea accompany us after the disappearance of all the recent mammalia, as well as the fossil remains of living

* M. Marcel de Serres pointed out this fact to me when I visited Montpellier, July, 1828.

species of several other classes. We even find the skeletons of extinct quadrupeds in deposits wherein all the land and fresh-water shells are of recent species*.

Necessity of accurately determining species.—The reader will already perceive that the systematic arrangement of strata, so far as it rests on organic remains, must depend essentially on the accurate determination of *species*, and the geologist must therefore have recourse to the ablest naturalists, who have devoted their lives to the study of certain departments of organic nature. It is scarcely possible that they who are continually employed in laborious investigations in the field, and in ascertaining the relative position and characters of mineral masses, should have leisure to acquire a profound knowledge of fossil osteology, conchology, and other branches; but it is desirable that, in the latter science at least, they should become acquainted with the principles on which the specific characters are determined, and on which the habits of species are inferred from their peculiar forms. When the specimens are in an imperfect state of preservation, or the shells happen to belong to genera in which it is difficult to decide on the species, except when the inhabitant itself is present, or when any other grounds of ambiguity arise, we must reject, or lay small stress upon, the evidence, lest we vitiate our general results by false identifications and analogies. We cannot do better than consider the steps by which the science of botanical geography has reached its present stage of advancement, and endeavour to introduce the same severe comparison of the specific characters, in drawing all our geological inferences.

Tables of shells by M. Deshayes.—In the Appendix the reader will find a tabular view of the results obtained by the comparison of more than three thousand tertiary shells, with nearly five thousand living species, all of which, with few exceptions, are contained in the rich collection of M. Deshayes. Having enjoyed an opportunity of examining, again and again, the specimens on which this eminent conchologist has founded his identifica-

* See vol. i. chap. vi.

tions, and having been witness to the great time and labour devoted by him to this arduous work, I feel confidence in the results, so far as the data given in his list will carry us. It was necessary to compare nearly forty thousand specimens, in order to construct these tables, since not only the varieties of every species required examination, but the different individuals, also, belonging to each which had been found fossil in various localities. The correctness of the localities themselves was ascertained with scrupulous exactness, together with the relative position of the strata; and if any doubts existed on these questions, the specimens were discarded as of no geological value. A large proportion of the shells were procured, by M. Deshayes himself, from the Paris basin, many were contributed by different French geologists, and some were collected by myself from different parts of Europe.

It would have been impossible to give lists of more than three thousand fossil shells in a work not devoted exclusively to conchology; but we were desirous of presenting the reader with a catalogue of those fossils which M. Deshayes has been able to identify with living species, as also of those which are common to two distinct tertiary eras. By this means a comparison may be made of the testacea of each geological epoch, with the actual state of the organic creation, and, at the same time, the relations of different tertiary deposits to each other exhibited. The number of shells mentioned by name in the tables, in order to convey this information, is seven hundred and eighty-two, of which four hundred and twenty-six have been found both living and fossil, and three hundred and fifty-six fossil only, but in the deposits of more than one era. An exception, however, to the strictness of this rule, has been made in regard to the fossil shells common to the London and Paris basins, fifty-one of which have been enumerated by name, though these formations do not belong to different eras.

It has been more usual for geologists to give tables of characteristic shells; that is to say, of those found in the strata of one period and not common to any other. These typical species are certainly of the first importance, and some of them will be seen figured in the plates, illustrative of the different tertiary eras; but we were more anxious, in this work, to place in a clear light, a point of the greatest theoretical interest, which has been often overlooked or controverted, viz., the identity of many living and fossil species, as also the connexion of the zoological remains of deposits formed at successive periods.

The value of such extensive comparisons, as those of which the annexed tables of M. Deshayes give the results, depends greatly on the circumstance, that all the identifications have been made by the same naturalist. The amount of variation which ought to determine a species is, in cases where they approach near to each other, a question of the nicest discrimination, and requires a degree of judgment and tact that can hardly be possessed by different zoologists in exactly the same degree. The standard, therefore, by which differences are to be measured, can scarcely ever be perfectly invariable, and one great object to be sought for is, that, at least, it should be uniform. If the distinctions are all made by the same naturalist, and his knowledge and skill be considerable, the results may be relied on with sufficient confidence, as far as regards our geological conclusions.

If one conchologist should inform us that out of 1122 species of fossil testacea, discovered in the Paris basin, he has only been enabled to identify thirty-eight with recent species, while another should declare, that out of two hundred and twenty-six Sicilian fossil shells, no less than two hundred and sixteen belonged to living species, we might suspect that one of these observers allowed a greater degree of latitude to the variability of the specific character than the other; but when, in both instances, the conclusions are drawn by the same eminent conchologist, we are immediately satisfied that the relations of these two groups, to the existing state of the animate creation, are as distinct as are indicated by the numerical results.

It is not pretended that the tables, to which we refer, comprise all the known tertiary shells. In the museums of Italy there are magnificent collections, to which M. Deshayes had no access, and the additions to the recent species in the cabinets of conchologists in London, have been so great of late years, that in many extensive genera the number of species has been more than doubled. But as the greater part of these newly-discovered shells have been brought from the Pacific and other distant seas, it is probable that these accessions would not materially alter the results given in the tables, and it must, at all events, be remembered, that the only effect of such additional information would be, to increase the number of identifications of recent with fossil species, while the proportional number of analogues in the different periods might probably remain nearly the same.

SUBDIVISIONS OF THE TERTIARY EPOCH.

Recent formations.—We shall now proceed to consider the subdivisions of tertiary strata which may be founded on the results of a comparison of their respective fossils, and to give names to the periods to which they each belong. The tertiary epoch has been divided into three periods in the tables; we shall, however, endeavour to establish *four*, all distinct from the actual period, or that which has elapsed since the earth has been tenanted by man. To the events of this latter era, which we shall term the *recent*, we have exclusively confined ourselves in the two preceding volumes. All sedimentary deposits, all volcanic rocks, in a word, every geological monument, whether belonging to the animate or inanimate world, which appertains to this epoch, may be termed *recent*. Some *recent* species, therefore, are found *fossil* in various tertiary periods, and, on the other hand, others, like the Dodo, may be *extinct*, for it is sufficient that they should once have coexisted with man, to make them referrible to this era.

Some authors apply the term *contemporaneous* to all the formations which have originated during the human epoch; but as the word is so frequently in use to express the synchronous origin of distinct formations, it would be a source of great inconvenience and ambiguity, if we were to attach to it a technical sense.

We may sometimes prove, that certain strata belong to the recent period by aid of historical evidence, as parts of the delta of the Po, Rhone, and Nile, for example; at other times, by discovering imbedded remains of man or his works; but when we have no evidence of this kind, and we hesitate whether to ascribe a particular deposit to the recent era, or that immediately preceding, we must generally incline to refer it to the latter, for it will appear in the sequel, that the changes of the historical era are quite insignificant when contrasted with those even of the newest tertiary period.

Newer Pliocene period.—This most modern of the four subdivisions of the whole tertiary epoch, we propose to call the *Newer Pliocene*, which, together with the *Older* Pliocene, constitute one group in the annexed tables of M. Deshayes.

We derive the term Pliocene from *πλειων*, major, and *καινος*, recens, as the major part of the fossil testacea of this epoch are referrible to recent species*. Whether in all cases there may hereafter prove to be an absolute preponderance of recent species, in every group of strata assigned to this period in the tables, is very doubtful; but the proportion of living species, where least considerable, usually approaches to one-half of the total number, and appears always to exceed a third; and as our acquaintance with the testacea of the Mediterranean, and some other seas, increases, it is probable that a greater proportion will be identified.

* In the terms Pliocene, Miocene, and Eocene, the Greek diphthongs *ei* and *ai* are changed into the vowels *i* and *e*, in conformity with the idiom of our language. Thus we have Encenia, an inaugural ceremony, derived from *εν* and *καινος*, recens; and as examples of the conversion of *ei* into *i*, we have icosahedron.

I have been much indebted to my friend, the Rev. W. Whewell, for assisting me in inventing and anglicizing these terms, and I sincerely wish that the numerous foreign diphthongs, barbarous terminations, and Latin plurals, which have been so plentifully introduced of late years into our scientific language, had been avoided as successfully as they are by French Naturalists, and as they were by the earlier English writers, when our language was more flexible than it is now. But while I commend the French for accommodating foreign terms to the structure of their own language, I must confess that no naturalists have been more unscholar-like in their mode of fabricating Greek derivatives and compounds, many of the latter being a bastard offspring of Greek and Latin.

The newer Pliocene formations, before alluded to, pass insensibly into those of the *Recent* epoch, and contain an immense preponderance of recent species. It will be seen that of two hundred and twenty-six species, found in the Sicilian beds, only ten are of extinct or unknown species, although the antiquity of these tertiary deposits, as contrasted with our most remote historical eras, is immensely great. In the volcanic and sedimentary strata of the district round Naples, the proportion appears to be even still smaller.

Older Pliocene period.—These formations, therefore, and others wherein the plurality of living species is so very decided, we shall term the *Newer* Pliocene, while those of the tertiary period immediately preceding may be called the *Older* Pliocene. To the latter belong the formations of Tuscany, and of the Subapennine hills in the north of Italy, as also the English Crag.

It appears that in the period last mentioned, the proportion of recent species varies from upwards of a third to somewhat more than half of the entire number; but it must be recollected, that this relation to the recent epoch is only *one* of its zoological characters, and that certain *peculiar species* of testacea also distinguish its deposits from all other strata. The relative position of the beds referrible to this era has been explained in diagrams Nos. 3 and 4, letter *f*, chapter II.

Miocene period.—The next antecedent tertiary epoch we shall name Miocene, from μειων, minor, and καινος, recens, a minority only of fossil shells imbedded in the formations of this period, being of recent species. The total number of Miocene shells, referred to in the annexed tables, amounts to 1021, of which one hundred and seventy-six only are recent, being in the proportion of rather less than eighteen in one hundred. Of species common to this period, and to the two divisions of the Pliocene epoch before alluded to, there are one hundred and ninety-six, whereof one hundred and fourteen are living, and the remaining eighty-two extinct, or only known as fossil.

As there are a certain number of fossil species which are characteristic of the Pliocene strata before described, so also there are many shells exclusively confined to the Miocene period. We have already stated, that in Touraine and in the South of France near Bordeaux, in Piedmont, in the basin of Vienna, and other localities, these Miocene formations are largely developed, and their relative position has been shown in diagrams Nos. 3 and 4, letter *e*, chapter II.

Eocene period.—The period next antecedent we shall call Eocene, from ἠως, aurora, and καινος, recens, because the extremely small proportion of living species contained in these strata, indicates what may be considered the first commencement, or *dawn,* of the existing state of the animate creation. To this era the formations first called tertiary, of the Paris and London basins, are referrible. Their position is shown in the diagrams Nos. 3 and 4, letter *d*, in the second chapter.

The total number of fossil shells of this period already known, is one thousand two hundred and thirty-eight, of which number forty-two only are living species, being nearly in the proportion of three and a half in one hundred. Of fossil species, not known as recent, forty-two are common to the Eocene and Miocene epochs. In the Paris basin alone, 1122 species have been found fossil, of which thirty-eight only are still living.

The geographical distribution of those recent species which are found fossil in formations of such high antiquity as those of the Paris and London basins, is a subject of the highest interest.

It will be seen by reference to the tables, that in the more modern formations, where so large a proportion of the fossil shells belong to species still living, they also belong, for the most part, to species now inhabiting the seas immediately adjoining the countries where they occur fossil; whereas the recent species, found in the older tertiary strata, are frequently inhabitants of distant latitudes, and usually of warmer climates. Of the forty-two Eocene species, which occur fossil in England, France, and Belgium, and which are still living, about half now inhabit within, or near the tropics, and almost all the rest are denizens of the more southern parts of Europe. If some

Eocene species still flourish in the same latitudes where they are found fossil, they are species which, like *Lucina divaricata,* are now found in many seas, even those of different quarters of the globe, and this wide geographical range indicates a capacity of enduring a variety of external circumstances, which may enable a species to survive considerable changes of climate and other revolutions of the earth's surface. One fluviatile species (*Melania inquinata*), fossil in the Paris basin, is now only known in the Philippine islands, and during the lowering of the temperature of the earth's surface, may perhaps have escaped destruction by transportation to the south. We have pointed out in the second volume (chap. vii.), how rapidly the eggs of freshwater species might, by the instrumentality of water-fowl, be transported from one region to another. Other Eocene species, which still survive and range from the temperate zone to the equator, may formerly have extended from the pole to the temperate zone, and what was once the southern limit of their range may now be the most northern.

Even if we had not established several remarkable facts in attestation of the longevity of certain tertiary species, we might still have anticipated that the duration of the living species of aquatic and terrestrial testacea would be very unequal. For it is clear that those which now inhabit many different regions and climates, may survive the influence of destroying causes, which might extirpate the greater part of the species now living. We might expect, therefore, some species to survive several successive states of the organic world, just as Nestor was said to have outlived three generations of men.

The distinctness of periods may indicate our imperfect information.—In regard to distinct zoological periods, the reader will understand, from our observations in the third chapter, that we consider the wide lines of demarcation that sometimes separate different tertiary epochs, as quite unconnected with extraordinary revolutions of the surface of the globe, and as arising, partly, like chasms in the history of nations, out of the present imperfect state of our information, and partly from the irregular manner in which geological memorials are preserved, as already explained. We have little doubt that it will be necessary hereafter to intercalate other periods, and that many of the deposits, now referred to a single era, will be found to have been formed at very distinct periods of time, so that, notwithstanding our separation of tertiary strata into four groups, we shall continue to use the term *contemporaneous* with a great deal of latitude.

We throw out these hints, because we are apprehensive lest zoological periods in Geology, like artificial divisions in other branches of Natural History, should acquire too much importance, from being supposed to be founded on some great interruptions in the regular series of events in the organic world, whereas, like the genera and orders in zoology and botany, we ought to regard them as invented for the convenience of systematic arrangement, always expecting to discover intermediate gradations between the boundary lines that we have first drawn.

In Natural History we select a certain species as a generic type, and then arrange all its congeners in a series, according to the degrees of their deviation from that type, or according as they approach to the characters of the genus which precedes or follows. In like manner, we may select certain Geological formations as typical of particular epochs; and having accomplished this step, we may then arrange the groups referred to the same period in chronological order, according as they deviate in their organic contents from the *normal* groups, or according as they approximate to the type of an antecedent or subsequent epoch.

If intermediate formations shall hereafter be found between the Eocene and Miocene, and between those of the last period and the Pliocene, we may still find an appropriate place for all, by forming subdivisions on the same principle as that which has determined us to separate the lower from the upper Pliocene groups. Thus, for example, we might have three divisions of the Eocene epoch,—the older, middle, and newer; and three similar subdivisions, both of the Miocene and Pliocene epochs. In that case, the formations of the middle period must be considered as the types from which the assemblage of organic remains in the groups immediately antecedent or subsequent will diverge.

The Recent strata form a common point of departure in all countries.—We derive one great advantage from beginning our classification of formations by a comparison of the fossils of the more recent strata with the species now living, namely, the acquisition of a common point of departure in every region of the globe. Thus, for example, if strata should be discovered in India or South America, containing the same small proportion of recent shells as are found in the Paris basin, *they* also might be termed Eocene, and, on analogous data, an approximation might be made to the relative dates of strata placed in the arctic and tropical regions, or the comparative age ascertained of European deposits, and those which are trodden by our antipodes.

There might be no species common to the two groups; yet we might infer their synchronous origin from the common relation which they bear to the existing state of the animate creation. We may afterwards avail ourselves of the dates thus established, as eras to which the monuments of preceding periods may be referred.

Numerical proportion of recent shells in the different Tertiary periods.—There are seventeen species of shells discovered, which are common to all the tertiary periods, thirteen of which are still living, while four are extinct, or only known as fossil*. These seventeen species show a connexion between all these geological epochs, whilst we have seen that a much greater number are common to the Eocene and Miocene periods, and a still greater to the Miocene and Pliocene.

We have already stated, that in the older tertiary formations, we find a very small proportion of fossil species identical with

* See the Tables of M. Deshayes in Appendix I.

those now living, and that, as we approach the superior and newer sets of strata, we find the remains of existing animals and plants in greater abundance. It is almost as difficult to find an unknown species in some of the newer Pliocene deposits, although very ancient and elevated at great heights above the level of the sea, as to meet with recent species in the Eocene strata.

This increase of existing species, and gradual disappearance of the extinct, as we trace the series of formations from the older to the newer, is strictly analogous, as we before observed, to the fluctuations of a population such as might be recorded at successive periods, from the time when the oldest of the individuals now living was born to the present moment. The disappearance of persons who never were contemporaries of the greater part of the present generation, would be seen to have kept pace with the birth of those who now rank amongst the oldest men living, just as the Eocene and Miocene species are observed to have given place to those Pliocene testacea which are now contemporary with man.

In reference to the organic remains of the different groups which we have named, we may say that about a thirtieth part of the Eocene shells are of recent species, about one-fifth of the Miocene, more than a third, and often more than half, of the older Pliocene, and nine-tenths of the newer Pliocene.

Mammiferous remains of the successive tertiary eras.—But although a thirtieth part of the Eocene testacea have been identified with species now living, none of the associated mammiferous remains belong to species which now exist, either in Europe or elsewhere. Some of these equalled the horse, and others the rhinoceros, in size, and they could not possibly have escaped observation, had they survived down to our time. More than forty of these Eocene mammifers are referrible to a division of the order Pachydermata, which has now only four living representatives on the globe. Of these, not only the species but the genera are distinct from any of those which have been established for the classification of living animals.

In the Miocene mammalia we find a few of the generic forms most frequent in the Eocene strata associated with some of those now existing, and in the Pliocene we find an intermixture of extinct and recent species of quadrupeds. There is, therefore, a considerable degree of accordance between the results deducible from an examination of the fossil testacea, and those derived from the mammiferous fossils. But although the latter are more important in respect to the unequivocal evidence afforded by them of the extinction of species, yet, for reasons before explained, they are of comparatively small value in the general classification of strata in Geology.

It will appear evident, from what we have said in the last volume respecting the fossilization of terrestrial species, that the imbedding of their remains depends on rare casualties, and that they are, for the most part, preserved in detached alluvions covering the emerged land, or in osseous breccias and stalagmites formed in caverns and fissures, or in isolated lacustrine formations. These fissures and caves may sometimes remain open during successive geological periods, and the alluvions, spread over the surface, may be disturbed, again and again, until the mammalia of successive epochs are mingled and confounded together. Hence we must be careful, when we endeavour to refer the remains of mammalia to certain tertiary periods, that we ascertain, not only their association with testacea of which the date is known, but, also, that the remains were intermixed in such a manner as to leave no doubt of the former coexistence of the species.

In the next page will be found a Synoptical Table of the Recent and Tertiary formations alluded to in this chapter.

N.B. By aid of this table, the reader will be able to refer almost all the localities of the Pliocene formations enumerated in the Tables of M. Deshayes (Appendix I.) to the newer or older division of the Pliocene period established in the foregoing chapter.

Synoptical Table of Recent and Tertiary Formations.

PERIODS.		Character of Formations.	Localities of the different Formations.
I. RECENT.	. . . . . .	Marine.	Coral formations of Pacific. Delta of Po, Ganges, &c.
		Freshwater.	Modern deposits in Lake Superior—Lake of Geneva—Marl lakes of Scotland—Italian travertin, &c.
		Volcanic.	Jorullo — Monte Nuovo — Modern lavas of Iceland, Etna, Vesuvius, &c.
II. TERTIARY.	1. Newer Pliocene.	Marine.	Strata of the Val di Noto in Sicily. Ischia, Morea? Uddevalla.
		Freshwater.	Valley of the Elsa around Colle in Tuscany.
		Volcanic.	Older parts of Vesuvius, Etna, and Ischia—Volcanic rocks of the Val di Noto in Sicily.
	2. Older Pliocene.	Marine.	Northern Subapennine formations, as at Parma, Asti, Sienna, Perpignan, Nice—English Crag.
		Freshwater.	Alternating with marine beds near the town of Sienna.
		Volcanic.	Volcanos of Tuscany and Campagna di Roma.
	3. Miocene.	Marine.	Strata of Touraine, Bordeaux, Valley of the Bormida, and the Superga near Turin—Basin of Vienna.
		Freshwater.	Alternating with marine at Saucats, twelve miles south of Bordeaux.
		Volcanic.	Hungarian and Transylvanian volcanic rocks. Part of the volcanos of Auvergne, Cantal, and Velay?
	4. Eocene.	Marine.	Paris and London Basins.
		Freshwater.	Alternating with marine in Paris basin—Isle of Wight—purely lacustrine in Auvergne, Cantal, and Velay.
		Volcanic.	Oldest part of volcanic rocks of Auvergne.

10

Reprinted from *London Edinburgh Philos. Mag.*, ser. 3, **7**(37):46–50 (1835)

On the Silurian System of Rocks. By Roderick Impey Murchison, *F.R.S., Vice-President of the Geological and Royal Geographical Societies, &c. &c.**

GEOLOGISTS having long felt that the older sedimentary deposits required a systematic examination, I have devoted the last five years to the study of this class of rocks, hoping thereby to fill up certain pages which were wanting in the chronology of the science†. A table published last year was the first attempt to convey to the geological student a correct view of the thickness, variety of strata, and fossil organic contents of a vast system, which, though arranged by nature in the most lucid order of succession, had not previously been pointed out. These rocks, rising from beneath the old red sandstone in Herefordshire, Shropshire, Radnorshire, Brecknockshire, Monmouthshire, and Caermarthenshire, and each distinguished by separate and *peculiar organic remains*, were respectively named after those localities where each of them could be best studied, and their places in the series most clearly established. I have no change to an-

* Communicated by the Author.

† See Lond. and Edinb. Phil. Mag. 1832 to 1835, in the Proceedings of the Geological Society.

nounce in the order detailed in that table (see Lond. and Edinb. Phil. Mag., vol. iv. p. 370), but I wish to simplify it by the abandonment of double names, as applied to any one formation, and by the adoption of the names of those places only where the respective rocky masses lie in juxtaposition.

The names finally adopted, and which will be incorporated in a work now in preparation on this subject, are,

1. *Ludlow rocks*, divided into upper and lower Ludlow rocks, with a central zone of limestone: in this formation no change of name is proposed.

2. *Wenlock limestone* and shale (*equivalent, Dudley*).

3. *Caradoc sandstones.* This name, supplying the place of the Horderley and May Hill rocks, has been derived from the striking and well-known ridge of Caer Caradoc, on the eastern flanks of which, and lying between it and the Wenlock Edge, are exhibited those peculiar strata which are the equivalents of the shelly sandstones of Tortworth.

4. *Llandeilo flags* (preferred to "Builth and Llandeilo"). When this table is reprinted, there will naturally be found many additions to the organic remains, some identifications of British with foreign species, and numerous corrections.

Notwithstanding the adoption of these names, there was still required a comprehensive term by which the whole group could be designated, and at once distinguished from the *old red sandstone* above, and the *slaty rocks* below. Without such a collective name for the group, I found it impracticable to proceed with the work which I had engaged to complete, it being essential to the clear exposition of the subject, no longer to speak of these deposits as "transition rocks" or "fossiliferous grauwacké." The term 'transition' might indeed, have been retained, if for no other reason than to impress upon foreign geologists, (the Germans particularly,) how vast a difference exists between the geological horizon of the mountain or *carboniferous* limestone and that of the limestones of Ludlow and Wenlock, which are not only separated by many thousand feet of strata from the limestone of the carboniferous system, but, further, contain an entirely distinct class of organic remains. It was, however, utterly hopeless to use the word 'transition' in any definite sense as applied to these lower deposits, seeing the extent to which it had been abused. By some it was confined to those older rocks in which the earliest traces of organic remains were supposed to be observed, whilst others had more recently so expanded the meaning as to comprehend in it the whole of the carboniferous series! Thus at a period when, from the rapid advances of the science, it had become indispensable to define the boundaries of groups naturally distinct from each other, dissimilar things were still confounded under one common name! and hence every geologist with whom I am acquainted had been for some time agreed upon the expediency of obliterating the term. The name 'transition' is, in truth, not applicable to any one class of stratified deposits in preference to another. Thus, for example, within the area of a map now preparing for publication and embracing parts of ten counties only, I shall be able to show *transitions* into every formation, beginning with the inferior oolite and terminating in descending order with the Llandeilo flags, many thousand feet below the old red sandstone; whilst the latter overlie other fossiliferous masses, the relative ages of which yet remain to be worked out! In various memoirs read before the Geological Society I have described these rocks as "fossiliferous grauwacké," but this term is in reality a misnomer, as the group contains few if any strata of the true grauwacké of German mineralogists. But whilst this system contains no such beds, it is underlaid and sometimes in discordant stratification, by a vast series of slaty rocks, in which much genuine grauwacké *is* exhibited. It was therefore manifest that if used at all in geological nomenclature, the term 'grauwacké' must be rejected as inapplicable to the first great system below the old red sandstone, and restricted to rocks which were *now* proved to be of much higher antiquity. My friend Professor Sedgwick will doubtless soon dispel the obscurity which hangs over these grauwacké rocks, with which his labours in Wales and Cumberland have so well enabled him successfully to grapple.

To return, however, to the system under review, I was urged by leading geologists both at home and abroad to propound an entirely new name for it. In consonance, therefore, with those views which have rendered the names used by English geologists so current throughout the world, I venture to suggest, that as the great mass of rocks in question, trending from south-west to north-east, traverses the kingdom of our ancestors the Silures, the term "Silurian system" should be adopted as expressive of the deposits which lie between the old red sandstone and the slaty rocks of Wales, including, as above detailed, the Ludlow, Wenlock, Caradoc, and Llandeilo formations. One of the largest of these formations, to which, indeed, the Llandeilo flags are frequently subordinate, has been named after the bold and picturesque ridge of Caer *Caradoc* in Shropshire.

I further propose that the system be subdivided into "Upper" and "Lower Silurian rocks," the former embracing the deposits of "Ludlow" and "Wenlock," the latter those of "Caradoc" and "Llandeilo." By this arrangement the observer will not be embarrassed when he finds that certain typical strata have disappeared. Thus, for instance, when the zones of limestone thin out, by which the formations of Ludlow and Wenlock are subdivided, it is no longer practicable to mark lines of separation between them. Under such circumstances the united mass will simply be described as "*Upper Silurian rocks*," whilst, wherever similar causes prevent the separation of the "Caradoc sandstones" from the "Llandeilo flags," the two will be termed "*Lower Silurian rocks*."

A wood-cut is annexed, explanatory of the manner in which these deposits are connected with the old red sandstone above them, with each other, and with the slaty grauwacké rocks beneath them. Although the lower Silurian rocks are frequently unconformable to the schistose grauwacké, as represented in the wood-cut, there are situations in which a passage from the one to the other may be detected.

Having alluded to a work which is in preparation upon the geology of the country of the Silures, and of several of the surrounding counties, I may take this opportunity of acquainting those friends who have fostered the undertaking, that the author is unceasingly occupied in promoting its completion. Geologists are not, I am sure, among those who feel surprise at the delay; for they well know that so large an area of country, and one so little previously examined, could not have been adequately described without several years of hard labour. In the mean time I may state that a map is now constructing, being a precise reduction of the Ordnance Maps, extending over the greater part of ten counties, all of which have been geologically coloured in the field; together with about twenty plates of engravings of fossils, many of which are of new species, and a vast number of coloured sections, vignettes, &c., all of which will, I trust, be completed so as to enable me to lay the work before the public in a few months.

[*Editors' Note:* Material has been omitted at this point.]

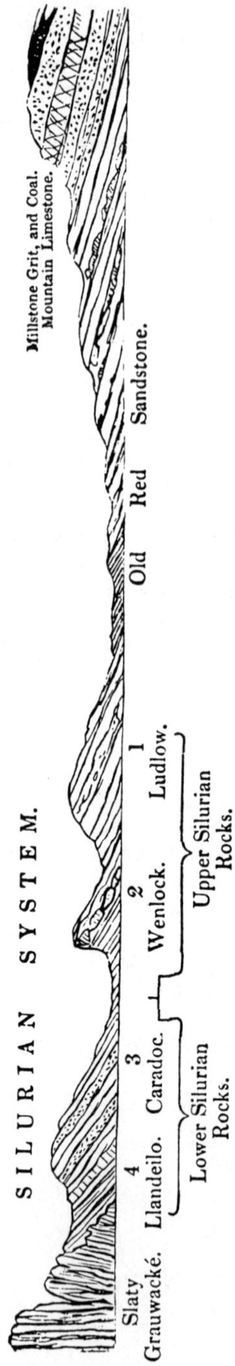
SILURIAN SYSTEM.
Slaty Grauwacké.
4 Llandeilo.
3 Caradoc.
Lower Silurian Rocks.
2 Wenlock.
1 Ludlow.
Upper Silurian Rocks.
Old Red Sandstone.
Millstone Grit, and Coal.
Mountain Limestone.

11

Reprinted from *Geol. Soc. London Trans.*, ser. 2, **5**:215–218, 221 (1836)

Remarks on a Section of the Upper Lias and Marlstone of Yorkshire, showing the limited vertical range of the Species of Ammonites, and other Testacea, with their value as Geological Tests.

By LOUIS HUNTON, Esq.

Communicated by J. F. Royle, Esq., F.G.S.

[Read May 25, 1836.]

ON presenting a stratigraphical arrangement of the fossils in the lias formation on the coast of Yorkshire, I wish to offer the following remarks relative to their geological distribution.

The section (p. 221.) refers to that part of the coast between Whitby and Redcar, in which the Loftus and Boulby Alum Works are situated, and was named by General Mudge "Easington Heights." Its greatest elevation is 681 feet. It includes the inferior with a part of the great oolite series; and the lias is better developed in this place than in any other part of the coast, the upper shale having a thickness of 200 feet, and the marlstone of 160; and there are about 150 of the lower shale also exposed.

With regard to the correctness and universality of the order given to the fossils, the position of no one species was fixed till after several had been extracted from the situation indicated in the section, as well as in various different localities from the Alum Hills, fronting the vale of Thirsk, to the peak near Whitby; and in the majority of instances, from thousands of specimens afforded by the quarrying of the lias shale in the manufacture of alum. Though it would have been easy to have added many more species, the geological situation of which is determined, yet wherever the limits of distribution have not been clearly defined, they have been omitted.

Upper Lias Shale.—Commencing at the surface of this deposit, the first fossil present in any quantity, is the *Nucula ovum* (see section, p. 221.); which, after descending a few feet and entering what is technically called the "hard or cement stone seam," occurs in great abundance; but below that bed it becomes gradually scarcer and smaller, so that after a descent of 60 feet, it is rarely met with, and I believe has never been found so low as the jet rock. The *Orbicula reflexa,* and the small, but delicate and beautiful *Plagiostoma pectenoideum* are associated with the Nucula, and similarly distributed. It may be worthy of observation, that the *Nucula ovum* is singularly characteristic of that portion of the upper shale used in the manufacture of alum, and that wherever the shale occurs without this shell, it has been found unfit for the remunerative production of the salt*. This is owing to the nature of the shale, the upper part, where calcined, being richest in sulphate of alumina; not, as might be supposed, from containing originally *more sulphur*, but a *less proportion*, the lower strata having such an excess of pyrites and bitumen, that the contents of a kiln made from such schistus, invariably overroasts. A large Pallustra, or Amphidesma, (the *Amphidesma donaciforme* of Phillips) is also plentiful in the upper part; but it is occasionally found in the lower, and a similar shell occurs in the Marlstone.

But of all organic remains, the Ammonites afford the most beautiful illustration of the subdivision of strata, for they appear to have been the least able, of all the Lias genera, to conform to a change of external circumstances. Of the species, most plentiful in the upper part, scarcely any are found in the lower; except a few doubtful individuals of *A. communis* (the prevailing species in the "hard seam") which are now and then seen so low as the jet rock; and though *A. fimbriatus* and *A. heterophyllus* are found in that stratum, they are always dwarfish, seldom attaining three inches in diameter, whereas in the upper beds, they are met with twenty inches in diameter.

The portion of the series mentioned before as the "Jet Rock," is a very compact and highly bituminous bed, from 20 to 30 feet thick and 150 feet below the surface of the formation. It contains many flattened, pyritous nodules, which generally inclose organic remains. The rock itself is also so very sulphureous, that when a small heap of it was calcined at the Loftus works some time ago, the melted bitumen and sulphur flowed in flaming streams. The whole of the jet used in the manufacture of ornaments, is extracted from this stratum; for when occasionally met with higher up, it is like coal in its fracture, and too brittle to work well; and in the inferior strata, as the marlstone, it forms merely a thin coat, enveloping a lump of fossil wood.

* The Alum Works, of which lias shale is characterized by the *Nucula ovum,* are the Peak, on the extreme east, and Stow Brow, Godeland Banks, Sandsend, Kettleness, Boulby, Loftus, Selby Hogg, and Osmotherly, on the extreme west. Failures have occurred to different works in attempting the calcination of the lower portions of the Upper Lias, or those strata undistinguished by the *N. ovum.*

The seams are sometimes 20 feet long by 6 feet broad, but they are often smaller, and their greatest thickness is 3 inches.

The Jet Rock is also peculiar for containing the remains of the Teleosaurus; for though Ichthyosauri and Plesiosauri appear pretty equally distributed through the upper shale, this (fluviatile?) reptile is rarely found but in, or immediately above, the Jet Rock. Its bones are generally scattered in separate nodules, the upper jaw being scarcely ever united to the lower. With the exception of the specimen in the Whitby Museum, I am not aware of any instance of this animal having been found entire, whereas the other saurians seldom want more than the head or paddles.

On inspection of the section, it will be perceived, that the Jet Rock has its peculiar suite of Ammonites, and here, as was before observed, *A. heterophyllus* and *A. fimbriatus* make their first appearance, being about the size of a half-crown, while in the "hard seam" they attain the enormous magnitude of 1 and 2 feet in diameter. The *Ammonites Crassulus,* (*n.s.*) common to the whole upper lias, on the other hand, has its greatest development in the Jet Rock, gradually diminishing from $1\frac{1}{2}$ inches to $\frac{1}{2}$ an inch in the overlying beds. An interesting, though perhaps it may appear trifling, circumstance attending this fossil is, that as it decreases in size it becomes more numerous, many hundreds being then found together.

The lowest bed of the upper shale is hard, compact and sandy, and is singular for its great barrenness of fossils, occurring as it does immediately beneath one so prolific as the Jet Rock. This scarcity, combined with the arenaceous nature of the stratum, may perhaps afford some hints for elucidating, hereafter, the anomaly of fluviatile reptiles being mixed up with pelagian shells and fishes. It is a plausible speculation, that during the formation of the upper lias, the bottom of the sea in which it was deposited, was gradually settling. This conjecture is supported, by the greater prevalence of vegetable matter and fluviatile (?) reptiles in the lower part, and the increase in number of the larger species of Ammonites, Nautili, and Belemnites in the upper. This conjecture is further supported by the mineralogical nature of the schistus, silex being more abundant in the inferior parts, and alumina in the superior, an arrangement which would naturally take place, if the coast during that epoch gradually sunk, and the land, the detritus of which afforded the material for the lias, as gradually receded; the finer argillaceous sediment being consequently carried into the deeper and more and more distant parts.

Marlstone.—On entering this series, we at once perceive a striking contrast to the upper lias, for independently of the decided mineralogical difference, not only the species and genera, but the *order* of Testacea have undergone a total revolution. Instead of a preponderance of Cephalopods and a scarcity of bivalves, we now have an overwhelming majority of the latter, some of the beds being almost constituted of Cardia and Gryphytes, which have also an extensive vertical distribution. *Avicula Cygnipes, A. inæquivalvis* and *Pullastra antiqua* are equally abundant in the whole series; but other shells, as *Cardium truncatum,* occur more plentifully in the inferior strata, whilst *Terebratula triplicata* and *T. trilineata* crowd the superior.

The species of Ammonites, though few in number, are, however, highly characteristic; thus we find *A. vittatus* about the centre of the series, confined to a very small range, associated in nodules with the *Cardium multicostatum, Turbo undulatus,* and *Pecten planus*; but the two latter occur in other parts of the formation. The *A. maculatus,* is constantly found at the junction of the *marlstone* with the *lower lias,* which here pass so gradually into each other, that it is impossible to determine where the sandstones end and the blue shale begins. I have long sought for *A. maculatus* in the upper and central portions of the *marlstone,* but have never found it many feet above the junction beds; and though this and other Ammonites from unequal geographical distribution, may be more abundant in one place than in another (*A. maculatus* is in greatest number at Staithes, *A. Hawskerensis* at Hawskerbottoms), yet they constantly maintain an invariable relative position. *A. Hawskerensis* I believe to be also a junction Ammonite, but from its rarity in other situations, than that mentioned, it is not entered in the table.

The above description may not, in some instances, exactly accord with previous statements, but one great source of error has hitherto been, the collecting of specimens from the debris of the whole formation, accumulated at the foot of cliffs or other similar situations, where they have long laid, and the inferring of their position from the nature of the matrix. A strong temptation to this method arises from the facility it affords of obtaining finer fossils than can possibly be procured from newly quarried nodules, the hard nature of which renders it very difficult to extract the fossil entire, without long exposure to air and moisture.

Annexed is a systematic list of the fossils found in the upper lias and marlstone, and noticed in the section.

[*Editors' Note:* Material has been omitted at this point.]

Section of the Upper Lias and Marlstone at Rockliff, near Whitby (Easington Heights, Mudge's Survey), showing the limited Distribution of certain Species of Testacea, and the Order Cephalopoda in particular, Local Dip S.W. One Foot in Ten Feet.

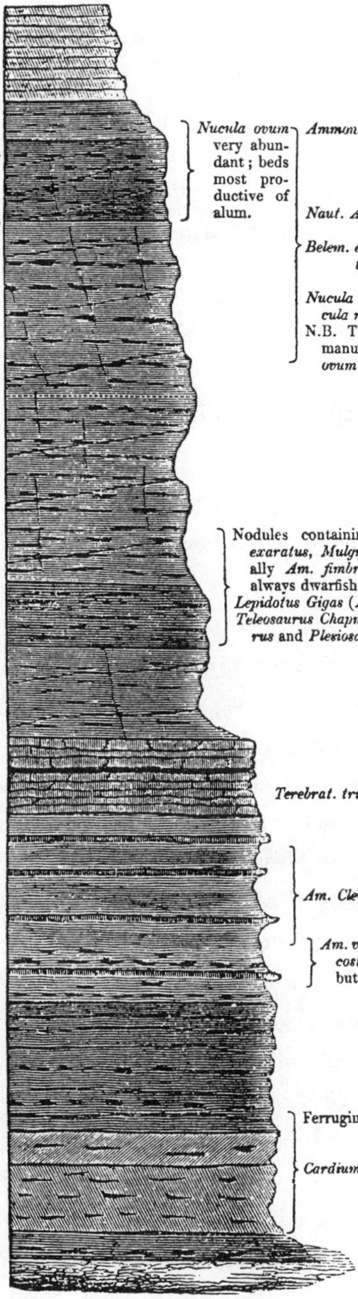

12

Reprinted from *Geol. Soc. London Trans.*, ser. 2, **4**(Pt.II):103-105, 328-334, Plate Xb (1836)

Observations on some of the Strata between the Chalk and the Oxford Oolite, in the South-east of England.

By WILLIAM HENRY FITTON, M.D., F.R.S. P.G.S., &c.

[Read June 15, 1827.]

(1.) IN a paper published in the Annals of Philosophy for November and December 1824*, I gave an account of the order and characters of the strata which occur beneath the chalk on the coast of part of the Isle of Wight and of Dorsetshire, and stated some reasons for supposing that a similar arrangement would be found to exist in the interior of England.

The principal objects of that paper were; *First,* to distinguish as a separate group, the series of strata now called the Lower Green-sand;—which had previously been confounded either with the beds containing green particles immediately below the chalk, or with the sandy and ferruginous strata conspicuously exhibited on the coast at Hastings, and then called "Iron-sand." *Secondly,* to indicate more clearly than had been done before, the peculiar characters of the group, which succeeds in a descending order to that just mentioned, and is remarkably distinguished by its fossils from the strata immediately in apposition with it, both above and below. For this latter group, which includes the Weald clay, the sand of Hastings, and the Purbeck limestone, and is well entitled to a separate denomination, I have adopted the name of "*Wealden,*" proposed by Mr. Martin, in his valuable memoir on the West of Sussex†.

(2.) The objects of the inquiries which have produced the following pages were, to compare some portions of the series of strata between the chalk and the Oxford oolite, in different parts of the South-east of England; to ascertain the existence of the Wealden in the interior; and, if possible, to determine its boundaries.

* Annals of Philosophy; New Series, 1824, vol. viii. pp. 365, 458, &c.

† "Memoir on a Part of Western Sussex": 4to, London, 1828.

In stating the result of this investigation, I shall give a series of sections of the strata below the chalk, at places which I have myself examined *, beginning on the coast near Folkstone, and following the outline of the chalk thence to the sea on the N. West of Norfolk. The relative situation of these places is shown in the map annexed to this paper†. The intermediate country, in general, I have not examined in detail, and some points of importance I have never seen; but those who may have opportunities of continuing the inquiry, will find, I hope, no difficulty in connecting their observations with mine.

(3.) The series of strata about to be described extends from the Chalk down to the Oxford oolite or coral-rag, and is composed of alternating but irregularly distributed beds of Sand, Clay, and Stone.

Among the sandy strata, it is important to discriminate not only between the Lower green-sand and that of Hastings, but to distinguish both from a third group, consisting principally of sand abounding in green particles, which lies beneath the Portland stone.

Clay, of several varieties, occurs in all parts of this series; but three groups, constituting the Gault, the Weald-clay, and the Kimmeridge-clay, derive peculiar importance from their generally occupying valleys, or depressions, at the foot of the escarpments of the Chalk, the Lower green-sand, and the Portland stone, respectively, and thus producing conspicuous natural features in the tracts where this succession is observable.

The stone of the tracts under consideration is either limestone; indurated sand-rock; chert; or siliceous matter intimately mixed with carbonate of lime, in the form of grit,—which has commonly a concretional structure, and seems to pass into continuous beds only by the approach and ultimate union of the concretions.

But the most remarkable distinction, in the suite described in this paper, arises from the great difference of character in the organized remains which the principal groups include. The fossils of the chalk and green-sands and those of the Portland stone, are all marine, and the species numerous. But in the *Wealden*, between the lower green-sand and the Portland stone, although the fossils are abundant as to quantity, the species are comparatively few, and by far the greater part of them belongs to fresh water. The whole of the phenomena, in short, presented by this remarkable assemblage of beds, are such as to accord with the hypothesis of their having been deposited in fresh water communicating with the sea.

* Plates X. a, and X. b.

† Plate IX.—The detail, upon a larger scale, will appear in the new edition of Mr. Greenough's Geological Map of England.

(4.) Combining these sources of distinction, the following arrangement and subdivisions may be adopted, the nomenclature of which, however exposed to criticism, is now probably too well established to be changed without inconvenience*.

CHALK....	*Upper.* *Lower.* *Marly.*	Fossils, marine: species numerous.
GREEN-SAND	*Upper green-sand.* *Gault.* *Lower green-sand.*	
WEALDEN..	*Weald-clay.* *Hastings sands.* *Purbeck strata.*	Fossils, for the greater part, of fresh water: species few.
Part of the OOLITIC SERIES	*Portland stone.* *Portland sand.* *Kimmeridge and Weymouth clay and sand.* *Oxford oolite* (Coral rag).	Fossils, marine: species numerous.

(5.) I proceed now to describe the Sections of these strata, beginning with the coast of Kent; and I shall connect with each section a list of such fossils as I have either found myself, or obtained on good authority from the places mentioned. The whole of the shells in these lists have been examined and named by Mr. James Sowerby, by whom also the drawings of the supposed new species were made, and the annexed engravings executed†. It is right to mention this explicitly, both that I may take the opportunity of expressing my acknowledgments for Mr. Sowerby's valuable and assiduous cooperation,—and that, being myself but very slightly acquainted with Conchology, I may place the portion of the following pages which relates to that subject on better authority than my own.

[*Editors' Note:* Material has been omitted at this point.]

* I have stated in another place (Annals of Philosophy, vol. viii. pp. 461, 462,) some objections to the employment of names for geological strata, which refer to characters not essentially connected with the structure or position of the objects to be designated. But the term green-sand, however faulty, besides the universal use of it in England, has been adopted both in Germany and France;—where, however, it may be regarded, in some measure, as free from the disadvantages of a significant name.

† Plates XI., *et seq.*

(168.) *Marine deposits coeval with the Wealden.*—It is obvious that, during a period of time sufficient for the accumulation of the Wealden, the deposition of matter in the adjacent seas could not have been inconsiderable; so that we might expect to find, interposed between the strata which then formed the

* "....que l'argile Veldienne qui contient les coquilles d'eau douce, n'est qu'une couche "particulière de terrains glauco-ferrugineux d'Angleterre."—p. 256.

† Passy, p. 272; and Plates I. and II.

‡ Ibid., p. 262, 265.

bottom of the sea, and the Lower green-sand, a series of beds coeval with the Wealden in point of date, but differing from it in possessing the characters of a marine deposit, and including marine shells and other productions of salt water;—with which, near the shore, the productions of the land, or even the freshwater shells of the rivers, might be occasionaly intermixed. And if the Portland strata constituted at that epoch, both the dry land, and the bottom of the sea, and were afterwards submerged, we ought now to find the Lower greensand, in some places immediately in contact with the Portland,—in others with the Wealden,—and in others again with the marine equivalent of this latter group.

Two results would probably attend the state of things here supposed, which are deserving of notice: 1st, That the Wealden and its marine equivalent could not both be found in the same place; and consequently, (since we have the former in England), that the *marine* beds of that date are not to be expected generally in this country: 2ndly, That the marine fossils of the beds cotemporaneous with the Wealden would probably be distinct, both from those of the Portland group beneath, and of the Green-sands above them; a consideration which gives peculiar interest to the fossils of this intermediate group.

The strata between the chalk and the oolitic system on the continent of Europe have not yet been sufficiently examined, to furnish all the evidence that may be expected upon this subject; but indications of such an equivalent to our Wealden as has been mentioned, have been already found in so many detached points, that its occurrence in other places, or even the existence of a continuous marine deposit of that age, is by no means improbable. Mr. De la Beche* has brought together evidence which shows that such a group exists in the department of the Haute Saône, in France; at Candern, in the Brisgau; near Aarau; in Poland; and on the confines of Silesia. To these may be added the Isle of Bornholm in the Baltic, and the vicinity of Helsingburg, in Scania, which have afforded specimens of fossil plants resembling those of our Wealden, along with marine shells; but at Bornholm, although the shells are also marine, they are generally such as may be supposed to have inhabited, either the estuary of a large river, or the seas immediately adjacent to the coast.

In most of the cases mentioned by Mr. De la Beche, a group of strata between the chalk and the oolitic system is found to contain pisiform iron ore; but the fossils which accompany that mineral are marine. A very extensive deposit of this kind in Poland and Silesia is described by Professor Pusch, which includes also argillaceous iron ore; and among its fossils are the genera Ammonites,

---

* Geological Manual, (1833.), p. 309.

Cardium, Venus, Trigonia, Sanguinolaria. The pisiform ore of the Haute Saône appears from the statements of M. Thirria to be above the Portland strata; but its relations to the superior formations are less distinct. It contains Ammonites, Hamites, Nerinea, Cirrus, Terebratula, and Pentacrinites, with teeth of fishes and of Saurians: and it is observable that the uppermost stratum in the group supposed by M. Thirria to represent the Portland stone, is "a compact conchoidal "limestone, a little tubercular, which contains a species of Paludina? near to Vivipara fluviorum "of Sowerby;"—a description which might very well be applied to some of our Purbeck strata.

M. Walchner has ascertained that the pisiform and reniform iron ores of Candern, belong to two distinct formations, of very different date. The lower rests upon beds of "Jura limestone", which he refers to the Coral-rag or Portland-stone, and which are decidedly below the molasse of Switzerland. The second and more recent pisolitic deposit occupies eroded cavities on the surface of several different formations; it consists of transported matter containing fossil bones, and is not covered by any solid strata. M. Merian is cited by M. Walchner in support of this view of the relations of these deposits; but in the description given by the former, of the strata which immediately cover the pisiform ore of Aarau, there are features very like those of our Wealden itself: for they are mentioned as consisting of "grit and bituminous schist passing into lignite, and "containing a great number of petrifactions, among which, though commonly indistinct, are a "Planorbis, and other freshwater shells."

The doubt which obviously presents itself with respect to the evidence above mentioned, is, whether a formation thus contiguous to the green-sands, and so very nearly coeval with them, may not really be a part of that formation itself; the presence of some fossils distinct from ours, not being sufficient to prohibit this identification. But the peculiar character and abundance of the iron ore, are important circumstances of difference.

The evidence respecting Bornholm occurs in a paper read before the Geological Society by Dr. Beck of Copenhagen*, who states that the beds below the cretaceous group in that island, exhibit many of the characters of a coal formation, containing coal and numerous impressions of ferns, among which are several of the genus Pecopteris; and with these was found the seed vessel of a Restiacea, considered by Dr. Beck as identical with one from the Hastings-sand at Heathfield in Sussex, which he himself had seen in Mr. Mantell's collection. The existence in the vicinity of Hoer, of fossil plants *analogous* to those of the Wealden, but probably belonging to the oolitic series, had been previously pointed out by Mr. Adolphe Brongniart†: but from what has been very recently mentioned to me by Professor Nilsson of Lund, after he had examined some of our best collections of the Wealden fossils, it is probable that a deposit contemporaneous with that formation may be found at Helsingburg in Scania, between the chalk and green-sands which occur near the coast of the Baltic, and the Lias, known to exist on the north-west of that place

The observations of M. Boblaye in the north‡, of M. Desnoyers § in the north-west, and of M. Dufresnoy||; in the south-west of France; with those of M. Elie de Beaumont, on the boundary of the Paris and London Basins¶; of Mr. De la Beche**, M. de Caumont††, M. Desnoyers‡‡,

* "Proceedings" (1835-6) vol. ii. p. 217. † *Annales des Sciences Naturelles*, 1825.
‡ Ibid., 1820, t. xvii. p. 35. § Ibid., 1825, t. iv. p. 353. || Ibid., 1829, t. xvii. p. 192
¶ Ibid., 1829, t. xvii. p. 354. ** Geol. Trans. 2nd Series, vol. i. p. 13.
†† *Mém. de la Soc. Linn. du Calvados*, t. i. 1824, p. 49 et 67; t. ii. 1825. p. 447, &c.
‡‡ *Mém. de la Soc. d'Hist. Nat. de Paris*, t. ii. 1825.

M. Constant Prevost*, and M. Passy †, on the coast of Normandy and the adjacent district; and more recently of M. Thirria, in the department of the Haute Saône‡, have shown that many of the beds between the chalk and the oolitic system in the different provinces of France, agree with each other, and with those of England, perhaps as nearly as the more distant portions of the same formations are found to do in our country: the Portland strata, which form the lower boundary of the Wealden, occurring at several detached points along their course; while above them, the chalk affords a limit which cannot be mistaken. It may deserve the inquiry, therefore, of resident geologists, whether traces at least, of all the intermediate groups between the Portland and the chalk may not be detected generally along the line of connexion; and I am not without hope that the task of comparative examination may be assisted by the preceding pages.

Among the points deserving of notice by those who examine the beds below the chalk in districts imperfectly known, I would mention especially the characters of the *Portland Sand*;—from the facility with which that formation may be confounded with other sands, likewise charged with green particles. I have reason to believe that a neglect of the distinctions between these different sands, and a consequent ignorance of the distance by which they are separated in the series of strata, has produced difficulty, and led to much confusion in the arrangement of the groups between the chalk and the oolitic system.

(169.) *Beds below the Wealden.*—The groups next below the Wealden, in the tracts described in the preceding pages, are connected by several characters; the Portland sand forms a transition to the Kimmeridge clay; and the latter, at its lower part becomes charged with sand and calcareous matter, and passes insensibly into the Oxford oolite. The fossils, also, of these associated groups are very much allied, and there is no reason to suppose that the deposition of this part of the series was interrupted by any important change of circumstances.

Portland strata.—The space occupied by the Portland strata, so far as they have yet been discovered in England, is bounded on the west by a line passing from Portland to the outcrop of the Portland stone in the Vale of Wardour, and thence towards Swindon. On the east of that line the Purbeck and Portland formations are visible only in the remaining portions of the once-continuous sheet, which seems originally to have invested a great part of Oxfordshire and Buckinghamshire. Whether the Portland beds exist, or ever have existed, under the sands of the Isle of Wight, Sussex, and Kent, it is now impossible to say; though it is probable that they did so, from their occurrence on the opposite coast of France, and their thickness in that quarter.

The peculiar circumstance relating to the junction of this formation with the Purbeck beds, is the distinct character of superficial soil, which the bed in which the trees and Cycadeæ are found, exhibits in the Isle of Portland. Trees and plants have occurred in the upright position in many other places, and in other parts of the series of strata; but the soil in those cases had lost its recent character, and been in a great measure assimilated to the surrounding matter. Some other thin beds alternating with the lowest strata of the Purbeck limestone, have likewise more

* *Mém. de la Soc. d'Hist. Nat. de Paris*, t. ii. 1825. p. 389.
† *Sur le Département, &c.* 4to, Paris, 1833.
‡ *Mém. de la Soc. d'Hist. Nat. de Strasbourg*, 1830, t. i.

or less of the character of a mechanical agglomerate; but none of them approach to the gravel-like texture and composition of the "Black-dirt" of Portland and the adjacent coast.

The small extent which the Portland strata occupy with reference to the Kimmeridge clay, is a remarkable fact; showing either that the former, like the Coral-rag, was originally deposited in detached portions;—or that the portions of the Portland group that we now see, are only the remaining and eroded borders of one continuous coating originally lodged upon the clay.

It is remarkable, also, that the *Wealden* has not yet been found beyond the limits of the Portland group; never reposing, as the green-sands are frequently seen to do, on any of the formations beneath the Portland stone: but before this exclusion be adopted, it will be necessary to examine more completely the counties on the north of Buckinghamshire. Reciprocally, in all the places where the Portland beds have hitherto been found, one or more members of the Wealden group are connected with it. It is possible, therefore, that the Portland strata formed the only land, at the period when the freshwater beds of the Wealden group began to be deposited.

The *Kimmeridge clay* thus more widely diffused than the Portland and the Wealden, forms the continuation upwards, of a long series of alternate strata of oolitic limestones, sand, and clay. The extent, thickness, and development, of this formation, vary much in different places, the upper beds alone being visible at the point from whence the name has been derived. The relations of the groups are fully seen on the coast near Weymouth, and still more completely on the shore of the Boulonnois, and near Scarborough: but, although the clay extends, without interruption, from Dorsetshire to Norfolk, it is seldom seen to advantage in the interior of England. Near Oxford, the lower members, which form the transition from the bituminous clays of Kimmeridge to the Oxford oolite, are evidently wanting; and the surface of the calcareous reestone beneath the clay, is deeply worn and eroded.

(170.) *Composition of Strata.*—Among the circumstances relating to the composition of the strata above mentioned, below the chalk in the South-East of England, the following are deserving of attention.

1. Proofs of the formation of stone, in the midst of strata of gravel, sand, and clay, distinctly of mechanical origin. This fact, which is of universal occurrence throughout the series of secondary strata, is especially remarkable in the Lower green and the Hastings sands of Kent and Sussex: sex: the former containing masses of siliceous grit, chert, and chalcedony, and concretions of chert, evidently formed subsequently to the division of the beds which inclose them, yet traversed by the same lines of false stratification which pervade the looser matter:—(22.) and (23.)

2. The change, in the fossil trees of Portland, of the original woody substance into silex; and the contrast of their composition with that of the surrounding matter in the "Dirt-bed", in which fragments of soft limestone are abundant; the strata, both above and below, likewise, consisting principally of carbonate of lime. In the Lower green-sand also, at Woburn, the petrifying petrifying matter of the coniferous wood is silex*.

3. The shells preserved in the green-sands of Kent, consist of carbonate of lime; but in the sands of Blackdown, though shells occur in still greater numbers, the calcareous matter has almost entirely disappeared, and with very few exceptions has been replaced by chalcedony; and the surrounding sand, instead of effervescing with acids, imbibes them tranquilly.

---

* The fact remarked by Mr. Brown, that all the fossil wood hitherto discovered in the strata mentioned in these pages, is either monocotyledonous, coniferous, or cycadeous, has been already adverted to: (112.) p. 225.

4. The occurrence of oolite containing freshwater shells, near the bottom of the Purbeck strata at Combe Wood (144.), p. 275-6, is another fact deserving of notice, and may perhaps explain some doubtful appearances in the uppermost bed of the Portland Series (the *Roche*); which, although full of marine petrifactions, and commonly oolitic, has very much the aspect of freshwater limestone. It can easily be imagined that the bank, or land, consisting of the Portland strata, when first raised above the sea, was covered with marine remains, which might subsequently have been cemented together, at the bottom of a freshwater lake.

5. The existence of pisolitic (or granular) oxide of iron, in the Wealden, is probably not confined to the coast near Hastings, where I found it to be generally diffused; p. 166. The fact of its occurrence in this formation is the more deserving of notice, from the great abundance of that ore in the *marine* deposits supposed to be contemporaneous with our Wealden, in other parts of Europe: supra, (168.) p. 330.

(171.) *Similarity of Deposits of different Epochs.*—The mineralogical resemblance between some of the strata above and beneath the Wealden, is very remarkable.—The upper part of the Portland formation in the Vale of Wardour is so like chalk, that it bears that name in some of the quarries; and both there and upon the coast of Dorsetshire it contains flint nodules, not distinguishable from those of the chalk.—The Portland sand abounding in green particles, cannot, in itself, be distinguished from the green-sands above the chalk, nor from those below; the green matter in all is of the same nature; the calcareous concretions are like those of the "Kentish-rag": yet the deposition of these groups was separated by intervals of time sufficient, in the first two cases, for the accumulation of all the Wealden strata, and in the two latter, for that of the chalk.—The strong resemblance of the variegated sands and marl of the Wealden to those of the new red sandstone, and of both to the tea-coloured and variegated marls of the beds above the chalk in the Isle of Wight, has been already mentioned.—The Wealden, again, has many striking points of resemblance to the coal measures. The shale, sandstones, and clay iron-stone of the latter are not distinguishable from those of the Hastings-sands, which are, in fact, a coal formation: and in addition to the Unios and other freshwater shells long known in the coal formations, the recent discovery of Cypris, in freshwater limestone, among the coal measures in Scotland*, is another point of agreement.

These resemblances, of which other examples might easily be given, in deposits separated by great intervals of time, demonstrate the identity of the processes which are still going on with those of the more ancient epochs,

* See, however, upon this subject, the remark of Mr. Sowerby, as to the possible resemblance of the fossil crusts of the *marine* genus Cytherina, to those of Cypris: Appendix, p. 345, under the head *Cypris Valdensis.*

and prepare the mind for the reception of other proofs of that continuity of action, which has of late been so ably discussed.

(173.) *Fossils.*—A systematic list of all the fossils mentioned in the preceding pages* is given in the Appendix B: and I have connected with it a series of columns, exhibiting at one view the stratigraphic distribution of the several species. My collection is so incomplete, that many of the columns are unoccupied; and most of them afford very inadequate representations of the facts: but it will be advantageous to keep in view the completion of such a list, whenever sufficient materials shall be obtained to fill it up. The number of specimens collected from any given district or series of strata, depends on so many circumstances purely accidental,—the texture of the matrix, the number of open quarries, the relative skill and activity of collectors,—that the richest collections can very seldom be regarded as fairly representing the fossil contents of the groups from which they are obtained; and there still are large tracts, in every part of the English series, the fossils of which are unknown. The proportion, therefore, which the species we are acquainted with, bear to the whole number contained in the strata, is so far from being determined, that any general reasoning upon this subject at present would be of little value.

From the imperfect lists of fossils given above, it would seem that even short distances are attended with greater variation in the contents of the strata, than might have been expected; and that although some species pervade the whole range of the formations in which they are found, others occur in several of the tracts described which have not yet been obtained elsewhere.

I cannot conclude these pages without expressing my great obligation to Mr. Lonsdale, for his kind and unremitting assistance during their passage through the press; which has been doubly acceptable, from its having been conferred on me by one of my most valued personal friends.

W. H. F.

London, August 16, 1836.

* It may be proper to mention here, that the whole of the author's collection referred to in the preceding lists, has been presented to the Geological Society, and will be found in their Museum.

Fig. 13. Abstract section of the Series of Strata represented in this Plate (X.b.) and in Plate X.a.

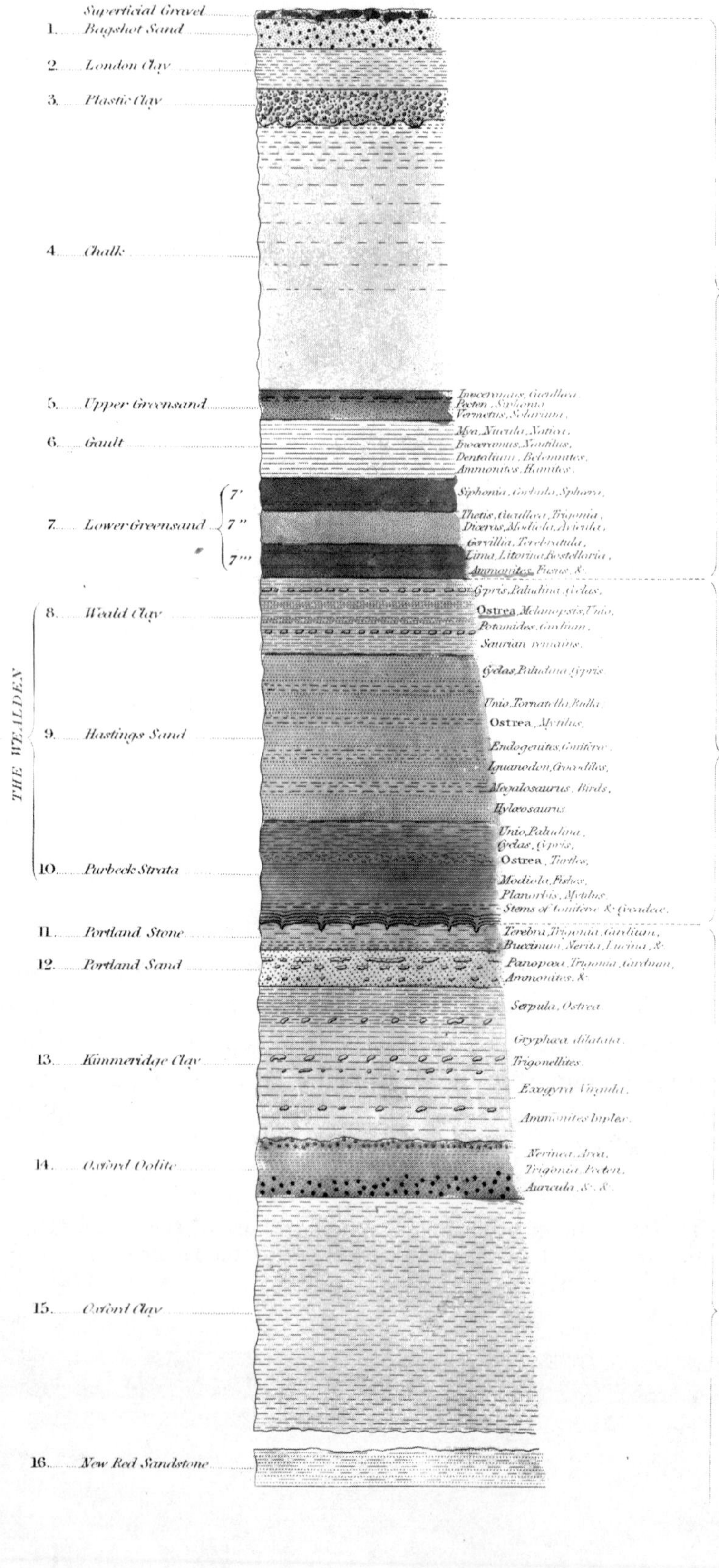

13A

GEOLOGIC OBSERVATIONS ON THE JURA OF SOLEURE [1838]

A. Gressly

These excerpts were translated expressly for this Benchmark volume by Barbara Conkin, University of Louisville, from Soc. Helv. Sci. Nat. Mem. ***2:****10–12, 20–22 (1838).*

In the regions I have studied, perhaps more than everywhere else, very varied modifications, whether petrographic, or paleontologic, interrupt, at every turn, the universal uniformity that has been attributed until now to the different terrains in different countries. They also reappear successively in several terrains and strike the geologist who wishes to study the nature of our Jurassic chains with astonishment.

Often, he will pause, with surprise, before formations that he believed to know well for a long time. Misled by a too confident faith in customary geologic dogma, which often generalizes from purely local facts, he perhaps mistakes characters he has seen so far to belong to such and such terrain and to such and such a division in particular; he will even tend completely to confound several terrains and will even come to doubt their existence.

Meanwhile, a new, more attentive, more prolonged and especially comparative examination will make him recognize . . . facts that will open a new route to geologic research and of which the scope will be calculable only when these will have attained their whole development and the results to which they lead have been appreciated for their proper worth.

My intention was to apply to the country where I live the geologic laws that M. Thurmann has verified with so much talent and success in the terrains of the neighboring Jura of Bern; but I found myself soon forced to modify these laws successively, according to the diverse regions that compose the Jura of Soleure, and the study of these diverse regions has necessitated on my part a system of research different than what one usually practices. In place of contenting myself with a certain number of vertical sections taken as descriptive types, I have followed each terrain in its horizontal extent, as far as possible, in order to study all its modifications.

I have succeeded, in this manner, in recognizing in the horizontal dimension of each terrain, diverse, very decided modifications that present constant peculiarities in their petrographic constitution as well as in the paleontologic characters of the assemblages of their fossils and that are governed by suitable and scarcely variable laws.

At once, two principal facts everywhere chracterize the assemblages of modifications that I call *facies* or *aspects of terrain:* The one consists in that a *given petrographic aspect of any terrain necessarily implies, wherever it occurs, the same paleontologic assemblage; the other is that a given paleontologic assemblage rigorously excludes fossil genera and species frequent in other facies.*

If it happens, by chance, that certain genera and species, peculiar to a facies, are found in another, it is a constant rule that the individuals of these genera and species are much more rare, much less developed and less characteristic than in the facies or general assemblage to which they characteristically belong. Likewise, it is always where the geognostic characters of the facies are more developed that the paleontological assemblages are more pronounced, that the genera and species are more numerous and the individuals more characterized, more developed and often in a perfect state of preservation. If the facies present intermediate modifications, of mixed geognostic characters, the fossils also partake of these blurred characteristics. They are then generally less frequent, badly preserved, and badly developed and belong to less numerous genera and species. Nevertheless, one sometimes finds there very identical fossils, which belong to species for the most part different from those of the principal facies or which are found only rarely in these latter.

I will support these curious facts in the description of each terrain we will study successively because they appear to me to repeat, on another scale and under other connections, a law similar to the law that rules, in general, the great geologic divisions. They play, with regard to the different aspects of a terrain, in its horizontal dimension, the role that certain fossils such as Ammonites noueuse, Gryphea arquee, Baculites, etc. fill, with regard to the great formations that, in the vertical dimension, characterize them in so precise a manner.

I think that the modifications, whether petrographic or paleontologic, that prove a terrain in its horizontal extent are produced by the different locations and other circumstances that so powerfully influence, nowadays too, the different genera and species of organized beings that populate the ocean and the present seas. At least I have often been astonished to find again in the distribution of our fossils the laws of organic associations, and in the assemblage of corresponding petrographic and geognostic characteristics the conditions of existence that rule in the undersea world.

[*Editors' Note:* Material has been omitted at this point.]

After having determined the principal facies that prevail in our Jurassic terrains, it remains to cast a glance over the laws that are basic and that rule their distribution, whether in the vertical or horizontal direction.

This first law results from facts that I indicated previously: *That each facies of whatever terrain presents very distinct characters, whether petrographic and geognostic, or paleontologic proper, which are in direct contrast to the characters that one generally assumes in the terrain and with characters of other facies of the same geologic horizon.* This law will serve to correct the classification of certain terrains and their subdivisions, in defining their position, and will thereby prevent grave errors in the determination of geologic horizon of localities separated by great massifs affecting particular characters.

A second law is intimately tied to the first: *Facies of the same petrographic and geognostic nature effect, in different terrains, very similar paleontologic characters and succeed each other very generally through a more or less numerous series of terrains superposed one on the other.* This law appears to me to present great interest on account of the zoologic determinations of fossils and the application one makes of it to characterize the terrains and their secondary divisions paleontologically. There is often a great resemblance in the organic forms of fossils of analogous facies, although they belong to very diverse terrains, and this resemblance is in large part the reason that a good number of fossils of different terrains have been identified. This resemblance has led to the rather accepted opinion that identical fossils are found not only in the subdivisions and distinct groups of terrains but also even in formations often separated from one another, in the vertical series of the geologic scale, by other very vast formations. . . . Without wishing to settle the question of knowing if fossils exist which pass from one terrain and even from one group to another, I believe I ought to call attention here to an important fact, little appreciated until now, based on this principle that the exterior surface of organic beings is always in intimate relation with the environment in which they live, as M. Agassiz has demonstrated with so much perspicacity in his lessons.

What I have said about the vertical succession of facies is, however, not without exception, and it is very natural that this law is modified according to the petrographic and geognostic aspect of rocks and terrains. Thus, one ought not be surprised to find in a rock of an oozy nature, superposed or subjacent to a coralline rock, fossils that live in ooze; but those ooze fossils do not indicate less than the corals, a deposit of shoals, or the littoral, although the rocks present a different type due to the circumstances of their deposition.

The distribution of facies, whether in the horizontal or vertical direction, is subjected again to other laws no less important.

Sometimes the diverse facies are limited sharply, in the horizontal direction; sometimes they pass one into the other by intermediate varieties, with poorly pronounced, vacillating characters, and by the effect of overthrusting that mixes the products of diverse facies and thus makes their study very difficult.

The sharp passages are noted particularly between the very pronounced coral facies and the pure oozes, above all when the coral banks are enclosed in subpelagic or pelagic deposits. At other times, this passage is made more slowly and in a much less evident way. This takes place particularly between the littoral facies, coralline and oozy, that most often are lost one in the other, as by an obliteration of their characters, from the centers or focus of rich development in organisms, as far as the periphery that shows only mutilated debris or a certain not very considerable number of badly developed and unimportant fossils.

The diversity of facies, in the vertical direction, increases in all the series taken up from base to top and diminishes very gradually in the reverse direction. This law presents some very curious phenomena; in the base, the pelagic facies of the oozy type predominates, and the others begin to be shown in a distinct way, starting only with the inferior oolite, which offers again characters neither pelagic nor littoral but an obscure mixture of all. We will soon see to what this phenomenon is due.

13B

GEOLOGIC OBSERVATIONS ON THE JURA OF SOLEURE [1840]

A. Gressly

This excerpt was translated expressly for this Benchmark volume by Barbara Conkin, University of Louisville, from Soc. Helv. Sci. Nat. Mem. ***4:****165 (1840).*

SUMMARY

After having reviewed the diverse terrains that compose the framework of our Jurassic mountains and their facies, sometimes littoral sometimes pelagic, it remains again to cast a last glance at the assemblage of these formations and at their distribution in the Jurassic regions between the Vosges and the Black Forest before we pass to the examination of orographic phenomena of our mountain chains. The sections and maps that are joined to this part of my memoir are designed to give a general idea as far as our knowledge permits, of the original state of the shores, beaches, bays, gulfs, shoals, and coralline banks of the basin that occupies us. I have, moreover, reunited in a general outline the principal characters of the formations and terrains that are successively deposited on the bottom of the Jurassic ocean; their different facies represent there under the petrographic, geognostic, and paleontologic reports, from the littoral near the Vosges and the Black Forest to the pelagic of the Swiss basin and the presumed oceanic of the Alps. We will see later that the deposits of the Cretaceous and molassic formation have followed the same laws of petrographic and biologic development that we have recognized in the Triassic and Jurassic formations.

13C

GEOLOGIC OBSERVATIONS ON THE JURA OF SOLEURE [1841]

A. Gressly

This excerpt was translated expressly for this Benchmark volume by Barbara Conkin, University of Louisville, Soc. Helv. Sci. Nat. Mem. ***5:****174–175 (1841).*

This general arrangement of the Jurassic terrains . . . leads us rather naturally to the following conclusions:

1. That there is a very marked difference in the petrographic, geognostic, and paleontologic composition of terrains, not only in the vertical but also in the horizontal direction;
2. That this difference results from local and general positions, according to whether the terrains are littoral, pelagic, or subpelagic or oceanic. There is a predominance of pure limestone, uniformity and simplicity of structure in the pelagic, more iron oxides and a greater variety of division and petrographic character in the littoral deposits;
3. That to this difference of horizontal position correspond paleontologic differences nearly as clear as in the vertical direction of successive formations;
4. That these geognostic, petrographic, and paleontologic characters will be able one day to serve to establish the shores, deeps, and other features of ancient oceans that in the past covered our continents and perhaps even, with the aid of a more systematic paleontology and based on a rigorous comparison of biologic laws that are observed in the undersea world today, to learn at least approximately the depth of their waters and the changes their depth has undergone in different epochs;
5. That the thickness of terrains can not scarcely serve to determine the time that passed during their deposition, considering that we have seen that pelagic deposits, thicker and more divided into numerous strata, were synchronous with very thin littoral deposits and, moreover, that the thickness of strata varies considerably from one place to another;
6. That if there is a means of estimating approximately the time that has passed during the deposition of a terrain, it is the study of the structure of banks of coral and mollusks in place, the successive layers of which can serve as so many chronologic dates. Meanwhile, it is not necessary to add to this method of estimation more worth than it actually has, considering that this structure is frequently subject to exceptions, and that there are some banks that have remained almost constantly in the same state; that some are formed only near the middle or at the end of deposition, while others have passed through all the phases of their development in a shorter span of time.

ELEMENTS

OF

G E O L O G Y.

BY

CHARLES LYELL, ESQ. F.R.S.

VICE-PRESIDENT OF THE GEOLOGICAL SOCIETY OF LONDON, ETC.;
AUTHOR OF "PRINCIPLES OF GEOLOGY."

"It is a philosophy which never rests—its law is progress: a point which yesterday was invisible is its goal to-day, and will be its starting post to-morrow."—EDIN. REV., No. 132. p. 83. July 1837.

LONDON:
JOHN MURRAY, ALBEMARLE STREET.
MDCCCXXXVIII.

Reprinted from the frontispiece and pages 271–279 of *Elements of Geology,* John Murray, London, 1838.

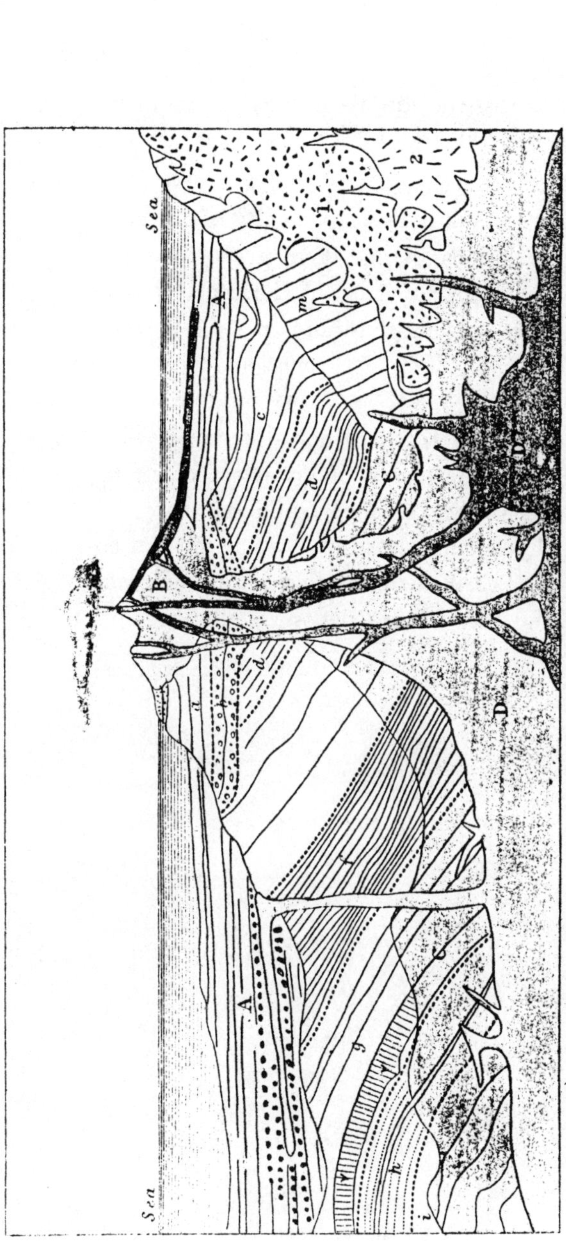

IDEAL SECTION of part of the Earth's crust explaining the theory of the contemporaneous origin of the four great classes of rocks.—see Chap. 1.

A Aqueous B Volcanic C Metamorphic (Gneiss, mica-schist &c.) D Plutonic (Granite &c.)

All the rocks older than A.B.C.D are left uncoloured.

ON THE DIFFERENT AGES OF THE AQUEOUS ROCKS

C. Lyell

[*Editors' Note:* The cross-section map on the previous page was the frontispiece in the original volume.]

In the last chapter I spoke generally of the chronological relations of the four great classes of rocks, and I shall now treat of the aqueous rocks in particular, or of the successive periods at which the different fossiliferous formations have been deposited.

Now there are three principal tests by which we determine the age of a given set of strata; first, superposition; secondly, mineral character; and, thirdly, organic remains. Some aid can occasionally be derived from a fourth kind of proof, namely, the fact of one deposit including in it fragments of a preexisting rock, which last may thus be shown, even in the absence of all other evidence, to be the older of the two.

Superposition.—The first and principal test of the age of one aqueous deposit, as compared to another, is relative position. It has been already stated, that where the strata are horizontal, the bed which lies uppermost is the newest of the whole, and that which lies at the bottom the most ancient. So, of a series of sedimentary formations, they are like volumes of history, in which each writer has recorded the annals of his own times, and then laid down the book, with the last written page uppermost, upon the volume in which the events of the era immediately preceding were commemorated. In this manner a lofty pile of chronicles is at length accumulated; and they are so arranged as to indicate, by their position alone, the order in which the events recorded in them have occurred.

In regard to the crust of the earth, however, there are some regions where, as the student has already been informed, the beds have been disturbed, and sometimes reversed. (See pp. 113, 114.) But the experienced geologist will not be deceived by these exceptional cases. When he finds that the strata are fractured, curved, inclined, or vertical, he knows that the original order of superposition must be doubtful, and he will endeavour to find sections in some neighbouring district where the strata are horizontal, or only slightly inclined. Here, it is impossible that they can have been extensively thrown over and turned upside down, for such a derangement cannot have taken place throughout a wide area without leaving manifest signs of displacement and dislocation.

Mineral character.—The same rocks may often be observed to retain for miles, or even hundreds of miles, the same mineral peculiarities, if we follow them in the direction of the planes of stratification. But this uniformity ceases almost immediately, if we pursue them in an opposite direction. In that case we can scarcely ever penetrate a stratified mass for a few hundred yards, much less several miles, without beholding a succession of extremely dissimilar calcareous, argillaceous, and siliceous rocks. These phenomena lead to the conclusion, that rivers and currents have dispersed the same sediment over wide areas at one period, but at successive periods have been charged, in the same region, with very different kinds of matter. The first observers were so astonished at the vast spaces over which they were able to follow the same homogeneous rocks in a horizontal direction, that they came hastily to the opinion, that the whole globe had been environed by a succession of distinct aqueous formations, disposed round the nucleus of the planet, like the concentric coats of an onion. But although, in fact, some formations may be continuous over districts as large as half of Europe, or even more, yet most of them either terminate wholly within narrower limits, or soon change their lithological character. Sometimes they thin out gradually, as if the supply of sediment had failed in that direction, or they come abruptly to an end, as if we had arrived at the borders of the ancient sea or lake which served as their receptacle. It no less frequently happens that they vary in mineral aspect and composition, as we pursue them horizontally. For example, we trace a limestone for a hundred miles, until it becomes more arenaceous, and finally passes into sand, or sandstone. We may then follow this sandstone, already proved by its continuity to be of the same age, throughout another district a hundred miles or more in length.

Organic remains.—This character must be used as a criterion of the age of a formation, or of the contemporaneous origin of two deposits in distant places, under very much the same restrictions as the test of mineral composition.

First, the same fossils may be traced over wide regions, if we examine strata in the direction of their planes, although by no means for indefinite distances. This might have been expected; for although many species of animals and plants have a wide geographical range, yet each species generally inhabits a small part only of the entire globe, and is often incapable of existing in other regions. But, in those cases where the fossils vary, the mineral character of the rock often remains constant; and, on the other hand, the fossils are sometimes uniform throughout spaces where the lithological nature of the rock is variable. In this manner we are frequently enabled to prove the contemporaneous origin of the same formation by one test, when the other fails.

Secondly, while the same fossils prevail in a particular set of strata for hundreds of miles in a horizontal direction, we seldom meet with the same remains for many fathoms, and scarcely ever for several hundred yards, in a vertical line, or a line transverse to the strata. This fact has now been verified in almost all parts of the globe, and has led to a conviction, that at successive periods of the past, the same area of land and water has been inhabited by species of animals and plants as distinct as those which now people the antipodes, or which now coexist in the arctic, temperate, and tropical zones. It appears, that from the remotest periods there has been ever a coming in of new organic forms, and an extinction of those which pre-existed on the earth; some species having endured for a longer, others for a shorter time; but none having ever reappeared after once dying out. The law which has governed the creation and extinction of species seems to be expressed in the verse of the poet,

> Natura il fece e poi ruppe la stampa. — *Ariosto.*
> Nature made it, and then broke the die.

And this circumstance it is, which confers on fossils their highest value as chronological tests, giving to each of them, in the eyes of the geologist, that authority which belongs to contemporary medals in history.

The same cannot be said of each peculiar variety of rock; for some of these, as red marl and red sandstone, for example, may occur at once at the top, bottom, and middle of the entire sedimentary series; exhibiting in each position so perfect an identity of mineral aspect as to be undistinguishable. Such exact repetitions, however, of the same mixtures of sediment have not often occurred, at distant periods, in precisely the same parts of the globe; and even where this has happened, we may usually avoid confounding together the monuments of remote eras, by the aid of fossils and relative position.

Test by included fragments of older rocks. — It was stated, that independent proof may sometimes be obtained of the relative date of two formations, by fragments of an older rock being included in a newer one. This evidence may sometimes be of great use, where a geologist is at a loss to determine the relative age of two formations, from want of clear sections exhibiting their true order of position, or because the strata of each group are vertical. In such cases we sometimes discover that the more modern rock has been in part derived from the degradation of the older. Thus, for example, we may find chalk with flints; and, in another part of the same country, a distinct series, consisting of alternations of clay, sand, and pebbles. If some of these pebbles consist of flints, with fossil shells of the same species as those in the chalk, we may confidently infer that the chalk is the oldest of the two formations.

The number of groups into which the fossiliferous strata may be separated, are more or less numerous, according to the views of classification which different geologists entertain; but when we have adopted a certain system of arrangement, we immediately find that a few only of the entire series of groups occur one upon the other in any single section or district.

The thinning out of individual strata was before described (p. 37.). But let the annexed diagram

Fig. 127.

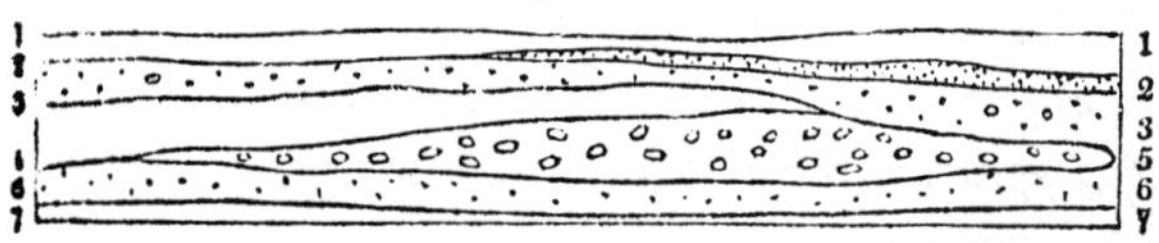

represent seven fossiliferous groups, instead of as many strata. It will then be seen that in the middle all the superimposed formations are present; but in consequence of some of them thinning out, No. 2. and No. 5. are absent at one extremity of the section, and No. 4. at the other.

If the reader consults the Frontispiece, he will see, that as the strata A rest unconformably upon the older groups, *a*, *b*, *c*, *e*, *f*, *g*, we should meet with a very different succession in a vertical section exposed at different places; in one spot A lying immediately on *c*, in another on *g*, and so forth. Now here the difference has been partly occasioned by denudation; the formations *a*, *b*, for instance, once extended much farther to the left, and but for denudation would have been everywhere interposed between A and the rocks *e*, *f*, *g*. In many instances the entire absence of one or more formations of intervening periods between two groups, such as A and *c*, (see Frontispiece,) arises, not from the destruction of what once existed, by denudation, but because no strata of an intermediate age were ever deposited on *c*. They were not formed at that place, either because the region was dry land during the interval, or because it was part of a sea or lake to which no sediment was carried.

In order, therefore, to establish a chronological

succession of fossiliferous groups, a geologist must begin with a single section, in which several sets of strata lie one upon the other. He must then trace these formations, by attention to their mineral character and fossils, continuously, as far as possible, from the starting point. As often as he meets with new groups, he must ascertain by superposition their age relatively to those first examined, and thus learn how to intercalate them in a tabular arrangement of the whole.

By this means the German, French, and English geologists have determined the succession of strata throughout a great part of Europe, and have adopted pretty generally the following groups, almost all of which have their representatives in the British Islands.

Groups of Fossiliferous Strata observed in Western Europe, arranged in what is termed a descending series, or beginning with the newest.

1. Newer Pliocene. 2. Older Pliocene. 3. Miocene. 4. Eocene.	Tertiary or Supracretaceous.*

* For tertiary, Mr. De la Beche has used the term "supracretaceous," a name implying that the strata so called are superior in position to the chalk.

[*Editors' Note:* Material has been omitted at this point.]

15A

Reprinted from pages xxvii–xxviii of *The Silurian System, Founded on Geological Researches in the Counties of Salop, Hereford, Radnor, Montgomery, Caermarthen, Brecon, Pembroke, Monmouth, Gloucester, Worcester, and Stafford; with Description of the Coal-fields and Overlying Formations,* Part I, John Murray, London, 1839, 586p.

DESCRIPTION OF THE MAP.

R. I. Murchison

The accompanying Map is intended to convey a clear general idea of the geological divisions of the surface of the tract illustrated, the details being explained by numerous coloured sections. It is a reduction of the Ordnance Survey, and therefore exhibits accurately the drainage and the relative positions of the towns and villages: the form of the hills is not delineated, but as many of the heights are marked by figures, and the names of the ridges are inserted in the line of their direction, it is hoped that the general features of the country will be understood.

In the selection of colours, I have represented each great system of strata under one common tone, and its sub-formations by modified tints of the same colour. The calcareous rocks of all ages appear under shades of blue or green; the formations to which they are subordinate being known by the colour which surrounds them. The rocks of igneous origin are all represented by bright red, a method which I have employed for some years[1], from having thought that masses, the varieties of which pass imperceptibly into each other, ought at once to strike upon the eye as being derived from the same source. But after all, as colours are fugitive, and details of high interest are often crowded together into small spaces, I have inserted *letters* at intervals (particularly on the sectional lines), the meaning of which is at once explained by reference to the table of superposition.

In addition to the tabular arrangement, usually employed by other authors, I have placed below the map on the right hand, a general transverse section, in which the sedimentary formations and their subdivisions are represented in the order in which they naturally overlie each other, where no volcanic rocks are present; whilst on the left, an ideal section is given of the submarine condition of these sediments before their elevation from beneath the sea, to mark the periods when volcanic action was most in play, either when giving rise to fine detritus of scoriæ and ashes, accumulated in alternating layers with sand and mud; or during the emission of those intrusive rocks, which, whether syenites, greenstones, porphyries or basalts, are all the products of fire. Many of these volcanic rocks being inseparably connected, even within small areas, and in the same ridges, their lithological distinctions can only be understood by consulting the explanations in the text; but the epochs at which they are supposed to have been erupted, are signified by different letters,—their relations to the sedimentary deposits being rendered more intelligible by reference to the plates of coloured sections (see Plates 29—37).

A small map of England and Wales has been engraved in the corner of the large map, for the double purpose of pointing out the geographical site of the Silurian Region, and of subdividing the kingdom geologically into eight sedimentary groups, each of which I conceive (in British classifi-

---

[1] In the Map of the Eastern Alps, Geol. Trans., vol. iii. Pl. 35., and in the coloured Ordnance Maps of the Silurian Region exhibited at various meetings of the British Association, and at the Geological Society from 1832 to 1837. I am glad to perceive that Mr. Griffiths has adopted this method in the new Geological Map of Ireland.

cation), to be entitled to be called "Systems," from their dimensions, lithological characters, and zoological contents.

In comparing, however, the rocks of our own land with those of foreign countries, it is not contended that each of these systems has a general or even an European range. The Old Red Sandstone, for example, so vast an accumulation in the British Isles, is scarcely apparent throughout a large part of the continent, while the deposits which I have termed the "New Red System," though thus arranged in England, may not satisfy some geologists of the French or German schools. The latter may contend that in their country, the "Keuper," "Muschelkalk" and "Bunter Sandstein," form a natural group (the Trias), which is separable from the underlying "Zechstein" and "Rothe-todte-liegende" by valid distinctions. In the region, however, which I describe, such distinctions do not exist. On the contrary, all the strata between the Lias and the coal measures are composed of red and green marls and sandstones, with conglomerates, both siliceous and calcareous, the lower members of which pass gradually downwards into the Carboniferous System. This subject is fully explained in the work; and I merely now allude to it, in order to show that in these tracts the natural order of this part of the series consists in the Carboniferous System being overlaid by one red system and underlaid by another. In regard to the subdivisions of the overlying red system, some foreign geologists are, I believe, of opinion that the fossils of the "Zechstein" and "Rothe-todte-liegende" agree better with those of the carboniferous epoch than with those of the overlying red sandstone, but this point is, I am led to believe, far from being satisfactorily adjusted; for although there may be a species or two of Mollusk common to the Magnesian Limestone and the Mountain or coal Limestone, the great mass of the fossils of the two formations are distinct. That the fishes are distinct has been clearly proved by M. Agassiz, while no researches have yet brought to light a Saurian in the coal measures, though several species of this class occur in the Magnesian Limestone. This last-mentioned zoological distinction is so strong that it seems to me to outweigh in value any conclusions drawn from the Mollusca. Again, if an identity between one or two species of Mollusca is to be made the groundwork for grouping together the Magnesian and Mountain Limestone, we may for the same reason connect the Magnesian Limestone with the overlying group (the Trias of the Germans), because the lily encrinite of the Muschelkalk, and, if I mistake not, a chambered shell of that rock have been discovered in the Magnesian Limestone.

As, however, the Silurian region exhibits passages from every formation into the overlying and underlying strata, so must of necessity the lines of demarcation be more or less arbitrarily drawn; and all that I seek to prove is, that the classification now put forth is best suited to this part of England, however little it may be adapted to other countries. In short, whatever may be the arrangement on the continent, British geologists have reason to claim for themselves the privilege of defining the upper and lower limits of the great Carboniferous System, seeing that it is more copious in our country than in any other part of Europe.

But to return to my own map. An effort has been made to separate the New Red Sandstone into three distinct bands. For the first time in English geology, the "Keuper Sandstone" is shown to lie in the upper red or saliferous marl[1], whilst beneath it is the ordinary New Red Sandstone which overlies the "Lower New Red Sandstone," the two last-mentioned deposits being usually separated by a calcareous conglomerate representing the Magnesian Limestone.[2]

[1] A memoir on the Keuper Sandstone of Worcestershire and Warwickshire by Mr. H. E. Strickland and myself is in the press, and will shortly appear in the Transactions of the Geological Society.

[2] In the map I have made use of three tints of light reddish colour to distinguish, 1st, the Red Marl and its

15B

Reprinted from pages 3-12 of *The Silurian System, Founded on Geological Researches in the Counties of Salop, Hereford, Radnor, Montgomery, Caermarthen, Brecon, Pembroke, Monmouth, Gloucester, Worcester, and Stafford; with Description of the Coal-fields and Overlying Formations,* Part I, John Murray, London, 1839, 586p.

INTRODUCTION.

R. I. Murchison

Design of the work.—Previous state of geological knowledge.—Order of the older fossiliferous strata.—Interval to be filled up, between the secondary deposits previously described, and the older slaty rocks.—Origin of this inquiry, and method pursued in following it out.—General view of the oldest fossiliferous deposits.

THE chief design of the present work is, to fill up an interval in geological history, by describing certain strata, which, although they occupy a considerable thickness in the crust of the globe, and connect the secondary deposits with the older slaty rocks, have never yet been adequately examined.

A few words will explain the previous condition of this subject.—When the materials of the earth's crust first became a subject of study, they were viewed principally with reference to their mineral characters; but the attention which was afterwards directed to the imbedded animals and plants, gradually produced a revolution in the science, and gave birth to what is now the largest and most important part of Geology. Since the period when Smith in England, and Cuvier and Brongniart in France, first identified strata by their fossils, a most rapid progress has been made in the application of this method of testing the age of rocks.

Sixteen years have now elapsed since Conybeare and Phillips, in their Outlines of the Geology of England and Wales, presented us with a connected view of the succession of the sedimentary British deposits, from the most superficial to those which support the carboniferous system; and, in the succeeding years, great and important additions have been contributed to our acquaintance with the youngest or tertiary deposits particularly by Mr. Lyell.

But this method has not been so extended as to carry the chronological succession *below* the Old Red Sandstone; partly, perhaps, from a preconceived opinion, that few organic remains were likely to be detected in these formations; partly from the belief, founded on just but inadequate observation, that the many mutations which these older rocks had undergone, must have nearly obliterated the evidences of their origin, whether consisting in a clear order of superposition, or in distinctness of zoological contents.

Though, undoubtedly, such reasons deterred many from grappling with this inquiry, it must not be supposed that the ancient strata have not been studied by enlightened observers. In Great Britain we may cite the names of MacCulloch, Greenough, and Sedgwick, as those of men prominently distinguished in throwing light upon them :— the first by long examination of the Scottish mountains and numerous writings upon them ; the second by his map of England and Wales ; the last by his exposition of the order of the rocks in the north of England, a portion of which task Professor Phillips has since followed out, by publishing a monograph of the carboniferous strata. But these authors had not the peculiar advantages which have fallen to my lot ; for in the regions which they studied, there is generally an abrupt boundary-line or break between the older and newer systems. For example, in Scotland there is no sequence of fossiliferous strata beneath and connected with the Old Red Sandstone ; neither is there such a sequence in Cumberland, Westmoreland, nor the adjacent tracts of Yorkshire, nor even in Devonshire, where some of the oldest masses exist. In fact, a perfect and unbroken series of links, connecting these older rocks with the younger deposits, does not occur in any portion of these islands which had been previously examined.

On the continent, where great attention had been bestowed upon the older and crystalline rocks, from the days of De Saussure and Werner to our own, the same belief was impressed on the minds of geologists, that the great dislocations to which these ancient rocks had been subjected, had entirely dissevered them from those fossiliferous strata with which we were well acquainted. In short, there existed no foreign work in which rocks of this age were classified according to a law, founded upon superposition and characteristic organic remains.

But to proceed to facts connected with our own country. No one was aware of the existence below the Old Red Sandstone, of a regular series of deposits, containing peculiar organic remains. For example, although it was supposed, that the limestone of Dudley was of greater antiquity than the Old Red Sandstone, no one had observed that those deposits were connected by an intermediate formation of very considerable dimensions, full of organic remains. It is this formation, now termed the "Ludlow Rocks," which seems to have most escaped attention, whilst, from its position, as will appear in the sequel, it is the key which accurately reveals to us the relations of the inferior masses to the overlying strata with which we formerly were acquainted.

Of the few memoirs which had been published in the Geological Transactions, relating to parts of the region illustrated, the earliest are by Mr. A. Aikin. That accurate observer had formed a project of describing Shropshire in detail ; but having long abandoned his intention, he no sooner heard of the progress I was making in the present work, than he placed at my disposal his original notes and drawings, illustrative of certain tracts around the Wrekin and the Caradoc. In truth, at the early period when Mr. Aikin undertook the task, it was almost hopeless to attempt to unravel the structure of Shropshire; for that county not only contains every sedimentary formation from the lias to the slates inclusive, but is also rendered most complex by the numberless dislocations of the strata, through the agency of volcanic rocks. The only other modern writers who had touched upon the ancient rocks of this part of England, were Mr. Leonard Horner in a memoir on the "Mineralogy of the Malvern Hills," Mr. Weaver in one on the "Geology of the Tortworth District," and Mr. J. Yates in a short paper on "Parts of the central Counties." These essays which are included in the Geological Transactions, though constituting valuable additions to our knowledge at the periods of their publication, had reference to limited tracts only, in most of which anomalous and disrupted relations prevented the adoption of any general view of a succession of the strata.

In this condition of the subject, I first explored the borders of England and Wales in 1831. The order of succession seen in the ridges on the left bank of the Wye, between Hay in Herefordshire and Builth in Brecknockshire, where the Old Red Sandstone is distinctly underlaid by grey fossiliferous strata, first led me to suspect, that I had met with a district containing a good part of the evidence required to lead to a systematic study of our older formations; a surmise which was confirmed by following out these rocks upon their line of bearing to the neighbourhood of Ludlow and Wenlock, where I found them much expanded. I explained my view of the subject to the first meeting of the British Association, held at York, and afterwards to the Geological Society of London. Perceiving, however, that a subject so new and so large could not be really advanced, except through patient and repeated examination, I re-explored the same districts, in 1832, and submitted details of the new acquisitions to the Geological Society; accompanying my memoirs with a set of geologically coloured maps of the Ordnance Survey.

An effort to classify these deposits was then made; but it was not until the close of the summer of 1833, that I was enabled to publish a tolerably correct synopsis or table of the various formations in the extensive region, over which my observations had progressively extended. Seven years have since elapsed, during which my attention has been almost exclusively given to the development of this subject and its collateral branches.

During the summer in which my first observations were made, Professor Sedgwick commenced a general inquiry into the structure of North Wales, for which his previous acquaintance with the slaty and crystalline rocks of Cumberland eminently qualified him. He first endeavoured to connect the transition rocks of the age of Dudley with certain calcareous slates pointed out by Mr. Greenough in North Wales; but finding no concordance between them, he was, to use his own expression, "driven to a new base line," in other words, to work upwards from the central axis or oldest rocks of Wales. As soon as he perceived that I had observed the links which connect the Old Red Sandstone with some of the inferior masses of his region, he felt the importance of pushing the inquiry, and by his encouragement I was materially stimulated to do so. In speaking of the labours of my friend, I may truly say, that he not only shed an entirely new light on the crystalline arrangement or slaty cleavage of the North Welsh mountains, but also overcame what to most men would have proved insurmountable difficulties, in determining the order and relations of these very ancient strata amid scenes of vast dislocation. He further made several traverses across the region in which I was employed, and, sanctioning the arrangement I had adopted, he not only gave me confidence in its accuracy, but enhanced the value of my work by enabling me to unite it with his own ; and thus have our joint exertions led to a general view of the sequence of the older fossiliferous deposits.

In the mean time, some kind friends who had watched the progress of my labours, and thought they might lead to important results, requested me to prepare a separate treatise; and although I had been at first desirous of confining these views (as in the case of every memoir I had previously written) to the Transactions of the Geological Society, my objections were overcome by a very flattering requisition[1].

In obeying this call, I hope I shall, at all events, promulgate some new facts, and place before geologists the history of a system of deposits, authenticated by numerous good evidences in the shape of organic remains, by far the greater part of which have never been published in any country.

While maturing these views, I became convinced that, as this large and ancient group contained peculiar organic remains, and was marked by distinctness of physical features, lithological structure, and order of superposition, it was well entitled to be considered a separate *system*. It was not therefore enough that, in my efforts to develope them, I had termed these deposits, in their natural descending order, the "Ludlow," "Wenlock," "Caradoc," and "Llandeilo" formations ; without some *collective* name, no general view could be carried out, nor their relations to the whole series of deposits established. For example, the group could not be defined "Transition Rocks," because nearly every modern author had so extended the meaning of this term, as to embrace in it all the deposits, from the carboniferous series inclusive, down to the oldest slaty rocks in which organic remains cease to be perceptible ; whereas the object I had specially in view was, to point out the existence of intermediate rocks of great thickness, *essentially different both in structure and organic remains from the carboniferous strata*. In early communications to the Geological Society, adopting a provisional name, I called these rocks "fossiliferous greywacke" ; but this term was deemed objectionable, there being few beds in the group which can be recognised as the "grauwacke" of German mineralogists—while there are fossiliferous

[1] Originating with the Right Honourable Frankland Lewis and his son Mr. George Lewis, this requisition was soon subscribed by many resident nobility and gentry through the very friendly exertions of Lord Clive. (See List of Subscribers.)

rocks having distinctly that mineral character, both above and below those which I describe.

A geographical term was finally adopted, derived from the Silures, whose power extended over the region where these rocks are best displayed, and the name of whose illustrious chief, Caradoc (Caractacus), has been transmitted to us in a bold range of hills, composed of one of the most important formations of the system to be described. The term was no sooner proposed than sanctioned by geologists, both at home and abroad, as involving no theory, and as simply expressing the fact, that in the "Silurian region," *a complete succession of fossiliferous strata is interpolated between the Old Red Sandstone and the oldest slaty rocks.*

M. Elie de Beaumont, for example, warmly encouraged me to use this name, and with his coadjutor M. Dufrénoy has since materially contributed to give it currency on the Continent. Soon afterwards M. Boué and M. de Verneuil announced the diffusion of "Silurian" rocks in Servia and the adjacent parts of Turkey in Europe, while our countrymen Hamilton and Strickland extended their range to the Thracian Bosphorus. More recently, M. Forchhammer of Copenhagen has visited the "Silurian region" to endeavour to recognize in it the rocks of Scandinavia; and even whilst I write, MM. D'Omalius D'Halloy and Dumont are exploring it to establish a parallel between its deposits and those of Belgium. Lastly, I may state, that M. de Boblaye has honoured me by offering to translate this work into the French language.

Shortly after I proposed the name "Silurian," one or two European tracts were, indeed, pointed out as resembling that which I had selected as a type. One of the most remarkable is on the southern frontier of the Ardennes, where three formations were recognised by Dr. Buckland and Mr. Greenough as occupying the same geological position as in England and Wales; namely, between the carboniferous deposits and the older slaty rocks.

That similar deposits exist in many parts of Europe is evident from the works of Strangways, Brongniart, Von Buch, Hisinger, Dalman, Keilhau, and other writers, as well as from the fossils found in Norway, Sweden, Russia, Poland, Germany and France; though patient comparisons must be instituted before the peculiar relations of the Silurian Rocks of those countries can be established[1].

Their spread through distant regions will be again brought under consideration, in the concluding view of the general distribution of organic remains. In the mean time it

[1] I cannot attempt to enumerate the names of all the foreign authors who have written on Transition Rocks, though many of those who treat of the Organic Remains are alluded to hereafter. See Part II. Judging from the maps of Hisinger and Keilhau, I presume that both Upper and Lower Silurian Rocks exist in several districts of Sweden and Norway. It would appear that I may have been led into an error (p. 169 *note*) in supposing that the Old Red Sandstone occurs in these regions. I hope, however, at some future day, to examine personally the older rocks of Scandinavia and Russia, and to show to what extent they agree with those of our own country.

may be stated, that such deposits unquestionably exist in North America, not only as inferred from the publications of our countrymen Bigsby and Weaver, but also from the works of several American authors who have described organic remains (Dekay, Green, &c.), as well as from the letters of Mr. Featherstonhaugh and Professor Rogers. And although I have not yet seen a sufficient number of fossils to enable me to enter into details, the identity of certain species of trilobites, common to that continent and Great Britain, sufficiently sustains the accuracy of the inference. It is, indeed, probable that the Silurian strata are distributed throughout that quarter of the globe, since Mr. Charles Darwin has recently brought home from the Falkland Islands masses of rock, charged with fossils which can with difficulty be distinguished from specimens of the Caradoc sandstone,—while the existence of other Silurian Rocks in Southern Africa has been established by the evidence of organic remains collected by Dr. Smith, and first transmitted to me by Sir John Herschel.

We have, therefore, every reason to believe, that although the complete order of the foreign deposits of this age has not yet been pointed out, a little more labour will enable us to place them in parallel with our own.

But to return to England,—soon after I issued my first prospectus, Professor Sedgwick assigned to the older and contiguous rocks of Wales the name of Cambrian; and the reason is obvious; for the strata so designated are not only conterminous with the Silurian System, but are in several parts seen to rise from beneath its lowest beds, and fairly to unite with them. These names are not, however, propounded as immutable; they are simply offered as the best means we possess of clearly defining the objects to be attained.

To show that these systems, like other large groups of strata in Great Britain, occupied their true positions in the sea cliffs, both in relation to each other, and to the younger deposits, I terminated my survey, by following them into Pembrokeshire,—proving, that after a course of one hundred and sixty miles, they preserved the same relation to each other and to the overlying deposits, as in Shropshire and Herefordshire. Professor Sedgwick will complete this portion of the task, by explaining how the two systems range to the coast of North Wales, and how they are there related to the younger formations.

In tracing them to distant parts, the reader must not expect to find the Silurian rocks preserving an uniform lithological character, or conforming precisely throughout large spaces to the prevailing mineral types of the region described. This remark applies not merely to other countries of Europe, but even to distant parts of Britain: thus, although the calcareous flags of Llandeilo with their accompanying schists are considered to form the base of the Silurian System, their place is sometimes taken, often indeed they are underlaid, by sandstones of considerable thickness. Again, besides the frequent absence of beds of limestone in the Ludlow and Wenlock formations above alluded to, those deposits do not preserve the same lithological characters over wide tracts, their colour and composition in parts of Caermarthenshire and Pembrokeshire, and also at Tortworth, being very different from those of the same rocks in Shropshire and Herefordshire. And thus the lesson which has been already taught us by an examination of the younger deposits is repeated,—that the *zoological* contents of rocks, when coupled with their *order of superposition*, are the only safe criteria of their *age*. By such proofs we are enabled to distinguish the Silurian deposits from all others previously described, and through every lithological change we can thereby separate the system into upper and lower divisions.

Quitting the Silurian rocks, and taking an ascending prospect, I may observe, that although the stratigraphical position of the Old Red Sandstone and its general relations, both in England and Scotland, were well known, no complete account of its range, succession, and zoological contents, in Herefordshire and the adjacent counties, had yet appeared, though it there forms one of the largest of the stratified British groups[1]. Perceiving that besides this great thickness, the Old Red Sandstone had an individuality of mineral character, and peculiar organic remains, I ventured for the first time to designate it a system. This step was taken, not only to mark distinctly the vastness of the strata by which the Silurian deposits are *separated* from those of the carboniferous æra, but also to indicate at one glance, that the coal measures of our country are *included* between two great red coloured systems of rock—the New Red Sandstone above, the Old Red Sandstone below.

Had my labours terminated at this point, the Silurian Rocks, and their immediate relations to the overlying and underlying deposits, might have been long ago developed; but I felt that unless the structure of the whole region in question was made known, the map would have been imperfect, and the work would have borne an unfinished character. I therefore resolved to give sketches of all the coal fields included in the area examined, and further to explain their relations to the surrounding deposits. In adopting this resolution, however, I could scarcely estimate the additional exertions required; for, with the exception of the basins of the Forest of Dean, and of South Wales, none of these coal tracts, so important in the mineral economy of the nation, had been described.

But these subjects did not exclusively engross my attention. To render their history complete, it was necessary to examine minutely all the rocks of igneous or volcanic origin, which are interlaminated with, or penetrate the strata at so many points, as well as to follow out the lines of elevation and dislocation. Further, to arrive at clear

[1] The highest mountain in South Wales, called the "Fans" or points of Brecon, from its double summit, is 2862 *feet* above the sea, and is entirely composed of nearly horizontal beds of Old Red Sandstone. Through an error of the press, (p. 170) this mountain is stated to be 2500 feet high. The height of the Caermarthen Fans is accurately given as 2590 feet. Both these mountains are seen in the view facing p. 346, and the heights of both are correctly marked in that lithograph and on the map.

deductions, it was indispensable that all the overlying formations should be examined, and their boundaries laid down upon the map.

In searching the younger deposits I learnt, that the New Red System was separable into distinct formations, one of the uppermost being found to contain a band of sandstone with peculiar fossils. Though not previously noted, this rock most clearly represents the sandstone of the "keuper" or "marnes irrisées;" while the central masses of sandstone, both by geological position and the plants they contain, are shown to occupy the place of the "bunter sandstein" or "gres bigarré" of foreign geologists.

Finally, I ascertained, that within the area of the Silurian region there were natùral sections, exposing passages throughout the whole series, and thus connecting all the formations from the oolitic to the slaty Cambrian rocks; a most remarkable fact, and one perhaps almost without parallel in any European district of like extent.

I need hardly say, that these numerous branches of the subject have caused a much longer examination than was originally contemplated and have considerably retarded the appearance of these volumes. Geologists, indeed, require no apology for the delay, as they well know that the illustrations of such a work could not have been rapidly prepared; but from those friends who have not made our science their pursuit, and yet have countenanced my efforts, a larger measure of indulgence is claimed. With the aid, however, of the map, the numerous coloured sections, and pictorial views, it is hoped that this class of readers will not encounter many serious difficulties. As owners of the soil, I hope they may derive some use, not only from the efforts I have made to indicate where coal *may* be advantageously sought for, but from those which demonstrate where it *never can be found*; while a knowledge of the true nature of the sub-strata cannot fail to be of value to the agriculturist.

In deciphering the history of these rocks, I have worked from the upper or known formations to the lower or unknown, and the reader will therefore be led through a *descending* geological series. The younger stratified deposits which flank the Silurian territory on the east and north will be first described, and then the inferior sedimentary masses; while the rocks of volcanic origin, and the effects produced by their intrusion amid the strata, will be pointed out in each district in which the order of the beds has been previously established.

After being thus carried back to remote periods, during which the order of the strata has often been deranged by great dislocations and volcanic eruptions, the reader's attention will be called to the various sorts of superficial detritus of the region, which being formed out of all the solid formations, could not be intelligibly explained without a previous account of the rocks from which it has been derived. In this branch of our subject, the reasons for pursuing a descending order of inquiry no longer existing, the historical method is adopted. Beginning, therefore, with the consideration of the earliest deposited loose materials, the student will have successively placed before him each newer accumulation of gravel or sand, whether entirely marine, or formed during the transition from a submarine condition, into the æra when the newly-raised surface was first occupied by lakes and broad rivers. He will, in short, be conducted, step by step, from the time when large portions of our island were beneath the sea, to the consideration of the deposits which are in actual progress of formation above it; and thus he will embrace in one view the passage from the vast geological epochs into the comparatively brief period of history.

The first part of the work, including the descriptive geology of the region, will then be terminated, by a review of some striking results of the present inquiry.

The second part is exclusively devoted to Zoology, and contains the evidences on which many of the principal conclusions are founded; being a description of all the fossils of the Old Red and Silurian Rocks, prefaced by a general view of the distribution of animal life during the accumulation of those ancient strata.

Geologists will determine whether the results now offered to them are worthy of their approval. If it be acknowledged, that no sound general views of the early periods of the earth's history can be obtained, without a close investigation of the beds which connect the sedimentary deposits previously known with the oldest stratified rocks, then I may venture to hope, that the endeavour to clear up this difficult subject, will be considered a step, however small, in the foundation of the science.

We already know, that certain deposits with their organic remains, may be expected to show themselves (though sometimes under different aspects) in distant lands. Thus the tertiary and secondary strata have been identified over remote parts; while our own island is remarkable for having afforded in great measure the original types of the secondary age which directed that investigation. It appears highly probable, nay, it is even to a great extent already ascertained, that phenomena of the same kind prevail with respect to the system described in this volume; and that Siluria, with its accumulations of remote antiquity teeming with organic remains, presents a table made up of some of the earlier and hitherto undeciphered pages, which the history of the earth in other countries offers to our study.

But the Silurian, though ancient, are not, as before stated, *the most* ancient fossiliferous strata. They are in truth but the upper portion of a succession of early deposits which it may hereafter be found necessary to describe under one comprehensive name. For this purpose I venture to suggest the term "Protozoic Rocks," thereby to imply the first or lowest formations in which animals or vegetables appear.

That there is a limit in the descending scale of formations, beneath which no traces of life have been discovered, is now pretty generally recognised; and looking merely to this fact, geologists may agree to use the word "Protozoic," however they may differ in their interpretation of the phenomenon.

One class of observers believe, that life did not exist when the earliest deposits took place. They hold that the oldest crystalline strata (ancient gneiss, mica schist, &c.) were formed during a period of great heat: and the astronomer seems to strengthen this opinion, by showing that the planet must have been in a state of total or partial fusion when it assumed its present shape. Such reasoners are led to suppose, that the earliest accumulations occurred under conditions which forbade the possibility of any vital organization ; and they are further strengthened in their creed, when on examining the innermost folds of the earth's covering which have been extruded, they find in them no vestiges of life. They conclude therefore, that it was only after a long time, and when the surface had to a certain extent been cooled down by succeeding changes and the addition of fresh materials, that animals and vegetables were called into existence.

Other geologists contend, that as yet we gaze but dimly into the obscure vista of these early periods ; and that even if organized beings did live when the first crystalline rocks were formed, we cannot now hope to discover evident traces of their existence, owing to the great metamorphoses which subsequent agencies have produced in these masses,—metamorphoses which may be well supposed to have obliterated all traces of primeval creation.

Without here attempting to decide this question, I would merely observe, that the term "Protozoic," as above interpreted, may be used by the maintainers of either doctrine.

In conclusion, I repeat that my chief object is to develope the upper portion of this vast series ;—its lower divisions belong to the task which has fortunately fallen to Professor Sedgwick.

15c

Reprinted from pages 579–586 of *The Silurian System, Founded on Geological Researches in the Counties of Salop, Hereford, Radnor, Montgomery, Caermarthen, Brecon, Pembroke, Monmouth, Gloucester, Worcester, and Stafford; with Description of the Coal-fields and Overlying Formations*, Part II, John Murray, London, 1839, 586p.

Introductory View of the Distribution of Organic Remains in the Older Formations.

R. I. Murchison

GEOLOGY reveals to us the extraordinary fact, and without its aid the fact never could have been known, that as the globe passed from one condition to another, whole races of animals perished, and were succeeded by others with organizations adapted to the altered state of our planet.

On this phenomenon is based the fundamental principle of the identification of strata by their imbedded remains; the passage from one deposit to another being marked by a change in the animals which lived and died during the accumulation of each. Thus, although the fossils of any one great series of beds possess a common character, yet those which are found in the lowest and highest strata of a great formation are for the most part dissimilar in species, and often in genera.

This principle, so little adverted to in the early days of the science, and yet so vitally essential to its advance, has hitherto been chiefly tested by the examination of the tertiary and secondary deposits, and its complete development required, that succeeding observers should laboriously work out its application to the numerous older strata which are contained in the crust of the globe. In the progress of inquiry, this method of proof having been sometimes too minutely exercised, and at others too vaguely, has undeservedly incurred discredit. Some persons, for example, have endeavoured to show, that because within a limited area every succeeding stratum was charged with peculiar species, so we ought to find the same distinctions universally; while others, merely comparing lists of species in the works of authors of different countries, or erroneously identifying figures and descriptions of fossils, have inferred, that many species have a much wider range in the geological scale than experience confirms.

The only effective remedy for the scepticism engendered by such loose comparisons is to publish monographs, with figures of all the remains found in each group of deposits, the stratigraphical limits of which have been precisely defined by competent observers. This object having been in great measure accomplished in respect to the tertiary and secondary rocks, the inquiry has been recently carried down to the still lower strata of the Carboniferous System. We may, therefore, say that the chronological zoology of the science in England has been, on a general scale, systematically conducted to the verge of the ancient strata illustrated in this work; though from this step downwards in the series, it must be allowed, that the natural history of the fossils has never yet been directly connected with the order and arrangement of the rocks.

In the commencement of this volume, while sketching the previous state of our knowledge of these older formations, I pointed out the confusion which had been introduced by so applying the word "transition" as to embrace in one meaning, the Carboniferous, Old Red, Silurian and Cambrian Systems. If the Old Red Sandstone had been clearly separated from the Carboniferous System above and from the Silurian Rocks below it, the term "transition" (as formerly used by certain English geologists, and applied by them to the *lower* divisions only) might well have been retained. Seeing, however, that all the continental and most of our native writers had in the mean time extended the application of this word to the carboniferous deposits inclusively, there appeared to be no possibility of preventing confusion without the introduction of a distinct classification. In the previous pages it has I hope been clearly established, that the Old Red Sandstone and Silurian Rocks are completely distinct from the Carboniferous System, by order of superposition and lithological characters; and my present aim is to prove that they are equally distinct by their imbedded organic remains.

If this principle of classification, which is established in the arrangement of the tertiary and secondary deposits, be once admitted as respects the more ancient rocks of the British Isles, I have no doubt that in other parts of the world, where deposits of the same age occur, it will be found equally true. It is, in fact, only in countries like our own where all those strata exist, which rising from beneath a well-developed carboniferous system, connect it by unbroken links with the inferior slaty rocks, that such an inference can be sustained; and hence it follows, that no counter evidence can be received, when derived from tracts in which many of these links are wanting, or where the convulsions and alterations of the rocks have been so great, that the correct identification of their contents is impracticable. Still less can we admit the validity of arguments founded, either upon mere lists of fossils which may have often been erroneously identified, or simply upon the names attached to formations by geologists who have not studied the whole sequence of the deposits in question[1].

[1] On this point I may remark that a paper by Mr. Weaver on the south of Ireland, containing much valuable matter, has just appeared in the Transactions of the Geological Society, vol. iv. This memoir seems to warrant the necessity of the above caution. After separating the Carboniferous from what he continues to term

Looking, therefore, steadily to the object of this work, and restricting myself to that field of inquiry with which I am conversant, I distinctly maintain, what is asserted in the Introduction; that the fossils of the Silurian System here represented, and amounting in all to about 350 species, are, with the exception of a very few (chiefly doubtful casts), essentially distinct from any of the numerous and well-defined fossils of the Carboniferous System; and further, that the Old Red Sandstone which separates these two systems is also characterized by fossils peculiar to it. In testing this inference, I merely require, that if the organic remains of formations of dissimilar age are compared, they may be selected from well-known and clearly established *geological* types of such rocks. Having for a series of years collected fossils from every stratum of the Silurian Rocks, throughout a large region, in which the stratigraphical order is clear, I now present the results. Professor Phillips had previously completed a valuable monograph of the organic remains of the Carboniferous System. If the naturalist will compare the figures in these, the only two works yet published upon the older fossiliferous rocks, which combine geological description with zoological proofs, he will at once see the truth of my position.

Beginning with the *vertebrata*, are not the *fishes* of the Old Red Sandstone as distinct from those of the Carboniferous System on the one hand, as from those of the Silurian on the other? M. Agassiz has pronounced that they are so.

Are any of the *crustaceans*, so numerous and well-defined throughout the Silurian rocks, found also in the Carboniferous strata? I venture to reply, not one.

Are not the remarkable *cephalopodous mollusca*, the *Phragmoceras* and certain forms of *Lituites*, peculiar to the older system?

the "Transition Rocks," Mr. Weaver arrives at the conclusion, that a large number (between 60 and 70 species) of the fossils of these two systems are *identical*. I need not inform my readers, that my zoological data and inferences are completely at variance with those of Mr. Weaver.

Respecting the Irish case, it may be stated that Mr. James de C. Sowerby, whose authority is much cited in the memoir of Weaver, is of opinion that all the fossils therein enumerated, p. 21, as belonging to the *transition* limestone of Cork and neighbourhood, are, with one exception, characteristic fossils of the *carboniferous* limestone of England and Ireland; and therefore that they are of the *same* geological age as those which, in another part of the memoir, are described as *exclusively* belonging to the carboniferous strata.

Judging from the printed lists, Professor Phillips also thinks that the Cork limestone fossils are carboniferous.

Mr. Sowerby coincides with me in believing, that the only fossils alluded to by Mr. Weaver, *which really belong* to the more ancient rocks (Silurian, &c.) are those enumerated (pp. 10, 15 *et seq.*). Should this view be substantially correct, the south of Ireland, so far from offering evidence to contradict my classification, will eventually be found to support it.

From an inspection of Mr. Weaver's map alone, I cannot avoid surmising, that the localities where true Silurian fossils might occur, are those alone where such *have really been* detected, as the strata in those situations are separated from the carboniferous limestone by large masses of Old Red Sandstone. (See Smerwick Harbour, &c., on the coast of Kerry, and the east of Bon Mahon river, Waterford.) The Old Red Sandstone, that important feature of separation, being wanting in all the remainder of the country described, is it hazarding too much to suggest, that some of the limestones which there occur, and which are loaded with carboniferous fossils, may be outliers and remnants of the base of that system, which we know to be so vastly expanded and widely diffused throughout other parts of Ireland?

Is there one species of the *Crinoidea* figured in this work, known in the carboniferous strata?

Has the *Serpuloides longissimum*, or have those singular bodies the *Graptolites*, or in short any zoophytes of the Silurian System been detected in the well-examined carboniferous rocks?

And in regard to the *corals*, which are so abundant that they absolutely form large reefs, is not Mr. Lonsdale, who has assiduously compared multitudes of specimens from both systems, of opinion, that there is not more than one species common to the two epochs?

If, therefore, it should prove after all, that a *few* species of *conchifers* (an order of beings capable, perhaps, more than any other of enduring vicissitude), continued in existence, from the formation of the Silurian Rocks to the accumulation of the carboniferous limestone, how can their presence break down the individuality and separation of systems, established upon such a vast preponderance of direct zoological evidence in the other natural classes? Even should a few other *mollusca* in the two systems be considered identical, there is no doubt, that by far the greater number of them which truly belong to rocks rising from beneath the Old Red Sandstone, are distinct from those which inhabit the strata above that system.

Such evidences are therefore, as before mentioned, nothing more than additional supports of the important truth which geology has already established; that each great period of change, during which the surface of the planet was essentially modified, was also marked by the successive production and obliteration of certain races.

Let it not, however, be imagined that I wish to inculcate the doctrine, of every ancient formation having been tenanted by creatures absolutely peculiar to it. The large natural groups of strata only, or, so to speak, systems, can be thus distinguished. We have, indeed, ascertained to a great extent the distribution of organic life in the epochs anterior to our own, and we now know, that with each great increment of newly deposited matter, new animals appeared, and that while some vanished with the lapse of time, others unsuited to sudden changes were destroyed.

Among the secondary causes by which such results have been aided, the volcano and earthquake have assuredly been most prominent. We believe, therefore, that inasmuch as the outbursts of volcanic matter, in the region under review, have been numerous and often repeated, they must have materially operated in the destruction of animal life; but as we have shown that all the lines of eruption were *limited*, so we infer that no general destruction can have taken place; and hence it is, that this territory, when examined carefully and in all its parts, presents us with so many examples of a perfect "sequence" in the succession of the strata and the progressive development of their zoological contents.

Thus certain species, whether endowed with powers to resist vicissitude or living in those parts where few active agents of destruction were at work, continued to live through long epochs; while others of a higher structure, passed away in comparatively short periods.

There is yet, however, a phenomenon of the highest importance, connected with the distribution of organic remains in the older strata, which has not been adverted to; namely, that the same forms of crustaceans, mollusks and corals, are said to be found in rocks of the same age, not only in England, Norway, Russia, and various parts of Europe, but also in Southern Africa, and even at the Falkland Islands, the very antipodes of Britain. This fact accords, indeed, with what has been ascertained concerning the wide range of animal remains in deposits equivalent to our oolite and lias; for in the Himalaya Mountains, at Fernando Po, in the region north of the Cape of Good Hope, and in the Run of Cutch and other parts of Hindostan, fossils have been discovered, which, as far as the English naturalists who have seen them can determine, are undistinguishable from certain oolite and lias fossils of Europe[1].

Another remarkable fact illustrating this point of inquiry is, that although the older fossiliferous strata often contain *vast quantities* of organic remains, the *number of species is much smaller* than in more recent deposits. Judging from my own labours, I can affirm, that after seven years devoted to this inquiry, the number of species collected at the end of four years was little less than the aggregate now published. The above resemblance, therefore, (if not identity) of certain forms, out of this number, which are found in opposite hemispheres, is the more surprising.

[1] The largest collection of fossils from the Himalaya, with which I am acquainted, was brought home by Lady Sarah Amherst, and being examined by Mr. James Sowerby was found to contain several Ammonites undistinguishable from our English species of the lias.

The specimens from Fernando Po and the opposite coast of Africa were collected by Capt. Sir C. Bullen, R.N., and when placed on the table of the Geological Society every person present who was conversant with lias fossils, thought these samples had been merely sent up from Lyme Regis (*Ammonites communis, &c.*).

The fossils from the Orange River, six hundred miles north-east of Cape Town, including *Astarte orbicularis* and *Gervillia aviculoides* of the green sand, and *Trigonia clavellata* of the oolite, were collected by Dr. Smith, the enterprising explorer of Southern Africa.

The fossils collected by Captain Smee in the Run of Cutch, are very numerous, and several of them, such as *Gryphæa dilatata, Trigonia costata, &c.*, are identical with English oolitic species. (See Proceedings of the Geological Society, vol. ii. p. 77.)

The fossils from the Falkland Islands were discovered by Mr. C. Darwin, and they appear to me to belong to the Lower Silurian Rocks.

It is well known that certain species of Trilobites are common to the older rocks of North America and various parts of Europe; the same fact has been since established repecting South Africa, through the labours of Dr. Smith above alluded to. At Cedarberg, one hundred and fifty miles north of Cape Town, he found an assemblage of fossils evidently composing an Upper Silurian group, in which are some species which mark the passage from the Old Red Sandstone to the Ludlow Rocks and others which are abundant at Ludlow and Wenlock. These fossils were first forwarded to me by my distinguished friend Sir John Herschel, and among them we observe the *Homanolotus Herschelii* (Nob.), Pl. 7 *bis*. f. 2., with *Calymene Blumenbachii* and other trilobites, associated with casts of shells undistinguishable from the following species figured in this work. *Cucullæa ovata, Leptæna lata ?, Orbicula rugata, Nucula, Turbo, Turritella* and *Conularia quadrisulcata*.

But it has been said, that a species in these old rocks has been often decided by a mere cast, grounds upon which no conchologist can pretend to set up real distinctions. Now to this objection I would first reply, that sharply defined casts may sometimes suffice to prove the identity of forms to be compared, and secondly, that it is by no means on such evidences alone, that my conclusions have been drawn; for in numberless instances the fossils of the Silurian System preserve their shelly covering as perfectly as the testacea even of the youngest deposits.

On this point, however, of specific identity (particularly in the case of very distant deposits) let us not insist too strongly. It is enough to state, that on comparing the fossils from different localities of each Silurian formation we find the closest agreement; certain forms being found in one formation, certain other forms in another; and by this means we have been able to determine the age of the rocks, and to verify the justness of the division by order of superposition.

Extending this mode of investigation to fossils obtained from other parts of the world, and comparing casts with casts, an identity of form as well as of muscular and other impressions will be at once admitted. It is on these grounds we assert, that while the Silurian System was in the course of deposition, similar *mollusks*, *crustaceans* and *corals*, lived in very distant parts of the globe, and where climates of very different temperature now prevail.

In the meantime we may appeal to our records in the rocks, and if each succeeding system is found to contain different species of animals, there can be no difficulty in determining, whether their former geographical distribution was or was not similar to that which now prevails.

Examining the strata with this view, we find that in the ascending geological series the quantity of species increases considerably as we approach the younger deposits, and that in proportion to this increase, their geographical distribution harmonizes more and more with that of existing nature. For example, in the older Tertiary strata (Eocene), there are scarcely any examples of shells identical, being found in opposite hemispheres, and in the younger Tertiary (Pliocene) absolutely none; though in these deposits the number of species is at least tenfold greater than in the ancient rocks under consideration.

Now if this position be admitted, it is manifest that it cannot be explained by reference to the present distribution of animal life. For not only has each latitude, in our days, its different products, but even in distant parts of the same latitude (and in the same hemisphere), the amount of variation of species is often very great.

Again, reverting for an instant to our argument in the last chapter, we must recollect, that it does not merely depend upon the wide area over which the sediment containing these fossils was spread, but also upon the fact, that these ancient rocks are of vastly greater thickness than any of the succeeding deposits, and must therefore have occupied a much longer period during their formation.

Hence it is that the geologist, whose science consists in unfolding the records of lost epochs, is not manacled by present conditions, and if the phenomena to which he appeals convince him, that the former changes of the planet were of surpassing grandeur, he may also conclude, that some general cause must explain the distribution of the same species over distant regions. If the existence of formations, so nearly universal as respects the surface of the globe, be admitted, it would seem to be a fair inference, that however we may explain it, there must have *then* prevailed a generally equable temperature.

But let us pass from these general views to the immediate objects of this work.

The following pages and accompanying plates are intended to develop the amount of zoological mutation throughout a large portion of these ancient fossiliferous strata; and the inference we obtain is, that most of the groups which graduate into each other by lithological and geological characters, exhibit also a true zoological transition. Thus, in reviewing this ancient succession in ascending order, we perceive a certain number of species common to the Upper Formations of the Cambrian System and the Lower Silurian Rocks; and that while some of the Lower Silurian animals lived on to those days when the Upper Silurian beds were deposited, scarcely a single species which existed when the Silurian æra commenced, can be detected in the strata which mark its close.

Again, in proceeding upwards from the Silurian period into that of the Old Red Sandstone, we find in its lowest member, a partial intermixture of a few fossils, typical of the Upper Ludlow Rock or the last-formed beds of the Silurian System, associated with some new species; and ascending higher amid these red deposits, we meet with fishes of entirely distinct genera and species, as unlike those of the Silurian æra as they are to those of the overlying Carboniferous System. In the upper division, however, of the Old Red System (though of enormous thickness) scarcely the trace of fossil animals has yet been detected, and hence we are unprovided with zoological links to connect the whole series, though I have no doubt that such proofs will be hereafter discovered, and that we shall then see in them, as perfect evidences of a transition between the Old Red and Carboniferous rocks, as we now trace from the Cambrian, through the Silurian, into the Old Red System.

When, therefore, it is asserted, that the fossils of the Carboniferous æra are dissimilar from those of the Silurian, the reader must bear in mind, that the strata so broadly distinguished, are separated by accumulations of enormous thickness, and that the vast time occupied in their deposit, accounts satisfactorily for an almost entire change in the forms of animal life.

The same order and change of species is apparent throughout the overlying formations. Thus, although there is a slight community of character between the upper members of the Carboniferous rocks and the lower strata of the New Red System, particularly in the plants[1]; and although certain *mollusca* of the Magnesian Limestone

[1] I have adverted slightly in this volume, to the vegetables which may be found in the Silurian System; for although I am persuaded of the truth of the induction of M. Adolphe Brongniart and other fossil botanists, that each geological period, from the tertiary down to the carboniferous deposits, is marked by a distinct Flora,

differ little from those of the coal measures, we cannot examine even that formation, **still less pass onwards through the overlying red sandstone and marl without** perceiving, **that in the Sauroids of those days,** we are **surrounded by entirely new types of animal life, which conduct us by almost imperceptible gradations into the period of the** ***Ammonites*** **and** ***Belemnites*** **;** and as we enter among those groups, we take leave of many of the generic forms, such as *Goniatites*, *Orthoceratites* and *Producti*, which abound in the older periods. Lastly, in the tertiary system which links on the ancient deposits with our own, we entirely lose sight of many of the former types, and are introduced to a series of animal forms much resembling those which are now in existence.—Such is the vast succession with which geology has made us acquainted[1].

I have now only to observe, that in the following plates and their descriptions the organic remains are for the most part figured in the order in which they lie in the strata, a method by which the chronology of the subject is strictly maintained. Every fossil which occurs in more than one formation, is figured as belonging to that in which it is most abundant, and its repetition in other strata is explained in the text and lists. If naturalists should criticise this distribution, they will I hope excuse me when they consider, that my great object is to record perspicuously *a geological* succession.

Of the zoological department I have now to state, that the fishes are described by M. Agassiz; the mollusca and conchifers by Mr. J. de C. Sowerby; the corals by Mr. Lonsdale; the crustaceans by myself. I have further to mention with gratitude, that in various points of illustration I have been materially aided by Mr. Broderip, and that Professor Phillips, Mr. C. Stokes, Dr. Beck, Dr. Milne Edwards, and Mr. König have also contributed their assistance. The support, therefore, of such eminent naturalists must have greatly augmented the value of this division of the work.

the almost total absence of plants in the Silurian strata scarcely entitles me to enter upon this field of inquiry. It is, however, fair to state, that two or three plants, said to be species well known in the carboniferous strata, have recently been found in certain rocks of North Devon (considered to lie far below the culm or coal measures), and therefore as old, at least, as any portion of the Silurian System. If these data be eventually sustained, they will only compel us to suppose, that the changes in the physical state of the bottom of the sea, which produced the variations of its marine inhabitants, did not of necessity cause a change in the vegetable forms of the land; because the submarine operations may have exercised little or no influence upon aerial conditions or climate; and hence it may eventually be demonstrated, that the changes in animal and vegetable life do not continue to run parallel to each other beneath a certain limit in the order of the strata, or in other words during the earliest sedimentary accumulations. I confess that (judging from the analogy of the younger deposits) I at one time held a different opinion; but if the evidences above alluded to are substantiated, I have no difficulty in comprehending the rationale of the case.

[1] Although the stratigraphical order is strictly maintained in the collocation of the plates, the description sometimes embraces a class or family in the same chapter.—Thus the *Trilobites* of all the Silurian strata, described in Plates 7, 7 *bis*, 14, 23, 24 and 25, occupy one Chapter.

16

Reprinted from *Geol. Soc. London Trans.* **5**:633–634, 688–702 (1840)

On the Physical Structure of Devonshire, and on the Subdivisions and Geological Relations of its older stratified Deposits, &c.

By Rev. ADAM SEDGWICK, M.A., F.G.S., F.R.S., &c.
Woodwardian Professor of Geology in the University of Cambridge;

And RODERICK IMPEY MURCHISON, Esq., F.R.S., V.P.G.S.

[Read June 14th, 1837.]

PLATES L. to LVIII.

CONTENTS.

PART I.

INTRODUCTION, p. 633.

CHAP. I.—Description of the five geographical regions of Devonshire, p. 634.

CHAP. II.—Successive deposits between the north coast of Devonshire and the culmiferous series, p. 638.

CHAP. III.—Succession of deposits between Dartmoor and the south coast of Devon, p. 649.

1. Section from the east side of Dartmoor to Torbay, p. 650.
2. Section from Dartmoor to the Plymouth limestone, p. 653.
3. Section from the Plymouth limestone to Bolt Head, p. 656.
4. Section from Berry Head to Start Point, p. 659.

Conclusions, p. 662.

Slate-rocks of Cornwall, p. 664.

CHAP. IV.—Culmiferous series. Its relations to the other formations,—structure,—fossils, &c., p. 669.

Lower culm-measures, p. 670.

Upper culm-measures, p. 677.

Plants, p. 681.

CHAP. V.—On the granite of Dartmoor, p. 685.

PART II.

On the classification of the older stratified rocks of Devonshire and Cornwall, (Read April 24, 1839.) p. 688.

PART I.

INTRODUCTION.

IN a communication made to the British Association during its meeting at Bristol (1836), we first pointed out (by help of a section extending from the north coast of Devon to Dartmoor) the true geological position of the great culmiferous deposits, which occupy so large a portion of that county. In this paper our object is more extensive; for we not only attempt to describe the order of the successive formations north of Dartmoor, but also, as far as we are able, to bring them into comparison with the formations which are expanded from the south side of the Dartmoor granite to Start Point, and to the other headlands of the south coast of Devon. We believe that our present attempt is almost entirely new; for though some portions of the county have been described in considerable detail, yet the several parts have not been brought into any distinct comparison with one another; and in all the published geological maps, the greater part of the slate-rocks have been represented under one colour, without an attempt either to separate them into distinct formations or to mark their succession*. Still less has any attempt been made to bring them into a strict geological comparison with the other mountain groups of England and Wales. Our present object is to exhibit a synopsis of the evidence by which the successive formations are determined. But in future and longer communications, we hope to lay more ample details before the Society, and to describe a large number of organic remains, by help of which, the successive deposits may be referred still more precisely to their true place in the series of British formations, inferior to the carboniferous groups. We are aware of the great difficulty of our task, and if we can succeed in taking only the first step towards its accomplishment, our labours will be well rewarded†.

[*Editor's Note:* Material has been omitted at this point. Also, Plates L. to LVIII. have been omitted.]

* Mr. Greenough represents the slate-rocks under two colours. But the separation is founded on mineral characters and not on age; for even a portion of the culm series, altered near the granite, is represented of the same colour with the oldest slate-rocks of Cornwall.

Rev. J. J. Conybeare made also a separation of the Cornish and Devonian slates into two groups, partly on the same principle, and certainly with no reference to any ascertained order of superposition. His papers, which he did not live to finish, contain some excellent details. Mr. H. T. De la Beche, in the first edition of the Ordnance Map of Devon, (which had been published sometime before our examination of the county,) included (with the exception of the district of Start Point) the whole slate series and the culm-measures under one colour, not attempting any separation of them.

† The details alluded to in the *Introduction* are now given in Part II. of this paper. The whole text of Part I. is printed (with the exception of verbal corrections and an expansion of one or two mere mineral descriptions,) from the communication made by the authors in June 1837. A *few sentences* at the end of it, in which the oldest rocks of North and South Devon were placed provisionally in the Upper Cambrian and Lower Silurian system, are now omitted, as the authors, after a more mature examination of the fossils, are convinced that, so far, their first conclusions were erroneous. This omission will be alluded to again in the Introduction to Part II. (June 1839.)

PART II.

On the Classification of the older stratified Rocks of Devonshire and Cornwall, &c.

IN a former communication we described the mineral structure, and the order of superposition, of the successive groups of rocks which are crossed by a section from the north, to the south coast of Devonshire; and by following the beds (along their line of *strike*) into Cornwall, we were enabled, within certain limits, to bring the *killas* (or slate rocks of that county) into comparison with some of the lower Devonian groups. In this way, we ascertained the existence of a great culmiferous trough, superior to all the other groups in North Devon, and distinguished from them both by its physical structure and its organic remains. Hence we concluded, that the culmiferous series must be represented, for the future, in our geological maps by a distinct and appropriate colour. We had stated the same conclusion, and on precisely the same evidence, at the meeting of the British Association in 1836. We further concluded, that, at least, all the upper portion of this series was of the age of the great coal-formations of England; inasmuch as all its vegetable fossils, as far as they were ascertained, agreed specifically with undoubted coal plants. To this conclusion we still adhere; and we regard the culmiferous rocks as forming a true geological horizon, to which either the ascending or descending series of deposits may be referred.

Again, in the arrangement of the several groups below the culm-trough, as they appear on the North Devon section, we have no modifications to suggest; and the very illustrations which were before suspended (1837), to explain the actual succession of the deposits, will serve our purpose at this moment.

Since our first notice of this section in 1836, Mr. De la Beche has examined the country in great detail; and his arrangement of the several fossiliferous groups, inferior to the lower culm beds, agrees with our own *. The

* There is one unimportant exception to this remark which we have noticed in our former paper (see p. 642, note). It fortunately does not affect the position of the fossiliferous groups, or any question connected with their classification.

same may, we believe, be said generally of the arrangement given by Mr. Williams. We shall, therefore, for the future, refer to the successive groups in the North Devon section with entire confidence. The only questions which can now arise, must refer to their age and classification,—not to their relative position.

To our four sections, illustrating the structure of South Devon, we make no material change or addition ; with one exception, however, viz. the interpolation of a small band of culm measures between the Ashburton limestone and the granite—discovered by Mr. Austen since our visit in 1836, and represented (Pl. LI., fig. 8.) on his authority, as has indeed been already stated in a note to our former paper. Neither have we any material change to make in our comparison of the deposits of North and South Devon. Considered as a whole, we still think that they belong to one system of formation ; and that the lowest group of South Devon is lower than the lowest group of North Devon. But as for our comparison of the subordinate groups (though still the best that we can offer), we only throw it out as provisional, being chiefly founded on mineralogical characters, and those perhaps not of a very conclusive kind.

Lastly, we have no change to make in our description of the fossiliferous groups of Cornwall. They are very nearly on a parallel with groups exhibited in the sections of North and South Devon ; and the fossiliferous slates of both counties belong to one system. We do not encumber ourselves with any details respecting the more crystalline and (as we believe) *altered* slates ; but we have, in our former paper, shown that they are not, merely on account of their structure, to be regarded as older than the calcareous and arenaceous deposits.

After these preliminary remarks we proceed (agreeably to a promise given in our former paper) to re-examine the classification of the respective groups inferior to the culm measures ; and we begin by expressing our entire conviction that the true equivalents in the several disjointed sections can only be made out by zoological evidence.

We were at first led to the opinion, that the series of deposits we are describing, were to be placed below the Silurian system, partly from the resemblance of the highest group (No. 5 of the North Devon section) to the Caradoc sandstone; and perhaps still more from the mineral structure and slaty cleavage of many parts of the lower groups. At that time we had not sufficient boldness to adopt the only other alternative, and throw all the old slates of Devon, and the (supposed) primary slates of Cornwall, into the epoch of the old red sandstone. We were aware, indeed, that several of the South Devon fossils conformed nearly, perhaps specifically, to the type of the carboniferous limestone. But on the other hand, no one was able to define the descending range of such organic forms; especially as it was asserted, that they *reappeared* in the south of Ireland, in beds alternating with true greywacké—or rocks supposed to be older than the old red sandstone. Hence we placed, *provisionally,* but with no satisfaction to ourselves, or confidence in our evidence, all the older fossil groups of Devon and Cornwall in that great but ill-defined interval that exists between the Caradoc sandstone and the older slates of North Wales. Since 1836 we have received repeated proofs of the insecurity of our first hypothesis. Thus, Mr. De la Beche (1837) exhibited fossil plants from the quarries of Marwood (Pl. LI., sec. 2.) one of which Professor Lindley identified with *Stigmaria ficoides,* a very common coal-plant; and near the same place we had ourselves found a *Bellerophon,* not to be distinguished from *B. lobatus* of the old red sandstone. Again, Mr. Lonsdale, after an extensive examination of the fossils of South Devon, had pronounced them, more than a year since, to form *a group intermediate between those of the Carboniferous and Silurian systems,* and hence he concluded that the fossiliferous rocks of South Devon were subordinate to the old red sandstone. This view could be effectually maintained only by one who knew the fossils of the Silurian system. The work in which those fossils are described, proves that the fossils of the Silurian and carboniferous systems are quite distinct. It also proves that the fossils of the old red sandstone of Herefordshire are of a peculiar type; and, at the same time, it points out a want of zoological continuity in the development of organic forms between the epochs of the carboniferous and Silurian deposits. There is, indeed, a true zoological transition between the highest group of the Silurian system and the lower part of the old red sandstone; but between the middle and upper groups of the old red sandstone and any part of the carboniferous limestone no such transition had been discovered. These facts were well known to Mr. Lonsdale, and hence his conclusion respecting the age of the South Devon limestone was based on incomparably better evidence than the mere specific resemblance of some of its fossils to those of the carboniferous limestone: for he contended, as above stated, that the organic remains of South Devon formed a connecting link between the carboniferous and Silurian fossils; and thus supplied that zoological continuity which before had been sought for, but never found.

But this is not the whole of the evidence by which we were gradually led to distrust the views, we had first entertained respecting the age of the older stratified groups above described. For one of the authors, during the summer of 1838, re-examined a part of the great transverse section of Devonshire, and was more than ever convinced that, with some ambiguous exceptions, there was on each side of the great trough a true passage from the lower division of the culm-measures into the inferior fossiliferous group. Thus the section from Marwood across the culm series to the calcareous slates of South Petherwin seemed to be perfectly symmetrical (Pl. LI. fig. 2.), the lower beds of the culm-measures being repeated (as described in our previous paper) on both sides of the trough; from beneath which emerge on the respective sides, and in a regular uninterrupted sequence, the calcareous slates of Petherwin and Barnstaple. Now, if this view be correct, it follows, almost inevitably, from our own evidence, that the calcareous slates in question must be very near the upper limit of the old red sandstone; and this conclusion being once admitted, we should have little difficulty in classing, not merely the calcareous rocks of North Devon, but also the fossiliferous slates of South Devon and Cornwall, with the old red sandstone.

To the preceding facts, bearing on the classification of the older rocks of Devon and Cornwall, we might add some remarks on the mineral structure of the two counties—on the existence of red conglomerates among the beds of the Cornish *killas*—and on the large tracts, both of North and South Devon, filled with rocks identical in structure with the most ordinary forms of old red sandstone. But we forbear to dwell on such facts as these, wishing to build our present conclusions chiefly on zoological evidence.

Guided then by the evidence already stated, and the conclusions to which it seemed to point, we have re-examined the fossils obtained, either by our own labours or by the assistance of our friends, from the two counties; and the conclusion arrived at by Mr. Lonsdale respecting the age of the South Devon limestone, we now apply without reserve both to the five groups of our North Devon section, and to the fossiliferous slates of Cornwall.

This is undoubtedly the greatest change which has ever been attempted at one time, in the classification of British rocks. But all the facts brought to light by those, who have laboured in the same field with ourselves, were gradually leading towards the same point; and the establishment of the Silurian organic types must before long have made the change inevitable*.

So far from thinking ourselves rash and hasty in drawing the preceding

* The reasons for the new classification of those stratified rocks of Devon and Cornwall which are below the culm-measures, were briefly stated by the authors in a paper published in the Philosophical Magazine for April 1839.

conclusion, we think we may rather be accused of being over-cautious and tardy in accepting evidence, however opposed to commonly received opinions. But, on this score, we may state in our justification, that a considerable part of our fossil evidence has been but recently brought to light. Several of the organic remains, from the neighbourhood of New Quay, on the north coast of Cornwall, which we owe to the kindness of Mr. Carlyon, and many other good specimens given to us by our friend Mr. Pattison of Launceston, have been but very recently submitted to a careful examination; and the beautiful collection of Plymouth fossils committed to our care by Mr. Hennah, was only unpacked, in consequence of unavoidable delays, during the present spring. An examination of this collection by one of the authors, assisted by Mr. Lonsdale, Professor Phillips, and Mr. James Sowerby, confirmed the previous conclusions of Mr. Lonsdale, which were chiefly founded on Mr. Austen's collection; and, as before stated, led the way to our proposed classification of the Cornish and Devonian formations*.

At first, we did not intend to give any map among the illustrations of this paper; as it was thought that any of the published geological maps of England would enable the reader to follow our details. But it has now been thought better to give a reduced map of the country we describe, chiefly taken from the corrected Index geological map attached to Mr. De la Beche's Report on the Geology of Cornwall and Devon; in order that we may convey, at a single glance, the connexion between the structure of Devonshire and Cornwall, and the directions in which our several sections are drawn†. All details of structure are, however, intentionally avoided, and the stratified groups are all represented by two colours. In this way we mean to convey our belief, that the inferior groups all belong to one system, viz. that of the old red sandstone. There may, indeed, among the lowest rocks of the two counties, be some which belong to a still lower system. But if such there be, we are unable to point them out; and we think it best and safest, as far as regards their classification, to merge them all at present in one common colour.

Neither did we at first intend to publish any figures of the fossils derived from the several groups above described. But it was urged, that a paper proposing such great changes in the classification of our older rocks would not be well received, or even understood, without more graphic proofs than mere

* Since this paper was read, Mr. Williams has permitted us to examine his excellent collection of fossils from Cornwall and North Devon. We may allude to them generally as confirming our views; but none of them will appear among the illustrations of this paper, as their specific descriptions are, we trust, hereafter to be given by Professor Phillips.

† See Plate L.

verbal descriptions. In the subjoined plates (Pl. LII. to LVIII.) we have therefore figured some of our best fossils. The figures and determination of the fossil shells are all by Mr. James Sowerby: any notice we may take of the corals will be given on the authority of Mr. Lonsdale.

We might perhaps now content ourselves by simply giving a list of the fossils derived from the respective deposits, and then leave the reader to draw the proper conclusion from the facts before him; but we think it advisable, in the first place, to put our evidence in a more general and popular form. In our previous paper, the successive groups are described in the ascending order in which we first observed them, during our traverse from the north to the south coast of Devon. But in noticing the fossils, it is almost necessary (as our sections have no well-defined base) to take a reversed order, and commence with the upper division of the culm-measures.

1. *Culm Measures.*—In the first place, then, we have nothing to add to the evidence given by the vegetable fossils of this upper division of the culm-measures, or to the conclusions we drew from it. On this point, therefore, we merely refer to our former communication (*suprà*, p. 669, *et seq.*). The black limestone of the lower division contains many shells of the genus *Posidonia*, three or four species of *Goniatites*, and two or three other genera not yet well ascertained. Now the genus *Posidonia* is found abundantly both in the upper and lower limestone shales, of the true carboniferous series of England and Ireland: for example, in the *calp* of Ireland, which is in the place of the lower shales, and in the upper limestone shales of Northumberland, not far below the millstone grit. There is a species in Northumberland, which makes a very near approach to a species very abundant in the black culm limestone. Again, of the *Goniatites* of the culm limestone, there are three species very closely related to those of the mountain limestone*. Among them is one which Professor Phillips is unable to separate from the *G. Henslowi,* found in the great scar limestone of the Isle of Man; and it deserves remark that this species occurs also in the Petherwin group, below the culm-measures,—a fact which tends to prove the reality of a passage from the base of culm-measures into the next inferior group.

Where then shall we place the culm limestone, and carbonaceous shales that form a part of it? If we arrange them with the upper shales of the mountain limestone, then nearly the whole culm-trough of Cornwall and Devon will be the equivalent of the middle and upper divisions of our true carboniferous series. This view was adopted by Prof. Phillips when he published his index map of the British Isles. But if we identify the black, calcareous, culm shales

* Report on the Geology of Cornwall, &c. p. 117.

with the *calp* of Ireland, or the lower limestone shales of England, the culm series of Devon will, as a whole, be very nearly the equivalent of our entire carboniferous series. Considering, however, the enormous thickness of the whole culm-measures and the paucity of fossils in all the inferior beds of its lower division, we continue to place its base in a doubtful position. This is a measure of necessary caution, especially when we bear in mind the extreme difficulty, in the northern parts of our island, of drawing any precise line of demarcation between the base of the coal-measures and the old red sandstone*.

Before the sections of North Devon had been made out, the culm-measures were naturally included in the vast deposit of graywacké, of which we knew neither the beginning nor the end. But to merge them in that series now, is to reject all the direct evidence we possess, and all the analogies supplied by the corresponding parts of our island. Lastly, there seems now to be no difference of opinion as to the superficial distribution of the culm-measures; for the boundary line indicated roughly by us in 1836, and more in detail in our previous paper, agrees, as far as it goes, with that lately published by Mr. De la Beche.

2. *Fossils of Petherwin and Barnstaple.*—The fossiliferous slates in the neighbourhood of these two places rise immediately from beneath the base of the culm-measures. Their position is therefore symmetrical, and, as before stated, they appear on the whole to form a mineral passage into the overlying system. In this way, we establish a mineralogical connexion between the North Devon and Cornish sections. But the connexion is made incomparably more secure by help of the fossils. Those we have obtained from Barnstaple, Croyde Bay, Marwood, &c. (all of them places in the highest division, No. 5, of our North Devon section,) certainly agree as a group with those of Petherwin, and have many species in common, as will appear by our lists. It is enough for our present purpose to quote the following; *Spirifer attenuatus, S. bisulcatus, Orthis interlineata, Atrypa concen-*

* The difficulty we have pointed out does not arise from any doubt respecting a great principle of classification, but merely involves the adjustment of a boundary line. One of the best illustrations of this difficulty is seen on the east coast of Scotland, north of St. Abb's Head. The old red sandstone first appears as a conglomerate, resting transversely on the edges of the graywacké; and then passes so insensibly into the *red carboniferous rocks,* that, through a considerable extent of coast, it is hardly possible to draw any well-defined line between the two formations. We may, however, observe by the way, that there are several pretty extensive tracts coloured as old red sandstone in Macculloch's Geological Map of Scotland, which we believe every geologist who was guided by the English types would class with the lower division of the mountain limestone; and there are one or two places in Dumfries-shire, where we think he has confounded the new with the old red sandstone.

trica, A. decussata, A. striatula; also a smooth and a plaited species; *Leptæna caperata, L. scabricula*? *Pecten s. n. Pallastra s.n.** All these shells are found both among the calcareous slates of Petherwin and in the calcareous slates near Barnstaple; and several of the species common to both localities are extremely abundant. That there are several species among the Petherwin fossils which have not been yet found among those of Barnstaple, takes not at all from the force of our direct evidence. It deserves remark, that there are several species in these upper groups which do not appear in the lower; and among those we may notice the Goniatites and the Endosiphonites†, (Clymenia of Munster,) as they seem to point towards the types of the carboniferous period.

In the lower division of the Barnstaple calcareous slates occur the Marwood fossils, given in our lists, among which are several species of *Cucullæa.* These shells are associated with the genus *Bellerophon,* one of which, as before stated, cannot be distinguished from the *Bellerophon globatus* of the old red sandstone. Very nearly on the strike of the Marwood beds occur the fossil plants of the Sloly quarries. According to Professor Lindley, (*antè,* p. 683.) one of the species belongs either to a *Stigmaria* or a *Lepidodendron*; while others are new or incapable of determination. Professor Henslow, who has recently examined a good series of fossil plants from the same locality, informs us, that he is able out of them to distinguish three or four species, not one of which he believes to have been described in the published works of Sternberg, Brogniart, or Lindley. Either of these results would fall in with our present views. We undoubtedly consider the Sloly beds as out of the limits of the carboniferous series, and therefore we should expect that its fossils would conform to an older type. On the other hand, the appearance of a few species of the true carboniferous fossils in the upper division of the old red sandstone would be in perfect harmony with what we know of the geological distribution of organic forms. But the assumption that the ordinary species of coal-measure plants may exist in the *old graywacké* rocks of an entirely distinct period, takes for granted conditions which are not supported by the evidence of the Devonian sections, as we now interpret them.

We are unwilling to push our present remarks any further; but we can

* Some of the fossils quoted are of doubtful species; but the doubt does not apply to the identification of the individual specimens from the opposite sides of the culm-trough.

† The appropriate name of *Endosiphonite* had been given to the genus by Mr. Ansted before he was acquainted with the previous discovery of Count Munster. The name *Clymenia* is by no means so good; but having the priority, ought to be retained. We cannot but regret that the original name (*Planulites*) was rejected by Count Munster, as it harmonizes with the general nomenclature of fossil *Cephalopodes,* better than his present name *Clymenia.*

hardly avoid surmising that the peculiar plants of the "*terrain d'anthracite inférieur*" of Professor Voltz, and those of a supposed similar age in Brittany and other parts of France, may eventually be found to be on the same parallel with this portion of our North Devon series.

3. *Inferior Groups of the North Devon Section.*—The number of fossils hitherto collected is far too small to justify us in affirming to what extent the fossils of the Ilfracombe calcareous slates (North Devon Section, No. 3.) differ from those of Barnstaple. A few species are certainly common to both, among which we may quote two species of *Leptæna* and a *Pecten* (see the list). Corals abound in the Ilfracombe slates, though the species are not numerous. The *Favosites polymorpha* (never yet discovered in the mountain limestone, or below the upper Silurian groups,) is the most abundant. With it is another entirely new species, which also abounds in the Cornish fossiliferous strata beneath the culm-measures. On the whole we may state, on the authority of Mr. Lonsdale (from whom alone we have derived any accurate information on this head), that the Devonian corals are all either of species peculiar to these groups or such as occur only in the higher part of the Silurian system.

Immediately below the Ilfracombe calcareous slates, are some hard strong beds, with numerous impressions of a large cordiform bivalve, unlike anything we remarked in the overlying beds. From the lowest group of all (North Devon Section, No. 1.) we have very few well-preserved fossils : among them are many beautiful casts of a *Fenestella,* and among the bivalves is an Orthis of a new species, and a Spirifer identical with one found near Barnstaple (*Spirifer attenuatus* ?).

Such is the summary of the evidence given by the fossils of the North Devon section ; and imperfect as we acknowledge it to be, we think it sufficient to bear out our proposed classification ; especially as we know of no conflicting fossil evidence that is in any way opposed to it.

3. *South Devon and Cornwall.*—In the comparison of the mineralogical groups of North and South Devon, we can add nothing to what has been stated in a former paper. We could not bring the great upper South Devon limestone into direct comparison with the calcareous zone of Barnstaple ; because, in that case, the great slate formation, which overlies the Plymouth limestone, would have no representative in North Devon. We therefore provisionally placed the great calcareous group of South Devon on a parallel with the lower fossiliferous groups toward the northern extremity of our section. The fossils appear on the whole to bear out this classification.

First, then, as to the South Devon corals.—None of them have been found in the mountain limestone, several are new or not known to occur in the Silurian system, and some agree specifically with corals of the highest Silurian groups. On this point we rely on Mr. Lonsdale's determination; and we give it in his own words:

Favosites polymorpha, (Goldfuss, Pl. XXVII. f. 4. Silur. Syst., Pl. XV. f. 2.). Common in the Plymouth* and Ogwell* limestones, and it occurs also at Barton and Torquay, and in slate at New Quay, Cornwall.

——— *spongites* (Goldf., Pl. XXVIII. f. 1. Silur. Syst., Pl. XV. f. 8, 9.). Plymouth and Ogwell limestones.

——— ? *ramosa*, (Brassart). The Rev. D. Williams's cabinet. Near Chudleigh.

——— *Gothlandica*, (Goldf., Pl. XXVI. f. 3. Silur. Syst., Pl. XV. *bis*, f. 3, 4.). Plymouth limestone; apparently scarce.

——— *fibrosa*, (Goldf., Silur. Syst., Pl. XV. *bis*, f. 7.). A single imperfect specimen of this species is in Professor Sedgwick's collection from Fowey; it occurs also at Barton, near Mary Church.

Porites pyriformis †, (Ehrenberg, Goldf., Pl. XXI. f. 7. Silur. Syst., Pl. XVI. f. 2.). Common in the Ogwell limestone; it occurs also near Plymouth, and Mary Church. Pl. LVIII. fig. 4.

Cyathophyllum turbinatum, (Goldf., Pl. XVI. f. 8. Silur. Syst., Pl. XVI. f. 11.). A fragment in the Rev. R. Hennah's collection.

——— *cæspitosum* †, (Goldf., Pl. XIX. f. 2.). Very common near Plymouth and Ogwell; it occurs also near Torquay. Pl. LVIII. f. 8 to 8 d.

Astrea (*Favastrea*) *helianthoides*, (De Blainville, *Cyathophyl. helianthoides*, Goldf., Pl. XX. f. 2.). Two specimens in calcareous slate from near Plymouth in the Rev. R. Hennah's collection, and one in limestone in the Museum of the Geol. Soc.; presented by the late Mr. Harvey.

——— (———) *pentagona*†, (De Blainville, *Cyathophyl. pentagonum*, Goldf., Pl. XIX. f. 3.). Plymouth, Ogwell, and Torquay limestones; not very abundant. Pl. LVIII. f. 1, 1 a.

——— (*Siderastrea*, De Blainville,) *Hennahii* †, (sp. n.). Plymouth limestone and Barton Quarry near Mary Church. Pl. LVIII. f. 3 to 3 b.

Strombodes vermicularis, (*Cyathophyl. vermiculare*, Goldf., Pl. XVII. f. 4.). Common in the Plymouth and Ogwell limestones. Pl. LVIII. f. 7, 7 a.

Stromatopora polymorpha †, (Goldf., Pl. LXIV. f. 8.). Two specimens from Appleway and near Teignmouth, in the collection of Mr. D. Sharpe. Pl. LVIII. f. 2.

Coscinopora placenta ? †, (Goldf., Pl. IX. f. 18.). Common in the Ogwell and Plymouth limestones; it occurs also at Torquay, Whatcomb, Appleway, and Newton Bushel. Pl. LVIII. f. 5 to 5 d.

Turbinolia ? *celtica* †, (Lamouroux, Pl. LXXVIII. f. 7, 8.). Dinas Cove, Padstow, Berry Pomeroy, and Fowey. Pl. LVIII. f. 6.

Fenestella antiqua †, (*Gorgonia antiqua*, Goldf., Pl. XXXVI. f. 19. Silur. Syst., Pl. XV. f. 15–18.). Common at Fowey, Petherwin; in yellow rotten slate in the neighbourhood of Plymouth; in North Devon the fossil occurs at Saunton, Linton, ? and north of Barnstaple. Pl. LVIII. f. 10 to 10 b.

Cystiphyllum Damnoniense†, (sp. n.). Very abundant near Ogwell; it occurs also near Plymouth, but sparingly. Pl. LVIII. f. 11 to 11 b.

Scyphia turbinata†, (Goldf., Pl. II. f. 13.) Slate near Plymouth. Pl. LVIII. f. 9.

The most abundant of the above species in the limestones of South Devon are *Favosites polymorpha*, *Porites pyriformis*, *Cyathophyllum cæspitosum*, *Strombodes vermicularis*, *Coscinopora placenta*, and *Cystiphyllum Damnoniense*, only the three first of which have been noticed in the Silurian System of England. The limestones are further characterized by the total absence of *Catenipora escharoides*, and by the absence or extreme scarceness of many of the corals most abundant in the Wenlock and lower Silurian formations.

* The specimens from the Plymouth limestone, are principally in the cabinet of the Rev. R. Hennah; those from the Ogwell limestone in that of Mr. R. A. C. Austen; and those from Barton, near Mary Church, and Teignmouth, in the collection of Mr. Daniel Sharpe.

† The species thus distinguished are figured in Pl. LVIII., either on account of their being new, or characteristic of the Devonshire limestones.

Coupling this account with the near resemblance of some of the South Devon shells to known fossils of the mountain limestone, we at once establish our main position, that the South Devon fossils are of an intermediate type between those of the Silurian and carboniferous systems. It also follows from the same evidence, that the North and South Devon corals belong to rocks which are very nearly on the same parallel, as the most abundant corals of the lower North Devon groups agree specifically with the most abundant corals of the Plymouth and Tor Bay limestones.

As far as regards the fossil shells, we must refer to the accompanying lists. We may however remark, in our present summary, that we have shells in the Plymouth limestone which make a near approach to *Terebratula Wilsoni, Atrypa orbicularis, Orthis depressa, Pileopsis vetustatis, Nerita Haliotis*, and *Orthoceratites circularis*, of the upper Silurian rocks. Most of them, however, when closely examined, present some points, we think, of specific difference. On the other hand, this limestone contains some genera and species very nearly resembling certain fossils (such as *Pecten dissimilis* (Phillips), *P. plicatus* (Sowerby), *P. arenosus* (Phillips), *Terebratula acuminata* (Sowerby), *Producta analoga* (Phillips), &c. &c.) of the carboniferous limestone. Lastly, we may remark, that many species in the Plymouth limestone (and those, we believe, the most abundant) are quite distinct from the fossils both of the carboniferous and Silurian systems; among which we may point out a very large *Atrypa, Buccinum spinosum*, a large *Natica*, and several small univalves. (See the Lists.)

We have many fossils from the neighbourhood of Newton Bushel, chiefly, however, in the form of siliceous casts: we forbear, however, to dwell on them, as Professor Phillips and Mr. Austen are, we believe, both engaged in their description. Some of them agree specifically with shells of the Plymouth limestone; others are new; several of them nearly approach the forms of the mountain limestone (as we learn from the description of Mr. Austen, and the recently published "*Report*" of Mr. De la Beche); and the corals, at least the most abundant species, agree with corals of the Plymouth and Ilfracombe limestones.

Such is a brief summary of the results arising out of an examination of the South Devon fossils; and the conclusion to which they point appears to us absolutely inevitable, viz. that the South Devon series is intermediate between the carboniferous and Silurian systems, and consequently on the parallel of the old red sandstone.

After the details of our previous communication, it is evident that all the calcareous slates on the south-eastern side of Cornwall are of the age of the owest group of the South Devon section, and therefore probably in the lower portion of the old red sandstone. On the north-western coast of the county, we were preceded by Mr. De la Beche, and we must refer our readers to the many excellent details of his "Report." A few of our fossils from both coasts will appear in the appended lists; and, imperfect as they are, they appear to sanction our general views. The Petherwin, and we think also the highly crystalline Tintagel slates, appear to be of the age of the Barnstaple slates, and are therefore in the upper part of the old red sandstone. Such is a general summary of the evidence on which our proposed classification is founded; and the result is given in our accompanying map. (Pl. L.)

Concluding Remarks, &c.

Having stated the grounds on which we have endeavoured to establish the classification of the stratified rocks of Devonshire and Cornwall, it may be well, before we quit the subject, to notice certain objections that have been urged against our views.

1. In the first place, our evidence may be considered defective, from the absence of the carboniferous limestone, which in many parts of the British Isles has so striking a position between the coal measures and the old red sandstone. To this we might reply, by pointing out several instances in the central counties of England, where the carboniferous system rests on the lower formations, without the intervention of any bands of limestone. We would, however, rather appeal to a more general fact—that the mountain limestone, after forming in the north of England a great succession of beds, and rising into mountains of considerable elevation, gradually thins out in its prolongation towards the south; so that in South Wales it occupies but one well-defined band, and at length so entirely dies away in the western extremity of Pembrokeshire, that the overlying culm measures and hard grits rest, in one district, immediately on older rocks, without the intervention of any calcareous matter whatsoever. With such a fact before us, we have no reason to be surprised at the absence of the mountain limestone, where the culm measures reappear in Devonshire and Cornwall.

Considering the enormous thickness of the culm measures of these counties, it is by no means improbable that the mountain limestone is there represented by contemporaneous deposits of a different mineral type. For surely it is now needless for us to prove that rocks, identical in geological position, may entirely change their lithological structure, as they pass from one county to another. A change of structure, however, implies a change of conditions during the time of the deposit; and such a change might greatly modify the forms of animal life. In this way we might, perhaps, explain the absence of many of the best-known mountain-limestone fossils from the carbonaceous shales and sandstones of the lower culm measures of Devonshire.

2. Again, it has been objected, that the fossils of the Devonian groups, now classed with the old red sandstone, are not the same with the known fossils of that formation.

We allow that the fossil fish of the old red sandstone have not yet been discovered among the Devonian rocks. But how can we build any conclusion on such a mere negative fact? The same genera and species are not distributed through the whole extent of every formation. Moreover, who can tell that fossil fish will not hereafter be discovered among the calcareous slates of Devonshire? But there is a more positive objection to be noticed, viz. that the known fossil shells of the old red sandstone differ greatly from those we have obtained from the Devonian groups. To this we may reply, that the shells hitherto described from the old red sandstone are derived exclusively from its very lowest beds, which probably do not appear on any of our previous sections. These objections are, however, of little weight against the broad fact (which, in truth, forms the sum and substance of our evidence), that the fossils described in this paper are of a type intermediate between the types of the carboniferous and Silurian systems*.

3. The extreme change, among the formations we have described, of the ordinary mineral type, is another objection to our views. It has been noticed in various parts of our two memoirs, and considered at some length in our paper, of April last, in the Philosophical Magazine; we need not, therefore, dwell upon it now. But we may briefly state, that as the old red sandstone exhibits so many changes of structure, in its range from the Orkneys to the north side of the Bristol Channel, we have no difficulty in admitting still fur-

* Were it true, that species abounding in the mountain limestone, and disappearing in the descending sections, appeared again in still lower parts of the gray-wacké series; there would then be comparatively little weight in our evidence, or certainty in our conclusions. But we have no such antagonist fact now to contend with. In a paper, published (April 1839) in the Philosophical Magazine, we expressed our hopes that the views of classification we then threw out might cast some light on the structure of the south of Ireland; and before our paper was struck off, we became acquainted with the contents of a memoir by Mr. C. W. Hamilton, to which we alluded in a postscript. At the very meeting which followed that on which our present paper was read before the Geological Society, Mr. Griffith exhibited a copy of his Geological Map of Ireland, and read a memoir, illustrated by sections, and accompanied by fossils; by help of which he showed, we think, to demonstration, that there is no anomalous recurrence of the carboniferous fossils among the true gray-wacké rocks of the south of Ireland. His sections of the rocks immediately below the mountain limestone exhibited a broken, imperfect series; most instructive, however, in one respect, as showing an undoubted ascending passage into the limestone.

ther changes, if supported by the evidence of fact or rational analogy. What difficulty is there in believing that a marine formation should, in one part of the world, have gone forward so tranquilly as to allow all the ordinary operations of animal life; while, in another part, the deposits of the ocean were interrupted or modified by rude mechanical movements, incompatible with the existence of organized beings? We see no difficulty whatsoever in the mineral structure of what we now regard as the old red sandstone of Devonshire and Cornwall. The classification we adopt is derived from its own appropriate evidence; and it at the same time removes the greatest difficulties presented by the unexplained phenomena of our older formations; gives a natural sequence and meaning to their history; and, we may add, gives a symmetry to the disposition of the colours on our geological maps, which they never had before.

The strongest proofs are given in this and our previous paper, of the insufficiency of mineral character, taken alone, to determine the epoch of any doubtful formation: and as the term *gray-wacké* has done much disservice to geology, by inducing observers to merge, under one unmeaning name, deposits belonging to distinct periods in the history of the earth, we venture to hope that the word may henceforth be discarded as a term of classification. Used for the mere purpose of mineral description, we can of course make no similar objections to it.

4. Lastly, we may remark, that in all the descriptions, both of this and our previous paper, we have, for obvious reasons, drawn our analogies from the nearest formations of our own island. But we cannot forbear expressing our belief, that the series of rocks of Devonshire and Cornwall, above described, may serve to clear up some of the difficulties in the classification of the older formations of the continent of Europe, where the old red sandstone has seldom been found, at least in what has been hitherto regarded as the ordinary British type. But now that the old red sandstone has been exhibited in such a different form, on such a vast scale, and with its own series of organic remains, we have little doubt that its true equivalents will be found distributed through various parts of Europe. To obtain, as far as possible, satisfaction on some of these points, is the first task we propose to ourselves during the investigations of the approaching summer.

Though the term old red sandstone, when designating great groups of rocks like the Cornish *killas* and Devonian slates, should involve no error of classification, still it would, mineralogically, be most inappropriate. We purpose therefore, for the future, to designate these groups collectively by the name *Devonian system*, as involving no hypothesis, and being agreeable to analogy. Thus the terms Carboniferous system, Devonian system, Silurian system, and Cambrian system, will represent a vast and apparently uninterrupted sequence of deposits; each having, as a whole, zoological and (often also) mineralogical characters of its own, yet each passing into the system next in order by almost imperceptible shades of difference. The old *palæozoïc* rocks appear, therefore, notwithstanding their enormous scale, and the wide diffusion of the same organic types, to have been formed in subordination to the same laws which, in after periods, produced the more clear succession of our secondary formations.

Section from the Foreland on the Bristol Channel to the English Channel. For the details of the portion between the Foreland and Dartmoor, see Plate LI. fig. 1.

17

CONCLUSIONS RELATIVE TO THE SEPARATION OF GEOLOGIC STAGES, AND TO THE SPECIAL FAUNAS WHICH THEY ENCLOSE

A. D'Orbigny

This excerpt was translated expressly for this Benchmark volume by Barbara Conkin, University of Louisville, from pages 149–156 of Cours Elémentaire de Paléontologie et de Géologie Stratigraphiques, *vol. 1, Masson, Paris, 1849.*

#185. Some people have thought that discordance was indispensable for separating two ages of terrain. Observation, as well as reason, supports a contrary belief. In order for discordance to exist at all places at once, between the last beds of one stage and the first of the following stage, it is necessary to suppose that the dislocation at the end of an epoch is manifested with the same intensity in all parts of the globe at once; this is not probable. It is enough to see revolutions that have operated to attain, in their dislocations, a considerable portion of the sphere, of the extent, for example, of the chain of the Andes and the Pyrenees. When one studies the geology of plains and mountains, one discovers, on the contrary, that parallel beds, belonging to very different stages, succeed each other during a more or less considerable lapse of time. Thus, it is necessary to believe that if, even certainly, some parts of each stage have been dislocated during geologic revolutions, others remained with the upper and lower stages in a more or less concordant position.

Here is how we have found the separation of stages marked, in the quite special study that we have made of it, as one will be able to see in our fourth part, where all the partial facts will be established.

#186. *Denudation of a stage, marked by irregular lines of erosion,* at its point of contact with the stage that immediately covers it, is equivalent to a discordance because it is a certain sign of geologic movement. One sees, everywhere, some examples of these lines of separation.

Polishing, erosion, the corroded surface of one rock, before the rock which succeeds it is deposited, . . . are also as good limits between two geologic stages as is discordance, since they are again produced by powerful movements of dislocation, and of movement of waters, which are the immediate result of it.

It is likewise with the lack, in some places, of an intermediate stage, recognized everywhere else, as one finds so often in nature. The lack of the albien stage, which one notices in the Alps, from Castellane to Gap, between the aptien stage and the cenomanien stage, and the lack, on the contrary of the aptien stage, between the neocomien and albien, from Castellane to Grasse, when, on the other hand, these four stages succeed themselves regularly in the departments of Meuse, Haute-Marne, Aube, and Yonne, and in England, on the Isle of Wight, are, for us, the equivalent of a discordance; for these lacks of stages prove a movement of upheaval, of subsidence, or a great denudation at the places where they are not found in the midst of beds in their ordinary order of succession.

#187. When one sees the deposits of caverns, those of the pampas, as well as those of nearly all bone breccias such as those of Sicily, Nice, and Algeria, to be of a same reddish ferruginous color produced, evidently, by the washing of terrestrial parts of the surface of the continents, one must believe that this ferruginous color can, in certain cases, indicate the end of a geologic epoch or the beginning of another. It is, in effect, what we have noticed in a multitude of circumstances. . . .

#188. Great deposits of pebbles, or of coarse sediments, such as the little pebbles that are often found at the base of a stage, as at Thouars (Deux-Sevres) in the toarcien stage and at Carry (Bouches-du-Rhône) in the falunien stage, also frequently mark the end of a stage and can serve to distinguish them because they have been brought there during prolonged movement in the waters, at the beginning of a new horizon.

Thick unfossiliferous beds, which one sometimes finds at the base of a stage, also mark its beginning, before it had contained animals [*Editors' Note:* Material has been omitted at this point.]

#191. The line of separation between two stages is, let us say, marked by a discordance of stratification in the beds, by denudation, by polishing, by erosion of the surface of the older stage of the two, by ferruginous deposits, by beds of pebbles, by exterior inequalities of ground, and finally, by difference of color and of mineralogic composition of the succeeding rocks. It remains to know if these different signs, by which one recognizes the end of a stage or the beginning of another, always suffice to make the distinction very certainly and to give their relative age. Discordance, we have said, is a positive means of distinguishing two geologic epochs that succeed at whatever place. This discordance, however, which very certainly indicates that there is no identity of geologic age between these two stages, remains mute on the question of knowing if, deposited at this place one above the other, these stages are really those that nature has made to succeed in the general order of the geologic ensemble. One sees, in effect, in some places of our globe, and above all in France, numerous stages following regularly one after the other, in their relative age and in their true order of superposition, . . . but one finds also in yet more places the natural order reversed, in connection with denudations or differences of levels. [*Editors' Note:* Material has been omitted at this point.]

When discordance separates two stages that ought to succeed naturally, or when it separates stages that leave between them either terrains or a great number of stages at least, as in the places that we have just cited, it always only denotes a simple difference of epoch without signifying anything as to the relative mutual age of these epochs. It is thus certain that here the domain of *nonpaleontologic geology* ceases and gives way to the *exclusive domain of stratigraphic paleontology.*

If, in effect, in these cases, one applied only geology, one could suppose that the stages are in their natural order of superposition. One would not take, from then on, any consideration of the gaps that exist between them, and taking the stage that covers the other as the one that ought to succeed it, one will commit an error that would extend, afterward, onto all the other upper stages that one will place lower than they ought to be, of which we could cite frequent examples in the works of some authors. Geology has not, effectively, without paleontology, any certain means of recognizing the age of stages, when the natural order is interrupted. This is because the mineralogic characters are insufficient and can, on the contrary, lead to grave errors, since present circumstances and the study of terrestrial stages have made us discover (#87) the synchronism of beds of very different natures. When geologists who are not paleontologists have reported these gaps, they have always been led by paleontologic characters of the stage, the only ones that one can then invoke with certainty. . . .

#192. We have established (#163) that the end of each great geologic period had been marked by the annihilation of beings composing the fauna of each of these periods and that a new fauna was then manifested on the surface of the globe. The result is a succession of distinct faunas characteristic of each terrain, of each stage, as the special considerations of our third part will prove. In effect, whether a fauna appropriate to a stage be taken in India, America, or France; whether it be in contact with the fauna that preceded it, whether it be, on the contrary, separated by a more or less great gap, whether it be finally contained in a sand, in a clay, or in a limestone, it displays no less everywhere the same zoologic characteristics of the ensemble and the identical more characteristic species. Thus, in all geologic circumstances, stratigraphic paleontology will always give, for the sedimentary beds, with the help of general characters of animal forms and of some common species, comparatively with what one has observed previously in the midst of stages superposed in their natural order, a certain means of recognizing the relative age of a bed.

#193. In one paragraph of this chapter (#190), we have said that differences of mineralogic composition of concordant stages would proclaim the respective limits of these stages. However, synchronism of present deposits (#87) has shown us that sediments of diverse nature are deposited, at the same time, in the present seas, as they were deposited in the diverse geologic stages (#144). As a result, this difference in the mineralogic nature can receive only a very restrained local application. [*Editor's Note:* Material has been omitted at this point.]

#194. In summary, all the partial considerations of our third part, where we give the result of the strict discussion of the paleontology of the globe relative to vertebrate and annulate animals and to more than

18,000 species of molluscan and radiate animals,[1] and all geologic results of our fourth part, where we examine the stratigraphic and paleontologic characters of stages, lead us to the following conclusions:

Animals not showing any transition in their specific forms succeed each other on the surface of the globe, not by passage but by extinction of existing races and by the successive creation of species in each geologic epoch.

The animals are distributed by stages, according to geologic epoch. Each of these epochs presents, in effect, on the surface of the earth a distinct fauna, characterized by special forms, and by species indentically the same throughout. Also, the Silurian, Devonian, Carboniferous, Permian, Jurassic, Cretaceous stages and even the lower stages of tertiary terrains from all the geologic beds of the globe, on which we have certain data, present identical paleontologic characters—that is, the same *facies* of assemblage, the same generic forms and a more or less great number of identical species, common throughout, that prove their complete contemporaneity.

NOTE

[1]This ensemble of species, discussed as to their geologic age and to their partial and general zoologic characters in the paleontologic stratigraphy, ended in 1847, was printed, at this moment, under the title of *Prodrôme de paléontologie stratigraphique universelle des animaux mollusques et rayonnés*—two volumes of more than 500 pages each, fruit of a labor of five years.

18

GEOLOGIC SUMMARY ON THE CEPHALOPODS OF THE JURASSIC TERRAINS

A. D'Orbigny

This excerpt was translated expressly for this Benchmark volume by Barbara Conkin, University of Louisville, from pages 600–611 of Terrains Jurassiques, Vol. 1, Cephalopodes, *Paleontologie Française, Masson, Paris, 1850.*

I call *Jurassic terrains* all the succession of stages and beds that occupy the interval comprised between the Keuper, the last Triassic deposits, and the Neocomian stage, the first member of the Cretaceous terrains. In thus limiting the Jurassic terrains, as did MM. Dufrenoy and Elie de Beaumont in their geologic map of France, I reunite therein, as did these scholars, all the deposits after and including the Lias sandstone of authors, to the Portlandian stage inclusively. Circumscribed in this manner, the Jurassic terrains form a regular group cleanly separated from lower and upper terrains by all the stratigraphic and paleontologic considerations that offer, everywhere, the most perfect harmony in their general results.

DIVISION OF JURASSIC TERRAINS INTO STAGES

Many divisions have already been proposed for the Jurassic terrains—divisions deduced from the mineralogic characters of the beds, others based on the presence of such and such dominant fossils. I will not seek to discuss here the worth of divisions established by these methods; all, when they are due to direct observation and not to theoretical ideas, offer partial or general facts of great interest. But when it is a question of coordinating them, however, one finds oneself stopped from continuing. How can one group facts based on mineralogic composition alone when one has seen, by the study of present cases, that these limits are altogether illusory[1]. On the other hand, how can one dare to rely on nomenclature of the fossils mentioned in whatever series of beds when one sees the determination of these fossils so carelessly made by authors that it is necessary often to cut out half of it. It thus becomes impossible, in the present state of things, to establish a perfect concordance between the heterogeneous elements written in the annals of geologic science. Before these insurmountable difficulties I have found only one possible solution: to examine nature itself. From my first observations of the ground of France, I recognized everywhere the same succession of fossil beings confined in the same limits of geologic height,

[1]See *Cours élémentaire de Paléontologie et de Géologie Stratigraphiques,* p. 81 following.

whatever might be, moreover, the mineralogic composition of beds that enclosed them. I have equally recognized that the mineralogic character of beds served only to fool observers in sometimes making them see imaginary parallelisms. . . . I thus devoted myself to follow the paleontologic horizons everywhere where they were found, to assure myself if they depend on a marked epoch or on a simple local facies determined by the littoral or pelagian nature of the deposits that I previously studied.[2] After having found in all parts of France—in the north, south, east, and west, in Provence as in Normandy, in the Ardennes as in the Vendee—the same results, and having progressed during fifteen years only from confirmations to confirmations without finding a single contradictory fact, I have finally acquired the certainty that the Jurassic terrains are neatly divided into ten superposed zones or stages, limited as well by the respective faunas that they enclose as by the lines of stratigraphic demarcation noted everywhere. I have followed them one after another to the circumference of the basins in France and beyond; I have recognized that they are not confused in any place and that they truly represent so many distinct geologic epochs succeeding themselves one after the other in a constant and regular order. This fact secured, it was a matter then of being positively assured if these different stages set off on our soil were the result of a series of local circumstances appropriate to France or if they depended on general facts obvious in all parts of the globe at one time. I have happily some means of reaching this last conclusion.

Research carried out in Russia by MM. Murchison, de Verneuil, Keyserling, and Hommaire de Hell has put me in reach of recognizing, by the faunas enclosed in the Jurassic terrains of these countries, studied comparatively with those of France, that from the Crimea to Moscow and from this place to the north of the Urals, the more or less numerous members of the Jurassic terrains observed in Russia, all depend, by their respective faunas, on our French stages. It is the same in England and Germany where the continuation of the same ancient basins as in France is found. The fossils from the Jurassic terrains of South America, collected by MM. Darwin and Domeiko, have led me to the same results, as well as those from the province of Cutch in the East Indies. In effect, these isolated places, located at immense distances from one another, offer with our French stages not only common characters in the assemblage of their faunas but often moreover identical species that prove their perfect contemporaneity. These distant confirmations that, for the Jurassic terrains, happened to corroborate my observations, gave me at the same time the certainty that all the causes of separation of stages have been general. After these confirmations it was to a lesser extent possible for me to remain in doubt as to the worth of stages such as I envisaged them, since all intermediate places, in England and Germany, showed the exact continuation of them. I have had to adopt them for the double reason that they are not arbitrary and are, on the

[2]See *Cours élém. de Paléont. et de Géologie Stratigraphiques,* I.

contrary, the expression of the divisions that nature has traced with grand strokes on the entire globe.

These divisions, beginning with the lowest, are the following: Sinemurian, Liasian, Toarcian, Bajocian, Bathonian, Callovian, Oxfordian, Corallian, Kimmeridgian and Portlandian stages. One will see by the synonymy of each in particular that several have been perfectly perceived above all by English geologists, who have always taken more account of paleontologic characters in their divisions of terrains, while these divisions, often misappreciated formerly due to mineralogic preoccupation and given little of the value accorded to fossils, have led to many erroneous comparisons with the English sections or to many purely local divisions.

One will perhaps find these divisions too numerous, but as I have just said, they are the expression of the limits traced by nature and have nothing arbitrary. They have moreover all equal value and are all as important. It is necessary, then, either to adopt all without exception or to suppress them entirely, in order to not make from diverse epochs that have succeeded in the Jurassic terrains only a single whole, which would be too monstrous. It is certain that the stages such as rigorous superposition and the limits of the faunas they enclose present, are as clear cut in the Jurassic terrains as are, for example, the Silurian, Devonian, and Carboniferous stages in the Paleozoic terrains.

One will see by the nomenclature adopted in the terminology of names of stages that I have wished for the Jurassic terrains, as I have already done for the Cretaceous terrains, to take names drawn from the places where the stage is found best developed, in order to put an end to this tangled nomenclature drawn from the local mineralogic composition, so variable according to the localities, and from fossils dominant at one place, which can be lacking elsewhere. Here is, moreover, this synonymy for all the stages, beginning with the lowest.[3]

1st STAGE: SINEMURIAN. D'Orb.. I have derived this name from the town of Semur *(Sinemurium),* where the better type is found, a layer which I then regard as *standard*—that is, being always able to serve as a place of comparison. It is the zone of *Ostrea arcuata, Ammonites bisulcatus.* I know at present 175 characteristic species [*Editors' Note:* Material has been omitted at this point.].

2nd STAGE: LIASIAN, d'Orb. I have kept this name for naming that of the *lias* given first by the English and generally adopted. It is, with the uniform terminology adopted, a derivation analogous to that of *carboniferous,* of *falunien, etc.* The type in France is at Landes, at Vieux-Pont (lower beds only) (Calvados). . . . It is the zone of *Ostrea cymbium,* of *Ammonites margaritatus.* I know 285 characteristic species. [*Editors' Note:* Material has been omitted at this point.]

3rd STAGE: TOARCIAN, d'Orb. This name is derived from the town of Thouars *(Toarcium),* Deux-Sevres, where one finds only this stage on

[3]The limits of this work not permitting me to give some extension to special geologic considerations, I refer to details of stratification, for the geographic extent of the stages, to the fourth part of the *Cours élém. de Paléont. et de Géologie stratigraphiques.*

azoic rocks, where there is the most beautiful development in France that can be regarded as the type locality, standard locality. It is the zone of *Lima gigantea* and *Ammonites bifrons.* I know 287 characteristic species. [*Editors' Note:* Material has been omitted at this point.]

4th STAGE: BAJOCIAN, d'Orb. The name is derived from the town of Bayeux *(Bajoce),* Calvados, where the most beautiful French type is found, the standard locality, in all respects similar to that at Dundry, where the English type is found. It is the zone of *Trigonia costata,* of *Ammonites interruptus* . . . . I know in this stage 602 species. [*Editors' Note:* Material has been omitted at this point.]

5th STAGE: BATHONIAN, d'Omalius. M. d'Omalius d'Halloy established this name for the great oolite near Bath, in England. In France, I find the coastal type with *Ammonites bullatus* . . . : the submarine type is shown at Luc . . . .I know in this stage 532 species. [*Editors' Note:* Material has been omitted at this point.]

6th STAGE: CALLOVIAN, d'Orb. I derive this name from Kelloway *(Calloviensis),* which in England, was the first place where the stage had been well defined. The French type is at Dive. . . .It is the zone of *Ammonites jason* and *refractus,* of *Ostrea dilatata.* I know in the stage 278 species. [*Editors' Note:* Material has been omitted at this point.]

7th STAGE: OXFORDIAN, d'Orb. This name derives from the town of Oxford, in England, where the stage is found. In France, the best characterized type is found at Neuvizy (Ardennes) . . . . It is the zone of *Ammonites cordatus.* . . . I know in this stage 729 species. [*Editors' Note:* Material has been omitted at this point.]

8th STAGE: CORALLIAN, d'Orb. I have preserved this name in deriving it from *calcaire corallien,* of M. Thurmann, the name perfectly applied in France for a long time and on account of the polyps that are encountered there. The coastal type is found in France at Dompierre (Charente-Inférieure). The submarine type of deposits or of reef deposits, at the point of Ché near La Rochelle . . .It is the zone of *Ammonites altenensis,* of *Iceras arietina;* I have cited in this stage 638 species. [*Editors' Note:* Material has been omitted at this point.]

9th STAGE: KIMMERIDGIAN, d'Orb. I have derived this name from the town of Kimmeridge, in England, where the first type has been described. In France it is found at Tonnerre (Yonne). . . . It is the zone of *Ammonites Lallieri,* of *Ostrea deltoides* and *virgula.* I know around 200 species. [*Editors' Note:* Material has been omitted at this point.]

10th STAGE: PORTLANDIAN, d'Orb. I derived this name from the Isle of Portland, in England, where the first type has been described. The French type is found at Biney. . . . It is the zone of *Ammonites giganteus* and *Irius* of *Trigonia gibbosa.* I know at present only 61 species. [*Editors' Note:* Material has been omitted at this point.]

Such are the ten stages that I believe ought to be adopted as the expression of the diverse epochs that have succeeded each other in the Jurassic terrains. If there was for some species a contradiction between the age indicated in the text that follows and the text previously printed, one can take the following distribution because it is the result of new observations and local verifications made since the printing of the work.

19

GENERAL GEOLOGIC DEDUCTIONS DRAWN FROM FOSSIL SPECIES

A. D'Orbigny

This excerpt was translated expressly for this Benchmark volume by Barbara Conkin, University of Louisville, from pages 256–257 of Cours Elémentaire de Paléontologie et Géologie Stratigraphiques, *vol. 2, bk. 1, Masson, Paris, 1852.*

#1609. In summary, when one looks at our *Prodrôme de paléontologie stratigraphique,* and at the results detailed in the generalities or in the fourth part, one sees that the number of species common to two stages is a rare exception, which exists only for some stages separated without doubt by geologic perturbations of less value than those where no common species are shown up to the present and that even in stages where these common species exist, their presence is explained by the concordance of stratification of the stages, or when these are not separated by intermediate unfossiliferous, terrestrial, or marine beds. It is thus not necessary to take these common species for facts of passage, but one must regard them as mixtures, as exceptions for which one can easily account. It is urged also to guard against false ideas that one could take up in some exceptional places, in confounding them with the normal state. When one finds numerous mixtures, in the Falunian and Subapennine stages of Piémont, for example, where the stages are concordant and without intermediate beds, it is necessary to guard against concluding anything from it because these mixtures evidently depend on causes that we have indicated previously. If, moreover, one wishes to have proof of it, one will find it in the comparative study of these same stages well limited at other places such as Touraine, Bordeaux, and Dax for the Falunian stage and Perpignan, for the Subapennine stage where one encounters no common species between the two stages. . . . Thus, when there is a mixture at one place, it is necessary to have recourse to well-limited stages to explain them and not to confuse some local exceptions with the real limits of faunas characteristic of geologic stages.

#1610. Definitions of a geologic stage with respect to species. After all we have said about the separation of special faunas, we ought to make known what we intend by a stage. A stage, for us, is an epoch identical in all respects to the present epoch. It is a state of repose of past nature, during which existed, as in present nature, continents and seas, terrestrial plants and animals, and marine plants and animals and, in the seas, pelagic and littoral animals of all depth zones. For a stage to be complete, it must show an ensemble of terrestrial and marine beings, which could represent an entire epoch, analogous to the development we see presently on the earth. When one perceives only some parts of this ensemble that ought to exist at one time, it is because the others have been annihilated during geologic perturbations or that they are unknown. Therefore, we could not admit beds, even discordant, as a stage, whatever is their thickness, if they do not contain their characteristic faunas. We regard, thus, as completely false the idea that each stage ought to have few special species because it is evident that if (which we will demonstrate) each stage is an epoch similar to the present epoch, it must enclose a proportionate fauna and that if we know only a few species, it is that the trace of the greatest part of the species that it enclosed has been entirely annihilated or is unknown to us. We think also that the number of species proper to each stage discovered today ought to be considerably increased by research that remains for us to do in order to understand the diverse ages of the earth, in all parts of the globe, and we will not be astonished if this number, by the pursuit, becomes tripled.

#1611. There is relative to the limits of the stage a danger that must be avoided carefully: that of attaching too much importance to the local distribution of fossils in beds, before being assured if these details are the same in all parts of the earth. Most often, in effect, when in a geologic basin one finds that such and such beds enclose such and such series of species, one is naturally brought to regard this arrangement as an important fact of stratigraphy, as a special distinct epoch, when it is most often, as we have discovered by comparison, only a purely local arrangement, which nowhere allows generalization and which is related only to the composition of sediments or to local oscillations of the ground, which we will seek to demonstrate in stages in particular cases.

Part IV

OPPEL (1856) TO BUCKMAN (1902)

Editors' Comments on Papers 20 Through 28

20 OPPEL
Excerpts from *The Jurassic System of England, France and Southwestern Germany, Divided and Compared According to Its Individual Members*

21 QUENSTEDT
Excerpts from *Historical Notes*

22 BRONN
Excerpts from *On the Laws of Evolution of the Organic World During the Formation of the Crust of the Earth*

23 PHILLIPS
Excerpts from *Anniversary Address of the President*

24 LAPWORTH
On the Tripartite Classification of the Lower Palaeozoic Rocks

25 DELAPPARENT
Excerpt from *Paleontologic Method: Divisions of the Sedimentary Series*

26 BUCKMAN
Excerpts from *On the Bajocian of the Sherbourne District: Its Relation to Subjacent and Superjacent Strata*

27 BUCKMAN
Excerpt from *On the Grouping of some Divisions of so-called 'Jurassic' Time*

28 BUCKMAN
The Term 'Hemera'

By the middle 1800s, the foundation and broad framework of the science of stratigraphy had been constructed. As the stage and zone concepts were more fully developed and more details were filled in, the edifice took on an unanticipated complexity in its appearance. It became clear that the stratigraphic column would have to be subjected to more careful analysis if its true nature and meaning were to be appreciated and realized. According to Arkell (Paper 32, 15-17), Oppel (Paper 20) "took a decisive step beyond d'Orbigny" and "placed stratigraphy on a new footing and breathed new life into it." He emancipated the zones "from the thralls of both local facies, lithological and paleontological, and of cataclysmic annihilation, thus giving them an enormous extension and transforming them from mere local records of succession to correlation planes of much wider (theoretically universal) application." (D'Orbigny had, in fact, also envisioned stages as being universal.)

Oppel attempted to follow the precepts of his teacher, Quenstedt, who in his *Der Jura* (Paper 21, 23, 24) advocated the greatest possible precision in the collection and identification of fossils and in the detailed measuring of sections, still the best advice for any stratigrapher.

We have selected only a few passages from Oppel's massive *Die Juraformation* (Paper 20, 123, 124, 135) that explain his use of certain species to mark zones in the stages. Misunderstanding seems sometimes to have arisen with respect to Oppel's zones as a result of some workers' believing he meant that each of his zones had entirely distinct species not found in others. This is true for only some of the species; some are distinct, others are not. Phillips (Paper 8) held a similar view concerning the distribution of fossils in stratigraphic successions.

With regard to the naming of stages and zones, Oppel stated:

> I have named the individual zones each according to one of its more important species. For this I had precedence from a number of geognostic papers. Even though isolated concepts are hereby transfered to the whole essence of the formations that are to be distinguished and are used to name them, this would also occur with any other method of naming, because if we choose, for instance, locality names, we hereby transfer to our zones those characteristics attached to the local deposit. . . . In addition, there is probably hardly a city or locality in the vicinity of which only a single zone is formed. [1858, 247,248]

Oppel went on to explain that some of his thirty-three Jurassic zones "are to some extent hard to recognize at some localities because of a change of the predominent facies. Finally, some . . . could only be included in certain districts by their denying in other areas those features which there served us to distinguish them" (p.248). Some few of his zones, he added, might later be deleted.

Quenstedt's *Der Jura* was published in 1858 (Paper 21), the same year that Oppel completed his *Die Juraformation*. Quenstedt was extremely critical of d'Orbigny, partly in return for what he believed to be d'Orbigny's unjust criticism. Quenstedt's wounded outcry (pp. 17-21, 23, 24) is, as Arkell (Paper 32, 14) phrased it, "extremely telling," and we believe it is of interest to present excerpts from it; in this we learn also of Quenstedt's stratigraphic methodology.

Also published in 1858 was Bronn's memoir (Paper 22), translated in 1859 by Dallas, in which he showed the existence of transitional species and genera that bridged some of the gaps in the fossil record, thus aiding in the acceptance of Darwin's theory (Bronn translated the *Origin* into German as soon as it was published). In Bronn's conclusions (pp. 89, 90), his points 3, 4, and 9 are relevant here. Bronn (point 4) saw formations (meaning systems) as chronostratigraphic units, as had Lyell (Paper 9B) in his concept of the recent. In addition (point 9), Bronn recognized the existence of what later would be termed *disconformities* and *paraconformities,* even between thin strata; these subtler kinds of disconformities often go undetected with the result that different species appear to have lived simultaneously.

In Phillips's *Anniversary Address* (Paper 23), we have a summation of contemporary understanding of the distribution of marine fossils and of the vital role they play in stratigraphy. While Phillips acknowledged that the fossil record is *incomplete,* he found it not *"insufficient* to sustain just conclusions" (p. xxxiii). He stressed the point that "affinity of fossils is accepted as evidence of the approximate contemporaneity of the rocks" (p. xxxii); and that "the series of strata and of fossils, taken as a whole, is but one—a function of elapsed time; but the terms of the series apparent at any place depend on the physical history of the locality—a function of local conditions. In other words, the agreements depend on the concurrence of time; the differences on disagreeing local conditions" (pp. xliii, xliv).

Phillips (Paper 23, xliv) distinguished between the time term *period* (meaning a span of time) and *epoch* (meaning a point on a time line or a moment) and believed that only the former is generally ascertainable in stratigraphy. Phillips, after studying the fossils and their occurrences, as he catalogued the fossil collections of his uncle, William Smith, realized (p. xxxviii) that there are facies fossils, and he called for more study of this phenomenon. He noted the existence of "a poor origin, a rich development, and a subsequent decrease" in fossil zones and that zones are "distinguishable by the prevalence of a few types, and by the paucity or absence of others," (pp. xxxviii, xxxix) an observation which he had made in 1829 and which would come into play later in the concept of acme-zones and Buckman's concept of hemera (Paper 28).

Ramsay, in his presidential address of 1863, pointed to ten unconformities (both angular unconformities and disconformities) in the British Paleozoic strata, which strata Murchison had viewed (Paper 15) as a "vast yet . . . connected" succession (p. 574); Ramsay expressed the suspicion that more time was represented by missing beds than by existing formations, a proposition later dealt with by Barrell (1917).

Lapworth's (Paper 24) classic division of the Lower Paleozoic into three systems, by his introduction of the Ordovician, forestalled the possible continuation of grouping strata on the basis of similarities in faunas (as more and more fossils became known) into oversized general stratigraphic units. Lapworth discussed the pros and cons of alternative approaches to division of the Lower Paleozoic and concluded by demonstrating that the only solution that would satisfy from every valid viewpoint was to recognize the three systems. For this division, Lapworth relied on "the relative magnitude of zoological change" (p.3).

Lapworth (Paper 24, 10) made the excellent point that while an "unconformability" (angular unconformity) is useful to the field geologist as a base line in mapping and drawing sections, "it is far otherwise with the cautious systematist, who endeavours to found his systems in accordance with those of Nature herself, upon principles, not of local, but of universal application. . . . an unconformability . . . is . . . wholly misleading except in local application." Lapworth considered the belief in worldwide stratigraphical breaks between systems to be the "degenerate and impoverished survival" of the "doctrine of universal convulsion and simultaneous destruction of all life" (p. 11). For Lapworth, "the group of rocks holding each fauna forms for the geologist a convenient Procrustean bed to which to fit the tolerably synchronous deposits of other lands" (p. 10). He viewed a large angular unconformity as a convenient clue to the recognition of a system boundary since it carries with it a large amount of lost time resulting in two distinct faunas.

By 1880, most of the guiding principles of stratigraphy had been stated more or less precisely, and terminology was temporarily standardized into the dual nomenclature of stratigraphic and chronologic terms at the Second International Geological Congress in Bologna in 1881. Thereafter, according to Gignoux (1955, 21), it was mainly due to deLapparent's widely circulated *Traité de Géologie* (Paper 25) that general agreement was reached on zones and stages. DeLapparent reiterated the use of fossils for correlation and determination of relative age, providing one "substitutes for the idea of identical species that of *representative species*" (p. 705) (previously suggested by Phillips, Paper 23, xlv, and recognized in principle by Sedgwick and Murchison, 1839).

DeLapparent (Paper 25), realizing that there have always somewhere been interconnected oceans, believed that "paleontology furnishes the means of putting all the local series studied by stratigraphers in order" (p. 706). Similar faunas, to deLapparent (p. 707), did not necessarily mean the same instant in time, and he presented a diagram (Fig. 175) that compared three hypothetical stages in two regions to illustrate his view. He suggested that one might observe lateral continuity between deposits with fossils of two successive stages, a suggestion to which we cannot subscribe. We believe stratigraphers have sometimes become too dependent on paleontological evidence while neglecting significant physical evidence of change; this was well stated by Grabau in 1905.

DeLapparent (Paper 25) noted that "a boundary of the first order marked at one place by an accentuated discordance, as much stratigraphic as paleontologic, will be effaced little by little . . .in a certain direction, in order to make one continuous series of concordant beds" (p. 708). This so-called continuous series, we believe, would be broken by a disconformity of the magnitude of a paracontinuity (Conkin and Conkin, 1973).

DeLapparent advised (Paper 25, 708, 709) going "from the particular to the general and to lean above all on regional successions, well established in a great number of places" in drawing boundaries and grouping the smaller stratigraphic units. The hierarchy he proposed consists of local layers or horizons, grouped into assises composed of paleontologic zones; diverse synchronous assises, between two instants (disconformities) marked by terrestrial change at least in one region of notable extent, composed a stage.

DeLapparent echoed past workers but, at the same time, anticipated later authors (Chamberlin and Grabau) by suggesting (p. 709, 710) diastrophism as the basis for boundary determinations, but he believed the boundaries of systems would not be "susceptible of mathematic precision" (p. 710), the events defining them not having been instantaneous everywhere. Thus, he believed, "the corresponding period did not have to begin or end everywhere at the same instant" (p. 710).

Finally, deLapparent (p. 712) suggested using the littoral deposits of northern and central Europe to determine boundaries of stratigraphic units and the pelagic deposits of the Mediterranean region to determine groupings of these divisions, as previously suggested by Lavoisier in 1789.

From 1893 until 1902, Buckman (Papers 26–28), working with the Jurassic, attempted to introduce a more sophisticated refinement into stratigraphic concept, methodology, and principle by introducing the term *hemera* (Paper 26) for "the smallest consecutive division which the sequence of different species enables us to separate in the maximum development of strata" (pp. 481,482). Buckman used *zone* in the sense of Moore (1865-1866)—that is, as the exact horizon of a particular fossil. Buckman meant hemera to be the time of this kind of zone, not the time of the zones shown in his Table IV, which as Arkell (Paper 32, 20) noted, might include two or three hemerae. Buckman (Paper 28, 556) proposed *biozone* for "the range of organisms [species to phylum] in time as indicated by their entombment in the strata" and *faunizone* for "belts of strata, each of which is characterized by an assemblage of organic remains" (p. 557). Buckman added that "hemera . . . is the time term which corresponds to 'faunizone' as the stratigraphical term, just as 'age' corresponds to 'stage'" (p. 557). Subsequently the term *epibole* was proposed by Trueman (1923, 200) for the deposits accumulated during a hemera.

In 1893 (Paper 26), Buckman had defined hemera by saying it "is intended to mark the acme of development of one or more species" (p. 482), and he stated that "it is used in a chronological sense as a subdivision of an 'age'" (p. 518).

Buckman (Paper 27), presented his chronological system of nomenclature for the Jurassic, "based entirely on zoological phenomena," and proposed a "geological calendar." Here again he referred to the hemera as "the shortest geological time-division . . . that is the time during which a particular species . . .had dominant existence" (p. 443). He admitted (p. 444) that the practical application of his ideas would be difficult due to incomplete knowledge of faunas and to irregularity in faunal succession.

Once more, in 1902 (Paper 28), Buckman attempted to clarify his term *hemera* and its use. Alluding to the sunrise, he stated that a hemera "was designed to mark the time from species-rise to species-rise" (p. 555). In this paper he emphasized that a hemera always existed everywhere, while a zone (in the sense of Moore, 1865–1866) might be absent in particular areas due to nondeposition or erosion.

Arkell (Paper 32) had praise for Buckman's hemeral tables based on actual field work but added that "it is one of the greatest misfortunes for Jurassic geology that when increasing age and frailty prevented Buckman from continuing active field work, he lost sight of the distinction between results obtained with hammer, collecting bag and field notebook, and those arrived at by speculation and deduction from matrices at home" (p. 36). Nevertheless, Buckman's works and ideas spawned much thought and discussion that led to more practical zonal concepts.

Arkell (Paper 32) charmingly records an argument he had had with Buckman over Buckman's insertion of an excessive number of hemerae into the timetable of the Corallian rocks:

> I attempted to justify my view (which I still hold) that many of his hemeral indices lived side by side on the same sea-bed, by reminding him of Lyell's principle that the present is a key to the

> past and instancing the sporadic distribution of many modern sea-shells around our coasts. He caused considerable provocation by remarking with a smile 'Ah! So you have been reading Lyell: a most misleading book.' At the time there seemed no more to be said, so outrageous was the heresy. But the remark often recurred to me and I realized that in this particular connexion there was at least a germ of truth in it. It showed me that Buckman was not oblivious of ecological arguments . . . but he did not consider them valid. [pp. 27-28]

Arkell criticized Buckman's hemera for being based on the acme of a species, inasmuch as a species might reach its acme at different times in different places.

In 1899, Lapworth prophesied that "zonal work is probably destined to effect in the history of geological research a revolution as great and an advance as rapid as those brought about by the use of the microscope in the history of biology" (p. 520). If this is true, it is due to meticulous collecting and study of sections as has been advocated over and over since the time of Schlotheim (1813) and Phillips (Paper 8) and to the attention paid to the physical evidence of change as urged by Grabau (1905).

REFERENCES

Barrell, J., 1917, Rhythms and the Measurement of Geologic Time, *Geol. Soc. America Bull.* **28:**745-904.

Conkin, J. E., and B. M. Conkin, 1973, The Paracontinuity and the Determination of the Devonian-Mississippian Boundary in the Type Lower Mississippian Area of North America, *Univ. Louisville Studies Paleont. Stratigr.,* no. 1, 36p.

Gignoux, M., 1955, *Stratigraphic Geology, English Translation from the Fourth Edition, 1950,* by G. G. Woodford, W. H. Freeman, San Francisco.

Grabau, A. W., 1905, Physical Characters and History of Some New York Formations, *Science (n.s.)* **22**(565):528-535.

Lapworth, C., 1899, The Survey Memoir on the Southern Uplands: A Review, *Geol. Mag.* **6:**472-479, 510-520.

Moore, C., 1865-1866, *Somerset Archaeol. Soc. Proc.* **13**.

Oppel, A., 1858, The Jurassic System in England, France, and Southwestern Germany, *Wuerttemb. Naturwiss. Jahresh.,* Stuttgart, pp. 148-262.

Schlotheim, E. F., 1813, Taschenbuch für die gesammte Mineralogie (Leonhard), vol. 7, Hermannschen, Frankfurt am Main, pp. 2-134.

Sedgwick, A., and R. I. Murchison, 1839, Classification of the Older Stratified Rocks of Devonshire and Cornwall, *London Edinburgh Philos. Mag.* (ser. 3) **14**(89):241-260.

Trueman, A. E., 1923, Some Theoretical Aspects of Correlation, *Geol. Assoc. Proc.* **34:**193-206.

20

THE JURASSIC SYSTEM OF ENGLAND, FRANCE AND SOUTHWESTERN GERMANY, DIVIDED AND COMPARED ACCORDING TO ITS INDIVIDUAL MEMBERS

A. Oppel

These excerpts were translated expressly for this Benchmark volume by Ursula Barber, University of Louisville, from Wüerttemb. Naturwiss. Jahresh. *1856, pp. 123–124, 135.*

We do not have a division of the Jura formation according to its smallest component parts based on the evidence of the same in the various countries. Only whole groups of strata have been compared to one another, but it has not been shown that any horizon, which is marked at one place by a number of species constant to it, can also be found in the most distant area with the same certainty. This task is, no doubt, difficult, but only by accomplishing it can an exact correlation of whole systems be assured. Thereby, it becomes necessary, regardless of the mineralogical characteristic of the strata, to explore the vertical distribution of each individual species in the most diverse places and subsequently to emphasize those zones that, by the constant and unique occurrence of certain species, set themselves apart from the adjoining ones as definite horizons. Hereby one obtains an ideal profile of which the component parts of the same age in the various areas are always characterized by the same species. I have attempted such a division and have found it to be feasible with most stages; with others, I consider it as still unfinished. The difficulty in this is to be found mainly in the insufficient number of well-described species. The more accurately the species can be separated, the more exactly can the strata be divided. . . . We must often be satisfied with few well-defined species when marking a horizon because the remaining occurrences have not been sufficiently explored or have not been described. Further complicating is the changing of the faunas. To correlate with certainty a coral formation with a clay formation of the same age would often be almost impossible if one could not find, by means of a larger number of comparisons, some analogies.

After I had put together the sequence of the individual horizons and had thus produced the ideal profile, I found that the comparison of the local systems to one another could be executed with greater certainty. An example will explain this. The lower Lias can, minus the bone bed, be divided into seven and the middle into six successive zones. If we compare the division by Phillips, Marcou, d'Orbigny, and Quenstedt to one another, we find that the last two authors have the seven just-mentioned zones as the lower Lias. Phillips, however, adds the two above lying to it, while Marcou places only the four lowest zones into his lower Lias. If we wanted simply to place d'Orbigny's lower Lias with that by Phillips, Marcou, or Quenstedt, without considering the more detailed component parts, an accurate comparison definitely could not be achieved since the values, which the different authors assign to one and the same stage, often differ completely from one another. In my work, therefore, I place the sequence of the individual horizons first, substantiate these more accurately, and on the basis of this, present only then the more general view and comparison of the local systems.

Even if the individual horizons often mark themselves off more exactly from one another than whole stages from one another, I have still retained the grouping of the Jura formation by stages, especially since the grouping of the lesser known fossils is hereby made easier. At the end of the study of each stage, I mention those species on which the division and comparison of the strata are mostly based. At the same time, I state in each appendix the synonymy of the individual species as far as the intended briefness of the work permitted it.

To each of the individual zones are always assigned those species that distinctly characterize it and that have not yet been found in any other strata. That not all species of the lower Lias could here be exactly assigned goes without saying; after all, even whole stages have particular species in common with one another. Are not such crossovers more likely in two adjoining zones.

21

HISTORICAL NOTES

F. A. Quenstedt

This excerpt was translated expressly for this Benchmark volume by Ursula Barber, University of Louisville, from pages 17–24 of Der Jura, *Laupp, Tubingen, 1858.*

Unfortunately, even with this draft, I cannot outline in its entirety how the species are connected among one another. . . .That, I am reserving for a later larger work. For the time being, I only want to furnish the collector with a practical book. The expert will recognize rather quickly that the preciseness of our classification, which is based merely on self-observation, is substantiated completely by the nature of the matter. A person who is not capable of understanding us can indeed not collect. More minute divisions have probably been made, but we do not have a truer description of a great formation, which has traced step by step those creatures layered one on top of the other. In that, our Swabian strata shall serve others as a model. Indeed, we cannot in paleontology establish the species as positively as is the case with the plants and animals in the living world; the mutilation of the remains alone does not make this possible. To learn now, with just some certainty, what is identical to one another or different, we must know above all what lived together. The best works of modern time, as those by Goldfuss, Bronn, d'Orbigny, Davidson, etc., leave here far too noticeable gaps, and yet, no precise determination is possible when I do not know exactly which layer the creature was taken from. I have already taken great care to determine such points as much as possible in the *Handbook of Paleontology;* now, however, this becomes the main task. When already the Floezgebirge of Wuerttemberg has remained not without influence, even on the presentation of foreign regions, . . .the easier will, due to the drawings, the present description win friends. Only from one side, from which I would have least expected it, did I experience the most bitter and most unjust opposition: from Alcide d'Orbigny—not only in his *Paléontologie* but also in the most disparaging tone in *Prodrôme de paléontologie,* Paris 1850. I would have gladly taken his classification into consideration. However, a presentation, which because of carelessness and mistakes is so distorted, must raise doubts even where it may be correct. [*Editors' Note:* Material has been omitted at this point.] Now I ask: With such cardinal errors, of what value are the grand citations? And yet, d'Orbigny did have it much easier than we, because he presents the whole Lias in three sections: Sinemurian, Liasian, Toarcian, without any subdivision, while in Sinemurian . . . alone, we make five to six definite subdivisions . . . . Precisely this gradual sure probing, this unfolding of how one subdivision emerges from the other and all fossils that can be positively defined have a specific layer, constitutes the crucial point of all investigations. After all, I do not wish to be anything else but a geognost who nevertheless takes the liberty of also scrutinizing the forms—otherwise no thoroughness is possible. D'Orbigny, nevertheless, does not blush to tell me and Prof. Bronn (Prodrôme de paléontologie I, p. XVI): "Réunir les espèces les plus distinctes en négligeant de s'occuper de l'étage ou elles se sont montrées." On page xx in particular he attacks me. To begin with, a lamentation is set up that I do not stop at the simple name of the species but instead avail myself of several adjectives; that would be stepping a century backward and must impede progress. To that I reply: What the actual species is we do not know, but a more in depth study of the forms convinces us rather quickly that also among animals in the wild state, if only they are presented to us in sufficient quantity, similar races and varieties are found as among the domesticated. To see this properly emphasized one cannot but resort to adjectives. Otherwise, the true viewpoint is altered. [*Editors' Note:* Material has been omitted at this point.] If the layer had been stated exactly, as I "backward person" am in the habit of stating, then one could make a decision. [*Editors' Note:* Material has been omitted at this point.] The reproach that I do not separate distinctly is utterly unfounded, therefore unjust. However, the progress of science lies not only in separating but also in correctly connecting, and the latter, wherein d'Orbigny unfortunately shows too little skill, is decidedly the most difficult. This connecting is not done, however, according to principles of the last century but according to foregoing presentation and illumination of all characteristic marks. Who reproaches me with that has either not read my works or did not understand them. For 20 years I am speaking of the two-pointed abdomen lobes of the Lias ammonites and of the single-pointed lobes of the later ammonites. It is one of the most important and remarkable laws in the organism of these creatures; has d'Orbigny even benefited from this? At first sight it must of course seem daring when I maintain regarding Ammonites Baugieri d'Orbigny . . . that something is drawn on it that in nature could not at all take place . . . . Now, d'Orbigny may punish me for my boldness; he needs only to show at a public meeting in Paris the specimen with septa in the living chamber. Then I must admit that in our small Swabia things grew differently than in the large France. It is not at all necessary "prendre la diligence de Paris" to assure oneself that, for instance, the alveolus of Belemnites giganteus . . . is not reversed in its sheath; it can consequently not have been copied from nature. That is indeed bitter medicine if one has to be told such things by a man "who is 100 years behind in science." But criticism must be if science is to advance. To err is human, and Mr. d'Orbigny will find with such reproofs also many an approving word, which—coming from him regarding me—I have looked for in vain to this day [*Editors' Note:* Material has been omitted at this point.] Had d'Orbigny, the great "palaeontographe universelle," only looked at neighboring Germany as I at his France, then he might stop mockingly restricting me to our Wuerttemberg.

D'Orbigny has never been able to reproach me with any other thing but that I have incorrectly compared a figure. In most of the cases in which he claims this, it is either not true or the mistake is not mine. But even if this be the case, so has he made at least the same or usually worse mistakes. The reason for this is after all in the imperfection of the illustrations that we, being humans, cannot help. The rebuke, however, that I used only the first commonly used name in Germany, I repudiate for the following reasons: When we refer to earlier writers of the last century and the beginning of this century, so this can happen with few exceptions . . . because the earlier ones, Linné included, had a much broader conception of species than we do, particularly with the imperfections of their figures. [*Editors' Note:* Material has been omitted at this point.] Indeed, furthermore, a precise interpretation of even our best drawings will in the future only be possible in instances where at the same time the definite layer is determinable. Here, till now, remained the most noticeable gaps. Though I acknowledge with pleasure that d'Orbigny does many things better than his contemporaries, his classification into stages, nevertheless, constitutes still far too broad a frame, and even in this frame the worst mistakes are yet made. I, on the other hand, was first to have drawn the divisions with a preciseness that alone seems appropriate to the nature of the matter, and only on this basis should the understanding of the Jura species take a real step forward. For the sake of exactness, I had to attach a variety name often, but this one is usually understandable in itself and does not complicate the matter . . . . In this matter, I have sometimes gone perhaps a bit too far and must therefore be the more amazed at the contrary reproach by d'Orbigny. Our language should be chosen in such a way that it, wherever possible, aids the memory and perception. Since now, however, a species is positively established only when it has been well described and drawn—because without that, communication is not possible, and as regards fossils, one will in future also have to know the layer; consequently, there still remains, thank God, a latitude that is capable of maintaining a counterbalance to the empty vanity of being first in giving a name. [*Editors' Note:* Material has been omitted at this point.] Of what avail is it if a man has seen the whole world, and he does not understand aright the things that lie in front of his own doors? As much as it may be the task of geology to compare the far with the near and to span the whole globe, one cannot after all warn enough again of that hastiness to hurry through the countries and believing to have comprehended all of them. The person who with fitting circumspection becomes engrossed in his surroundings will soon realize that for the strength of one man, even a few square miles will be enough. We cannot manage this little. I am not able to compare exactly the small piece of earth in the land of the Hohenzollern with that of the Hohenstaufen. The further I extend my horizon, the more difficulties confront me. The insight into the history of the development of our earth can, under circumstances, be clouded by the extent of the horizon, and exactly that which d'Orbigny so heavily relies on is his star of misfortune. He wants to bring too much under one system. The next task we have to fulfill in connection with the Jurassic is to draw up sections as faithfully as possible in the different districts *of*

Central Europe. With such a slow advance from the bottom to the top will the transformation, which our earth in the course of time underwent, fully manifest itself. To compare two hand-high layers one above the other according to their true contents can have a more fruitful effect on the development of science than the registration of stages from the most distant areas of the world of which one must understand from the beginning that they are not correct after all.

I will now peacefully and to the best of my ability represent the matter as I understand it after an insight 19 years ago in Swabia. The faithful worker will discover the good, and in the end all derision and biting envy must give way to the truth.

22

Reprinted from *Ann. Mag. Nat. Hist.*, ser. 3, **4**(20):89, 90 (1858)

ON THE LAWS OF EVOLUTION OF THE ORGANIC WORLD DURING THE FORMATION OF THE CRUST OF THE EARTH

H. G. Bronn

[*Editors' Note:* In the original, material precedes this excerpt.]

II. ***Results of investigations regarding the gradation or simultaneousness of the appearance and disappearance of organized beings.***

[*Editors' Note:* Material has been omitted at this point.]

3. There are, consequently, no definite formations in the palæolithic sense of the word, no definite creations, no successive and well-marked floras or faunas, any more than there exists any formation which simultaneously maintains the same mineralogical characters, the same thickness, and the same lithological and palæontological characters in all parts of the world.

4. A geological formation, or a geological flora or fauna, is the totality of the sedimentary strata which have been formed upon the whole earth during a certain space of time, or the totality of the animals and plants which have lived during that space of time. It is of little consequence here, whether the lithological character, the thickness, and the limits of demarcation of these strata have been uniform over all the surface of the globe, or have varied in different places, assuming here one aspect, there another; it is of little consequence whether the various species of organisms belonging to this epoch may have lived from its commencement to its termination, only endured for a portion of this time, or passed the limits assigned to this formation.

[*Editors' Note:* Material has been omitted at this point.]

9. The average absolute duration of organisms was sufficiently long to give us no reason for astonishment at the important differences presented by species in this respect, although the history of these species is often told us only by strata of but slight thickness, so that it often happens that we regard as simultaneous, phænomena which have been separated by long periods of time.

23

Reprinted from *Geol. Soc. London Quart. Jour.* **16:**xxxii-xxxiii, xxxv-xli, xliii-xlv (1860)

ANNIVERSARY ADDRESS OF THE PRESIDENT

J. Phillips

[*Editors' Note:* In the original, material precedes this excerpt.]

What a contrast is offered by this mass of willing though laborious, and consentaneous though scattered efforts, all tending to enlarge knowledge and fortify the basis of theory, with the heap of crude conjectures which formerly took the place now firmly held by the true, however imperfect, natural history of the earth, which we have founded on sections of the strata and classifications of organic remains!

[*Editors' Note:* Material has been omitted at this point.]

§ *Palæontological Data.*

The communications made at our meetings so rarely touch on theoretical subjects, that, but for the discussions which ensue, it might seem as if we had ceased to doubt, or ceased to hope and aspire after higher generalizations. In tracing the history of the strata, whether in neighbouring or in distant localities, we employ, without fear, the evidence of organic remains to determine the place of the rocks which contain them in the general scale of strata. Even when comparisons are made between the groups of strata on opposite sides of the Atlantic—as between the Trenton Limestone of New York and that of Llandeilo in South Wales*, or between the Silurians of Europe and their Arctic representatives in Prince Patrick's Land†,—the affinity of the fossils is accepted as evidence of the approximate contemporaneity of the rocks.

[*Editors' Note:* Material has been omitted at this point.]

* See Rogers's Geol. Map of N. America; and Hall's Geology of New York.
† Haughton, Notice of Fossils found by M'Clintock.

An objection to the cogency of our conclusions in regard to the history of the changes which the surface of the earth has undergone is founded on the incompleteness of the record which we find in the organic remains. It is *incomplete*, in some departments more than in others; but we must not admit it to be *insufficient* to sustain just conclusions; for these cannot extend beyond the data.

The main subject of the record by fossils is the history of the sea and its bed; and for this the data appear sufficient. They are incomplete in regard to fresh water, the land, and the air, and must ever remain so. It is only by the intercalation of reliquiæ derived from these, in the series of marine deposits, that we can assign to them their right date; and the occurrence of land and freshwater remains among marine deposits is, and must always have been, comparatively rare, and, with reference to the history of the ocean, almost accidental.

[*Editors' Note:* Material has been omitted at this point, but the author goes on to say: "If we cannot thus reconstruct the whole animal and vegetable kingdom, we have numerous and instructive examples of both at certain important epochs in the three great geological periods" (p. xxxii). In reference to groups of "marine genera rich in species," Phillips states: "A surprising resemblance runs through all the groups, though, as in the case of *Terebratula,* the *facies* differs in each of the great periods, and, indeed, in the several systems in those periods" (p. xxxiv).]

§ *Provinces in Space.*

The conclusion adopted by Brown, Forbes, Godwin-Austen, and most naturalists in regard to the plants and animals now in existence, is that each species of this vast multitude came into being at a determinate place, and spread from thence by natural means over the neighbouring tracts, sometimes arriving at very great distances and appearing in unexpected situations*. Thus natural provinces are constituted, each including a considerable number of forms peculiar to itself, with others communicated from beyond its boundaries. These provinces are not necessarily the same for animals as for plants, nor for all the families either of plants or animals †. The unequal facility of distribution of different races comes in further to disguise the classification in provinces. Still, with Botanists and Zoologists almost universally, the principle of referring each species to a definite local centre, and of constituting provinces for many species having proximate centres, has been firmly supported as a true representation of nature.

Cases of very limited distribution occur, as in the Galapagos, visited by Mr. Darwin, where a unique group of life enjoys what seems to be a separate small province characterized by the herbivorous *Amblyrhynchus* and several peculiar species of birds, which, however, on the whole manifest a great analogy to other species on the American Continent, with which these isles may once have been connected. By admitting local displacements of land and sea, Forbes found in many cases an easy explanation of the occurrence of some species and the absence of others in the fauna and flora of islands, mountains, and branches of the sea.

Each of these species is conceived to be perpetuated in essential structure and habits of life, though subject to some variations, at birth and during life, by the influence of surrounding circumstances.

[*Editors' Note:* Material has been omitted at this point.]

* See, on questions of this kind treated in this manner, E. Forbes in Memoirs of the Geol. Survey, vol. i.; and Forbes and Godwin-Austen, Natural History of the European Seas.

† See Woodward's Treatise on Mollusca for a Map of the Distribution of this class of animals.

§ *Provinces in Time.*

Geologists have generally accepted these views, as they have done in regard to fixed truths of chemistry and mechanics; and they have added, from the evidence which their peculiar studies yield, a definite origin of each natural group in time, the approximate duration of the life of many, and the epochs of termination of several. Thus every specific form is conceived to have sprung into being at a certain point on the globe, at a definite epoch of time; its existence is traced through provinces of space and through periods of time, so that it has a real physical history.

For those who adopt this view, the course of reasoning on the succession of life on the globe is clear and convincing. It is, however, not universally adopted; but the hypotheses which have been framed to replace it (which always involve the idea of indefinite change of form, structure, and habits) would not, if adopted, materially affect the conclusions of geology, or change the practice of naturalists. If it is by the course of progressive change from older types that new specific forms have arisen, there must have been for each of these a time and a place when it began to manifest the new specific distinction. Geology needs not to discuss these hypotheses, sanctioned though they be by eminent names, amongst whom our Darwin is preeminent for powers of generalization operating on a large basis of personal observation*. None of them appear to be wholly without a foundation of fact, though none of them can be held to penetrate more than a small way into the mystery of the origin of species.

[*Editors' Note:* Material has been omitted at this point.]

Neither Zoology nor Botany nor Palæontology can countenance the supposition of many local origins for one specific form which can be increased and diffused by known processes, nor more than one epoch for its production and duration. In this, as in so many other instances,

"....Vestigia nulla retrorsum;"

the group once ended, like the period to which it belonged, returns no more.

[*Editors' Note:* Material has been omitted at this point.]

* The work of this author on Fossil Cirripedia is one of the most remarkable examples of his former careful discrimination of species (Palæont. Soc.); while his latest publication, 'On the Origin of Species,' is the most elaborate essay yet produced in favour of the descent of all known forms of life from a small number of originally created types.

§ *Zones of Life.*

To complete our knowledge of a geological province, we should remember that, as in dredging the sea-bed account is taken of the depth of water, the force of currents, and the quality of ground, so in geological investigations of precision we should expect similar attention to the separate collections from the coarse sands and pebbles of the shore, the drift sands, the muddy expansions, and the calcareous rocks. The general results to which we are accustomed, which sum the whole into one catalogue for a given period, furnish no doubt, the best means of contemplating the system of life of the period; but in local monographs, and in monographs of species o groups, the exact distinctive characters of the repository should, as far as possible, be recorded. In regard to depth of water, it is only by the record left in the structure of the stone, and succession of laminæ, that we can find an independent measure; and in this Mr. Sorby has greatly helped us. A measure of depth derived from the analogy of the fossil to the recent groups is seldom to be implicitly trusted. We may remark, further, that a series of fossils for any system, to be complete, should include three groups—from calcareous, arenaceous, and argillaceous repositories; for these three sets of deposits are really contained in each natural system of strata. One of the earliest and most profitable exercises of my life was the making a catalogue of the fossils in the collection of W. Smith, before the removal of it to the British Museum. When I had, following my great relative's guidance, enumerated the 720 species of fossils in his collection, he set me to compose those tabular synopses of the distribution of the more remarkable groups of fossils in the several zones of stratification of which the Table of Echinodermata in the 'Stratigraphical System of Organized Fossils' (1817) is an example. No sooner were these constructed for a few groups, than two ideas were strongly imprinted on my mind:—

1. That the groups of fossils had some real relation to the mineral and structural characters of deposits; so that, while smooth oysters, *Gryphææ*, &c., were frequent in clays, and Terebratulidæ rare in them, we were to look for Zoophyta and Echinodermata in the calcareous bands; Spongiadæ in some sands and some particular limestones; few fossils of any sort in peroxidated sediments, coarse conglomerates, or pure sands of any colour. It appears to me that this class of inquiry is still not enough followed out, either by explorers in local districts or by the palæontologist of the study. With so much the more pleasure, therefore, I welcome the extended example of such studies given by Dr. Bigsby in his comparison of the Silurian Fauna of North America and Europe.

2. That in each great natural zone of fossils there was to be traced a poor origin, a rich development, and a subsequent decrease,—thus apparently giving to each great natural period a zone of maximum fertility in species, in which, more than at any other time, the various orders of life acquire their full expression. With this idea in his mind, a palæontologist finds the whole Carboniferous period one, though in it are several zones distinguishable by the prevalence

of a few types, and by the paucity or absence of others—the maximum of marine life being about the top of the Scar Limestone, or the base of the Yoredale Series.

So in the Silurian series of the Malvern tracts, the maximum zone of life is in the Wenlock group, while from below (in the black shales) the numbers of species rise from almost zero to 176, and then dwindle again to zero in the Downton Sandstones, at the base of the Old Red. So in the Lias, the lowest beds yield very few shells, and no Ammonites or Belemnites; but from this the numbers swell rapidly to a maximum, which for many groups is in the Oolitic series of Bath (but for Ammonites and Belemnites is in the Lias), and decays to a minimum in the comparatively poor Oolites of Portland, Swindon, and Aylesbury.

In the Cretaceous strata we may perceive very clearly the influence of physical conditions on the zones of greatest fertility; for while the Mollusca generally are more abundant in the Greensands and Gault, the Amorphozoa, Foraminifera, and Bryozoa seem more to affect the Chalk, and specially the Upper Chalk, from which most of the Ammonites have disappeared. Fishes are most plentiful in the Upper Chalk.

Nor can we fail, whilst studying the distribution of the Cretaceous fauna, to perceive a distinction of north and south districts, if not provinces, within the limits of Britain. The northern chalk of Yorkshire is comparatively poor in fossils, and, instead of the *Belemnitella mucronata*, so common in the south, gives us in abundance *Belemnitella quadrata*, which the chalk of the Baltic exchanges for *B. mammillata*. Observations of this kind, well carried out, appear to justify the belief that in different or only occasionally connected basins the succession of forms might be somewhat different; and that groups might be analytically resolved into several stages in one tract, united together in a second, or complicated with new auxiliaries in a third.

The two most contrasted provinces of the middle Palæozoic strata (the Devonian and Welsh tracts) offer, in regard to their forms of life, a very marked difference: little but Fishes in the latter; abundance of Mollusca, Trilobites, Zoophyta, and Bryozoa in the former.

The *latter* is a tract of mostly peroxidated sediments; the *former* has a large proportion of protoxidated rocks, and in these principally are the treasures of fossil life. Sediments brought in different directions, with germs of local groups of contemporaneous life, are indicative of these facts, which, besides, suggest to us the necessity of reconsidering with care the relation of the truly limited Devonian to the greatly expanded Carboniferous and Silurian strata.

In the case just mentioned, then, there is often traceable a concurrence and coextension of certain given fossils with certain mineral deposits—the fossils coextensive with and limited to those particular deposits. So also the deposit of Aymestry Limestone is the favourite home of the conspicuous *Pentamerus Knightii*, which may be looked for in vain where the limestone dies out in contemporaneous muddy shales.

[*Editors' Note:* Material has been omitted at this point.]

This unequal diffusion of definite forms of life may often be ascribed to the progress of oceanic currents, which transported at once the germs of life and the sediments in which they were buried. If we trace by this means some of the ancient currents of the sea for any particular epoch, we shall find, with surprise, some neighbouring tracts to have been almost unconnected; while very distant regions manifest some effective communication. Thus, while the Trilobites of Bohemia differ almost in every species from those of Scandinavia*, while only a small proportion of the fossils of North Devon occur in South Devon, the American genus *Maclurea* unexpectedly appears in the oldest limestones of the extreme north-west of Scotland, and Ammonites like those of Kelloway Bridge in Wiltshire are collected in the centre of Russia and at the mouth of the Indus. So in existing nature, when we find *Spirula* in so many distant basins of the sea, between which now are no connecting channels, we must appeal to earlier distributions of land and sea for the means of intercourse which no longer exist. Thus Nature in some material aspects retains, in the arrangements of life as well as in the form of land and the peculiarities of physical geography, traces of the history of an earlier time.

[*Editors' Note:* Material has been omitted at this point.]

I may also refer to the still unsettled questions regarding the triassic or liassic affinities of the bone-bed and the lowest Lias shales, sands, and calcareous layers,—on which a valuable communication has reached us from Dr. Wright. This active palæontologist is proceeding with his careful survey of the several divisions of the Lias in the south of England, and accumulating data for the further discussion of the boundary (if it be necessary to draw as a line what nature marks as a zone) between the Upper Lias and the Inferior Oolite.

§ *Distribution of Ammonites.*

Throughout the Lias and the whole Oolitic system, the evidence of Ammonites is acknowledged by all geologists to be of the highest value in determining the place of detached deposits on the general scale of ancient life and time. This arises in a great degree from the circumstances that the whole group of true Ammonites is limited in time between the Trias and the Tertiaries, and that the species are very numerous, very definitely marked, traceable from youth to age, and grouped naturally into distinct assemblages, whose place in the succession is constant. On these grounds Dr. Wright may well be justified in attaching to them the same importance which was assigned to them by W. Smith, whose unpublished Table of the Distribution of Ammonites, drawn up by my own hands in 1817, is now laid before the Meeting. It forms one of a series of such attempts, of which an example has been printed, viz. the Table of *Echini*, in the 'Stratigraphical System of Organized Fossils' (1817). Von Buch's valuable labours on the Ammonitidæ are well known.

[*Editors' Note:* Material has been omitted at this point.]

* Barrande and Murchison.

§ *Palæontological Periods.*

When tracts of the old sea-bed so far apart as Wales from Lisbon, Bohemia, Sweden, or New York, the Arctic zone, the Himalaya range, or China are compared, and in all these countries the lowest strata are found to be filled with many forms of life remarkably similar*, not only in the mass of the rocks, but also often in the several groups of strata one above another, taken in the same order, geologists conclude that these strata, which, taken separately or taken in mass, have such remarkable agreement, were produced in the same period of the earth's history. The agreement of the fossils being for the most part due to the deposition of the strata in one oceanic basin, or to the free communication between one such basin and another, so that the species of marine animals might be diffused over parts of each, we have no reason to doubt that similar successions of deposits—lying in the same part of a similar general series of strata, and containing similar kinds of fossils grouped in a similar manner—were of approximately contemporaneous origin. In several of these cases they have a similar base of metamorphic rocks, and a similar cover of Devonian or Carboniferous rocks,—all circumstances giving independent testimony to the truth of our general conclusion. We might further confirm it by the evidence of displacements of the strata, at definite epochs and through definite groups of the strata—movements which have raised the Silurians before the date of the coal, and remodelled the whole area of the land and sea before the Trias and after the Chalk.

It has, indeed, been objected that this conclusion is illogical—that identity of time cannot be inferred from similarity of conditions—that the successive conditions of Silurianism, Devonianism, &c., may be due to a necessary sequence in the order of natural phenomena, and may have commenced *at any time*, round any land which, by rising from the deeps of the ocean, came within the zone of light and life in the waters. The objection is not valid. The dates of the uprising of land are, indeed, as here suggested, various; but according to the period when each came under the influence of the conditions requisite for life are the deposits and the fossils which surround it. An elevation which can be proved to have happened after the Coal is followed by shore- and sea-deposits having the fossils of the period, not the primeval fossils: the land which is raised bears the vegetation of the period, not the primeval vegetation. Thus the series of strata and of fossils, taken as a whole, is but one —a function of the elapsed time; but the terms of the series appa-

* *Heliolites porosa* occurs in the limestone of North Somerset as well as in the limestones of Devon and the Niagara Limestone of New York. *Atrypa aspera*, which was found by Capt. Maclure at Princess Royal Islands, occurs also in the limestones of the Eifel and Devonshire, and in the Niagara group of the New World.

rent at any place depend on the physical history of the locality—a function of local conditions. In other words, the agreements depend on concurrence of time; the differences on disagreeing local conditions.

The correspondence in time here indicated by the term contemporaneity may be understood as having only so much of definiteness as belongs to the deposition of a stratified rock. Whatever phenomena occurred within the limits of the duration marked by the beginning and ending of such a deposit may be reckoned as of *one period*, though they may not all be assignable to one precise or momentary *epoch*. There are good reasons for doubting the exact *epochal* correspondence of distant parts of the same continuous sedimentary deposits (as, for example, in beds of obliquely laminated oolite and millstone-grit, in which the successive deposition of parts of the bed by currents flowing in ascertained directions is certainly traced), but none which forbid our using these deposits to mark a *period* having a certain place in a series of periods*.

Geology has thus obtained a true scale of relative time on which to register all the events in the earth's history which fall within the wide compass of her inquiry. Starting with a scale of strata exactly determined, and rendered as perfect as possible in one basin, we acquire a general series or scale of deposits, in the order of their deposition, suited to a certain area of the ancient bed of the sea. By examination of these deposits, we find in each of them fossil marks of an intelligible kind, by which the place of any given step in the series can be determined for the area in question. By comparison of one large basin of strata with another, it is found that the fossils taken in allied groups have larger ranges than the mineral or structural characters of the strata, and indicate that the basins were occasionally, or frequently, in communication, so that some corresponding forms of life are found to occur in both, in several or all of the deposits, and in the same order of succession.

By thus comparing basin with basin, it is found that hardly one of the stratified systems of rocks was so insulated by the circumstances of its formation as to show no conformity with another in the sequence of ancient life, however different may be the appearance and however unequal the completeness of the series of the groups of strata. If there were any seas entirely separated (as now we see the Caspian and other seas), the separation of them must have taken place at some definite period, previous to which they were influenced more or less by the general order of life in the other neighbouring parts of the ocean. It may be taken as a general result of all this inquiry, that there is but one general series of life represented in a fossil state,—each term of this series corresponding to a geological period, and, taken in a large sense, preserving one main or general character, amidst many local variations, over all the areas yet investigated.

* Mr. Godwin-Austen has bestowed much attention on this subject. See also, in my Treatise on Geology in Cab. Cycl. vol. ii. chap. vi., a discussion on this subject, and a diagram in illustration.

The conformity of which we here speak is sometimes of such a kind as to be expressed by the term "identity of species," more frequently by the near resemblance of forms in the same genus, for which the name of "representative species" has been employed by E. Forbes. The term species is and must remain somewhat ambiguous in palæontology while the observers of different regions prefer the inglorious task of inventing new names for their native fossils, instead of the useful labour of a strict comparison of them with sufficient examples of the species already described by foreign naturalists. But there is seldom any ambiguity in a generic term; and in each region we can at least *count* the specific forms, or what are considered to be such. This being done, we gain at once a power of numerical expression for the several groups which compose the system of life of each period in each region, or what is preserved of each system, and a representation of the associated life, independent of the particular determinations of the specific elements.

[*Editors' Note:* Material has been omitted at this point.]

24

Reprinted from *Geol. Mag., n.s.,* **6**(1):1-15 (1879)

On the Tripartite Classification of the Lower Palæozoic Rocks.

By Charles Lapworth, F.G.S.

By those accustomed to the hopeless confusion of the *Greywacké* of the earlier geologists, the publication of Murchison's grand work on the "Silurian System" was hailed with feelings of the most profound relief and satisfaction. His clear and brilliant presentation of the physical and palæontological proofs of an orderly sequence among the Palæozoic Rocks below the Old Red Sandstone, as originally set forth in all their force and harmony in his magnificent volumes, naturally astonished and dazzled the majority of his scientific contemporaries, and secured for his nomenclature of these ancient deposits an almost universal acceptance. His subsequent abuse of this advantage to strengthen and consolidate his own system at the expense of that of his equally-illustrious co-worker—the less fortunate but more cautious Sedgwick—was a gallant but unscrupulous defence of this original nomenclature, which by that time he must have felt himself almost powerless to disturb. His later extension downward of the limits of his System, till it embraced all the rocks between the supposed Azoics and the Old Red Sandstone—though, in a measure, forced upon him from without—ought perhaps to be regarded in part as a very natural return to the ideas of his early teachers, who had always held the practical unity of the rocks of the *Transitional* period. In this way, however, Murchison unwittingly destroyed many of the most beneficial results of his own labours; in a sense, spending his old age in the attempted re-erection of the very edifice it had been the pride of his manhood to destroy—the early years of his scientific career being devoted to the worthy task of proving the marvellous variety of the Lower Palæozoics; his later years to demonstrating their integrity, unity, and indivisibility.

At the present day it would be wholly superfluous to enter upon the discussion of the vexed question of the respective claims o Sedgwick and Murchison to the *Middle* and *Lowest* Divisions of the Lower Palæozoic Rocks. We may, however, without fear of contradiction, concede to Sedgwick the credit of having been the first to determine the limits and sequence of their larger subdivisions, and to Murchison and his followers the honour of having been the first to assign them their distinctive fossils. Sedgwick worked out single-handed the true stratigraphical arrangement of the rocks of the Lower Palæozoics of Wales, from the Bangor Beds to the summit of the Bala Series, and divided them into several successive groups, the propriety and convenience of which subsequent research has served only to make more distinctly apparent. The unavoidable—but none the less vital—defect in his earlier work lay in his not publishing the characteristic fossils of these subdivisions, until after the appearance of the great work of his rival, in which the most conspicuous forms appear as characteristic of the subdivisions of a supposed overlying system. Murchison's work, on the other hand, depended not only upon mineralogical characters and sequence of formations, but also upon palæontological peculiarities. He failed signally, however, in strictly and correctly defining his lower groups, and in correlating some of his most typical beds, with the result of greatly confusing his lists of characteristic fossils.

The rigid conservatism of Murchison in his old age, and his systematic disregard of the facts and arguments adduced in support of the Cambrian System, brought about its inevitable re-action after his death. The campaign against the Murchisonian nomenclature, so brilliantly opened by Professor Sterry Hunt, in his masterly paper on the "History of the Names Cambrian and Silurian in Geology," has since assumed extraordinary proportions. The Cambridge School, headed by Professor Hughes, the present talented occupant of the Woodwardian Chair, supported by several earnest and industrious adherents, has revived the claims of Sedgwick in all their entirety; and presses them on the attention of geologists with an energy and persistence that threatens to lead to the formation of a body of workers, determined to force from posterity, in honour of the memory of Sedgwick, the rights he demanded, but of which during his life-time he was so unfairly deprived.

But, on the other hand, the Murchisonian nomenclature is embodied in the maps and publications of the National Survey. It is embalmed in the classic memoirs of the illustrious Barrande, and in the numerous works of the best-known geologists of Europe and America. It is still held, almost in its widest sense, by the more influential officers of the Geological Survey, and is taught to their students and subordinates with that complacent pride which has naturally been engendered by a quarter of a century of uninterrupted success. Even yet, its advocates have such an unfaltering faith in its intrinsic propriety and consequent impregnability, that the fact of the daily increasing number and ability of their opponents is either contemptuously ignored, or, at most, is deemed unworthy of a more respectful recognition than a passing smile.

The utter impossibility of reconciling the antagonistic claims of these opposing schools has led, of late years, to the formation of a third party, in which the best-known names are those of the late Sir Charles Lyell, and of Dr. Henry Hicks. These concede the light of Murchison to all the strata between the base of the Arenig and the summit of the Ludlow, but emphatically assign the Lingula Flags and Paradoxides Beds to the Cambrian. This school—if school it may be called—has been greatly aided by the wide publicity given to their views in the numerous memoirs in which they record their steady and cautious advance in working out the natural succession among the subordinate members of the Lower Palæozoic Rocks. It is just possible that, owing to the modesty of the claims it makes for Sedgwick, and to its retention of such a large proportion of the prevalent nomenclature, this view might gradually and insensibly have taken possession of much of the field, were it not for the persistent exertions of the Cambridge School, smarting under a sense of injustice, and determined to rest satisfied with nothing less than a complete redress of those historic grievances, which their affection for the honoured name of Sedgwick has led them to regard as little less than personal to themselves.

But the partial success that has already attended the earnest conscientiousness and perseverance of the members of the Sedgwickian party has, in truth, hastened the evil day. The result that their efforts have had in calling the attention of geologists to the salient points of the question at issue, is as fatal in its effects upon their own theory, as it is upon that of their opponents. By their recent adoption of the Lyell-Hicks line of demarcation at the base of the Lower Llandovery, they furnish, indeed, a thorough demonstration of the almost perfect palæontological distinctness of the faunas of the so-called *Lower Silurian* formations, and those of the true or Upper Silurian, and the consequent impossibility of combining them philosophically in one and the same system. But, in spite of all that is implied to the contrary, this course is, in effect, a distinct abandonment of Sedgwick's fundamental argument that these systems were necessarily distinct, from the fact that in the typical areas their beds were stratigraphically discordant. It amounts, on the other hand, to an implicit adoption of the only safe principle, that we have no reliable chronological scale in geology but such as is afforded by the relative magnitude of zoological change—in other words, that the geological duration and importance of any system is in strict proportion to the comparative magnitude and distinctness of its collective fauna. It appears to me that it is impossible for them to rest here, but that their next and inevitable step will be the further admission that the Lyell-Hicks division of *Cambrian* and *Lower Silurian* are as rightly entitled to the rank of separate systems as the true or *Upper Silurian* itself; and that, eventually, their rigid sense of fairness and justice will lead them so to discriminate them.

For, amid all the confusion incident to this controversy, one grand fact stands out clear and patent to the most superficial student of Palæozoic geology — namely :— the strata included between the horizon marking the advent of *Paradoxides*, and the provisional line presently drawn at the summit of the Ludlow, imbed *three distinct faunas*, as broadly marked in their characteristic features as any of those typical of the accepted systems of a later age.

The necessity for a tripartite grouping of the Lower Palæozoic Rocks and Fossils, in partial accordance with this fact, has been very generally acknowledged for the last thirty years. The keen-eyed and philosophic Barrande was the first to recognize this truth, and his addition of the "*Primordial*" to the *First* and *Second* Faunas of Murchison's original Silurian marked a geologic era equal in importance to the establishment of a new system. How keenly its enthusiastic discoverer watched over, and how zealously he promoted and encouraged, the gradual detection and elimination of his "*Primordial*" Fauna in Europe and America, are matters familiar and delightful to all earnest students of the history of discovery among the Lower Palæozoic Rocks. How the facts obtained by Phillips, Salter, and Hicks in Britain, forced even Murchison himself to adopt Barrande's views, and in his later years to become their keenest and most unsparing advocate, is equally well known. The subsequent development of the "*Primordial*" Fauna in Britain by Hicks, Salter, Belt, and others; in Sweden by Angelin, Nathorst Linnarsson, and Sjögren; and in America by Billings, Emmons, Hall, and Hartt, has progressed with marvellous rapidity. Every systematic geologist worthy of the name has, in his turn, been compelled to acknowledge the distinctness and first-rate importance of the "*Primordial*" Fauna.

Under one form or another, also, the difference in the facies of the more recent *First* and *Second* Faunas of Murchison has been universally admitted from the first, and the rock-groups formed by their including strata have been separated—at least as distinct *sub-systems*—in all parts of the world. It is indeed true that there have been, perhaps, as many diverse views held with respect to the proper position of the line of demarcation between them, as there have been separate areas of investigation; and it is only of late that geologists have reached something like a concensus of opinion in drawing it with Hicks at the base of the Lower Llandovery. Nevertheless, no honest investigator, either British or foreign, has ever dreamt of disputing the grand fact of the distinctness of these two faunas and the consequent need for the separation of their containing rock-groups in any natural and workable plan of classification.

Thus, it is hardly possible that any geologist, who is familiar with the rocks and fossils of the Lower Palæozoics, or who is even fairly versed in the literature of the subject, would at present venture to deny the proposition that between the base of the known fossiliferous series and that of the Old Red Sandstone there lie three successive rock-groups—each of which is characterized by a special fauna of first-rate geologic importance.

Further insistence upon this point is probably needless. But if the fact be once admitted, it follows of necessity that the interests of science demand that these three successive rock-systems shall be distinguished by three separate and unmistakable titles.

At this stage, however, we plunge into the very midst of the conflict of the schools. Friends, foes, and spectators, seem all fairly agreed as to the advisability of a triple division of the sediments.

The points around which the strife is at its keenest bear upon the question as to whether these three divisions are of equal classificatory value, and if so, which party has the best right to give them their names.

The strict Murchisonian of the present day claims for the Silurian all the fossiliferous strata that lie between the Archean and the Devonian, and arranges them in his three sub-systems of the *Primordial, Lower,* and *Upper Silurian.* If he is a palæontologist, he seizes at once upon the indisputable fact that the general facies of the fossils of these three divisions, when viewed in their collective aspect, has a marked character of its own, wholly distinct from that of the faunas of the overlying rock-groups. Profoundly impressed by this distinction, the less striking differences between the faunas of the three members of the Lower Palæozoic itself dwindle in his eyes into utter insignificance, and the slightest party bias is sufficient to lead him to regard them with Barrande as forming "one grand and indivisible triad which is the *Silurian System.*" As discovery progresses, gradually demonstrating the former presence of organic existences in strata far below the base-line laid down by the founder of his system, departing gradually in facies from his typical fauna (but, nevertheless, connected therewith by almost imperceptible gradations), his former admission, and the traditions of his school, compel him to keep pace with it, extending his system and fauna downwards step by step. The result is, that if he is consistent, he is at last driven to demand also, with Barrande, the inclusion of the beds which Murchison, even in his latest years, acknowledged to be the very basement rocks of a *pre-Silurian* system.

If, on the other hand, he is a stratigraphist, he instances the fact that in Britain and America no general stratigraphical discordance interrupts the vertical succession of formations between the Archean and the Carboniferous. He points to the Llandovery beds of Britain, and shows that the grandest stratigraphical break in the entire series in the typical area occurs in the heart of a group of beds that the founder of his school placed partly in one sub-system and partly in the other; but which he, in common with all scrupulous geologists, included in a single formation, whose essential unity he fearlessly challenges his opponents to deny. He calls attention to the *Colonies* of Bohemia to show that even where the palæontological distinction between the sub-systems is most abrupt, yet, according to the greatest of Silurian palæontologists, there is actually an alternation of the two faunas in the beds of passage. Or, he points triumphantly to the succession in Scandinavia, where the Lower Palæozoics are reduced to a collective thickness of a few hundreds of feet, and are occasionally folded up and entangled almost inextricably together in a single section, and asks how is it possible to doubt the unity of a System whose members are individually of such insignificant dimensions, and, physically, are so indissolubly united!

The moving principle of the Sedgwickian, on the contrary, is the demand for historic justice. With true British instinct, he recognizes the fact that the revered founder of his school was unfairly deprived of the natural fruits of the labours of a lifetime by the overwhelming forces of influence and circumstance; and he chivalrously devotes all his energies to the task of overturning history to the extent of bringing back matters to the point they would have reached, had the relative position of Sedgwick and Murchison been reversed. To this paramount consideration everything else is sacrificed. The Silurian is cut down so as to include the Upper Division only of Murchison's original System; while all the fossiliferous beds below are assigned to the Cambrian. That in this way he commits precisely the same scientific error as the Murchisonian, never seems to occur to his imagination. That every fact adduced in support of Sedgwick's claim to the rocks of the *Second fauna* can be met by one in Murchison's favour equally cogent, is forgotten. That every error committed by the latter to the destruction of his claims can be paralleled by one equally fatal to those of his opponent, is similarly ignored. He has long since convinced himself of the fact that the Silurian, as he restricts it, is quite large enough to form a system by itself, and that its fauna is grand enough and special enough to characterize one; but we never find him carry out this argument to its legitimate conclusion—that, if so, his own Cambrian is not *one*, but *two* systems, whose individuality he is, by his own principles, equally compelled to recognize. He seeks in all kinds of out-of-the-way spots for evidences of local unconformities between the *Balas* and the *Llandoveries* to satisfy his stratigraphical conscience that there is sometimes an actual physical break between them; when, without leaving his closet, he could assure himself of the fact that the two systems of the so-called Lower and Upper Silurian are already known to be stratigraphically concordant nearly all over the world. Where this argument fails, we find him insisting upon the presence of conglomerates and upon the sudden change in the character of the organic remains. But each and all the principles of classification implied in these distinctions are violated in his own procedure. The grandest zoological breaks in the whole Lower Palæozoic Series (those between the *Olenus* beds and the *Arenigs* of Britain, and between the *Canadian* and *Trentonian* of North America), and the thickest and most persistent conglomerates that antedate those of the Old Red Sandstone (viz. those of the *Lower Girvan* and *Quebec Groups*), all occur in the very heart of his own Cambrian System. Yet of these we hear little or nothing, but all the strata between the Archean and the Llandovery are piled up into a single system, for the sole reason that they happen to occur in association in the mountain-area of North Wales, and were very naturally lumped together by the first scientific man who conscientiously studied them.

The Lyellian is certainly more politic than his excited neighbours, but, from a common-sense point of view, his disinterested procedure is equally unfair. To him, the fact that Murchison described the Upper and Lower Silurian Rocks in his original Silurian System in such a way that they can, to a certain extent, be recognized and identified in Europe and America, is all-weighty. He calls special attention to the fact that Sedgwick's Upper Cambrian was ultimately found to possess the fossils of Murchison's Lower Silurian, but he forgets to add that it was Sedgwick, and not Murchison, who first gave the natural divisions of this group, placing them in their proper relations to each other, and defining their true limits above and below. He points out with emphasis the grand distinctions between the *Primordial* and *Second* Faunas, and the consequent impossibility of uniting the rocks they characterize in one and the same system; but the fact of the stratigraphical break at the base of the Mayhill Sandstone is, however, contemptuously dismissed as of no special classificatory value, and the two Llandoveries are joined in a single formation. Thus, in one stroke, Sedgwick is deprived of his grand argument of a physical break between his own and the overlying rocks; and the *Second* and *Third* Faunas are re-united to form what is termed the Silurian System. In effect, Murchison receives the lion's share, simply on the ground of possession; while Sedgwick is deprived of half his system because he had the misfortune, in the earlier stages of the controversy, not to command so numerous and influential a following as his more socially fortunate opponent.

At irregular intervals, also, we catch a momentary glimpse of a stray individual who refuses to identify himself with either of these great parties; preferring rather to temporize by definitively assigning the rocks of the *Middle Fauna* to neither claimant in particular. He refers to them under such makeshift titles as the *Cambro-Silurian* or *Siluro-Cambrian*, according as his otherwise unexpressed personal bias inclines him to one or other of the contending parties. Occasionally, indeed, we do find him possessed of a true estimate of the grand importance of the group, but he often leaves it to be understood that he regards it as forming a transitional series of second-rate geologic significance; and, in effect, belonging properly to both Cambrian and Silurian at once. He is at the same time so fully impressed with the consciousness of his own forlorn and isolated condition, as well as of the hopelessness of stemming the current of vulgar use and wont, that he generally contents himself with simply recording his protest in this manner, and timidly guards himself against possible ambiguity and misconception by prefixing the qualifying term *True* or *Upper* when he comes to speak of the undisputed Silurian.

But, in addition to the foregoing, there are innumerable outsiders like myself, who care nothing for schools, but everything for the facts. There is that great and ever-increasing body of students who are attracted to the study of geology because of the flood of light it casts upon the mysterious problems of life and its distribution. Above all, there are those foreign geologists, who naturally expect from British investigators an authoritative and unmistakable geologic scale to which to refer the results of their own researches. To all these the crying scandal of this interminable dispute is an annoyance and a positive encumbrance.

But whose procedure shall we follow? Shall we adopt the Murchisonian's convenient plan of carrying down the base-line for the Silurian System as far as a Trilobite has yet been detected, sinking it deeper and deeper into the earth as the progress of discovery reveals the evidence of the former presence of organisms in strata of yet older and older date, at the same time extending its highest boundary upwards into the supra-Ludlow formations as we detect the presence of an occasional fossil of a Ludlow type yet higher and higher in the more rapidly accumulated Red Sandstones above the *Bone beds*, till our unwieldy system includes half the fossiliferous sediments of the globe, and its very subdivisions are almost equal in classificatory importance to the accepted systems of a later date?

Or shall we adopt the methods of the traditional followers of Sedgwick, and, drawing a rigid line of demarcation at the base of the Lower Llandovery, imitate our opponents to the extent of erecting all the anterior fossiliferous strata into a gigantic system, on the ground that they were so combined by its founder, and in the delusive hope that we shall find at its base a universal unconformability, so that the name *Pre-Cambrian* will ever remain a synonym of the metamorphic and possibly azoic formations?

Or, with Lyell, shall we condone the past, and give a double share to the stronger party; consoling ourselves with the reflection that, after all, the question is merely a question of names, and not of principle; arguing that the injustice we tolerate did not originate with us, and is less the crime of a party than the inevitable result of untoward circumstance; and justifying our procedure in the eyes of the world by the implication that the general adoption of the larger portion of the Murchisonian nomenclature is already an accomplished fact, upon which it would be ridiculous to expect that any feeble efforts of ours would ever have the slightest influence?

Or, ought we rather to cast in our lot with the few who employ the term *Cambro-Silurian* or *Siluro-Cambrian* for the rocks of the *Second Fauna*, and try once again the oft-repeated and as oft-defeated experiment of reconciling the claims of both parties by allying the strata in dispute to two systems at once, in the use of titles which their very founders themselves abandoned as inconvenient and absurd?

Or, finally, standing aloof from all parties, shall we, in the name of science, claim the right of fully recognizing the systematic equality of the three Lower Palæozoic Faunas, by regarding the three successive rock-groups which contain them as individually entitled to the rank and denomination of a complete system?

It seems to me that to every unprejudiced mind it will be apparent that the adoption of this last course has now become an absolute necessity. Geologic truth and convenience imperatively demand a separate place and name for each of these systems. It only remains for us so to arrange their titles that no real injustice shall be committed.

Dr. Hicks's definition of the Cambrian system as including the *Paradoxides*- and *Olenus*-bearing beds, from the base of the Harlech Grits to the summit of the Lower Tremadoc is by far the best that has hitherto been proposed. Thus restricted, the title is synonymous with that of the *Rocks of the First or Primordial Fauna* of Barrande, and is certain to be ultimately accepted everywhere among geologists, from its naturalness, geologic distinctness and convenience of application, not only in Britain and Western Europe generally, but also among the ancient rocks of the continent of America.

In the same way the general restriction of the title Silurian to the strata that are comprehended between the line marking the base of the Lower Llandovery, and that denoting the commencement of the brackish or fresh-water conditions of the typical Old Red Sandstone, appears equally inevitable. It covers the whole of the rocks of Barrande's *Third Fauna,* which, as we have seen, must be erected into a separate system as a matter of geologic convenience. It is fortunate that the application of Murchison's title to them has never been disputed, even by his bitterest opponent.

The various titles at present in use for the intermediate system are all certain to be discarded by the geologist of the future. They are all more or less erroneous, ambiguous, or inconvenient. The retention of the designation *Lower Silurian* would be as systematically erroneous as it is historically unjust. To call it *Upper Cambrian* would be to allow the followers of Sedgwick to commit the very error they so emphatically condemn in the procedure of their opponents. The perpetuation of the Sedgwick-Murchison controversy, by the general adoption of such a title as the *Cambro-Silurian* or *Siluro-Cambrian*—even were it possible—would be, to say the least of it, excessively unwise. Neither party is likely to forego its claims when the object of contention is so conspicuously labelled with the names of both.

Before, however, we can take a single step to free ourselves from the present difficulty, we must dispose of two formidable objections, which, under the guise of universally accepted scientific principles, have grown grey in the service of prolonging this unfortunate controversy, and have, as yet, stubbornly barred the way to anything like a peaceful solution.

By those who still retain the Silurian System of the later days of Murchison in all its magnitude, the argument of their founder that there is no universal stratigraphical break to be detected among the Lower Palæozoics, at least as far down as the base of the Lingula Flags, is held to be an overwhelming reply to all objectors. Similarly, it has been the habit for their opponents, in their turn, to point triumphantly to the local breaks in Britain between the Mayhill and Bala beds, as affording in themselves a positive demonstration of the truth of their own view that these formations belong to wholly distinct systems. For a corresponding reason, also, the latter party claims for the Cambrian all the fossiliferous strata that underlie the Llandovery, from the fact that the physical succession among them is uninterrupted by a general physical break. Of such pre-eminent value is this principle considered, even by those who profess to stand aloof from this controversy, that a strong tendency is abroad to sacrifice in its favour the Old Red Sandstone itself.

To the field-geologist, pure and simple, who desires, above all things, an unmistakable base-line for his system, capable of being rigidly defined upon his maps and sections, the presence of a decided unconformability affords the very thing of which he stands most in need. The grouping founded upon stratigraphical breaks commends itself to his mind with a force that is practically irresistible. But it is far otherwise with the cautious systematist, who endeavours to found his systems in accordance with those of Nature herself, upon principles, not of local, but of universal application. Though fully cognizant of the value of an unconformability as affording him a fairly reliable horizon within a limited area, he soon learns that it is of all things most untrustworthy when it extends over regions of large diameter. It is at most a local phenomenon, wholly misleading except in local application.

It is surely a work of supererogation in these days to point out how the tendency of the entire course of geological discovery for the last fifty years has been to reduce to a mere shadow the magnitude of the miraculous and world-wide stratigraphical breaks that bounded the geologic systems of our forefathers. The doctrine of universal convulsion and the simultaneous destruction of all the life upon the earth at the end of each great epoch has so long since passed into the limbo of exploded hypotheses, that it would be highly amusing, were it not so painful, to see its degenerate and impoverished survival—the dogma of the necessity for general stratigraphical and palæontological breaks between our modern systems—dragging out its miserable and ridiculous existence, even in our midst, and claiming allegiance from men of standing in the science.

One concession, and one only, appears to be all that is needful to meet the real facts of the case. As a general rule, our British systems have been founded, less upon palæontological than upon mineralogical considerations, and it is more of the nature of a series of happy accidents, than a geologic necessity, that they happen to possess such distinctive faunas. In all cases, however, it is clear, both here and elsewhere, that the faunas that characterize our accepted rock-systems owe their distinctness—such as it is—to the fact that in the more typical areas there happened to be an absence of fossiliferous strata to unite them. Whether the time thus zoologically unrepresented was occupied in the upheaval and partial denudation of the rocks of the preceding system (as locally between the so-called *Lower* and *Upper Silurians* of Britain), or whether, on the other hand, it was filled by the deposition of barren strata (as between the corresponding systems of the United States), the result is precisely the same. The faunas of the consecutive systems differ to the extent of the progress made in the locally unrepresented interval; and the group of rocks holding each fauna forms for the geologist a convenient Procrustean bed to which to fit the tolerably synchronous deposits of other lands. The unconformability argument is worthless except from the point of view that the faunas of our typical British systems are likely to be the more distinct the longer these separating interregnums lasted. It is best, that is, simply as a matter of convenience and clearness of definition, to choose, if possible, the longest non-fossiliferous periods to divide them, and these are almost certain to occur where there is the greatest appearance of unconformability.

Nevertheless, the same effect may be owing to a cause in its nature diametrically opposite—the required palæontological break being due to a more than ordinary depression of the sea-bed, and the consequent cessation of almost all deposition in that area—a circumstance to my mind of equal importance from a classificatory point of view with an unconformability itself. An extraordinary regional depression of this character seems to have been the actual cause of the apparently sudden change in the facies of the Palæozoic fauna at the commencement of the Arenig period, both in Britain and Scandinavia, and when fully worked out will, in all probability, enable us to lay down a palæontological line of demarcation far more strictly synchronous throughout its geographical range than that which we shall be compelled to adopt at the base of the Lower Llandovery.

Nor is the venerable objection—that, owing to the established laws of scientific nomenclature, a moral obligation is binding upon us to adhere rigidly to the limits of each system as originally laid down by its founder—worthy of a whit more respect.

This is a claim whose absurdity verges upon the ridiculous when it is advanced by the Murchisonian in support of his contention that the *Paradoxides and Olenus* beds appertain to the Silurian, for they actually antedate all the strata of Murchison's original *Silurian System.* It is, therefore, only occasionally employed by him in a restricted sense in defence of his retention of the strata of the *Second Fauna.*

It crops up continually, however, in the writings and arguments of those belonging to the opposite party. It is urged again and again with a wearisome iteration, as if this conservative rule in geologic nomenclature were necessarily to over-ride every other scientific canon whatsoever. But even if we grant that Sedgwick, and not Murchison, first correctly defined and characterized the rock-group which yields the *Second Fauna,* this rule is equally inoperative in the face of our present recognition of the grand geological importance and distinctness of this *Fauna.* Of this fact Sedgwick was originally wholly unaware; nor does he ever appear to have estimated it at its true value. To us, however, who have watched the gradual elimination of the *Primordial Fauna,* the grand distinctness of the *Second Fauna* is so glaringly apparent, that it is impossible for us to conceive of the rock-group which it characterizes as a mere subdivision of the Cambrian.

It is all very well to plead for historic justice, and to demand, out of respect to the memory of a genius, the adoption of the nomenclature which the general geological world was, in a sense, deprived of the opportunity of accepting during his lifetime. But time and geological convenience will soon make short work of any scheme of nomenclature, however historically just, if it be not in all its parts the natural expression of the inter-relationships and mutual subordination of the facts it is its special aim to associate and systematize.

No amount of enthusiastic regard for the memory of a martyr will bolster up an unwieldy system for ever. The giant size upon which its weaker advocates pride themselves must in the end be the main cause of its inevitable dismemberment. We shall best promote the interest of the man whose memory we venerate, by modestly claiming for him as much, and no more, than truth and geological convenience will allow.

Thus, however reluctant we may be to interfere with the schemes of classification propounded by our great masters in the science, it appears to me that the time has now arrived when we can no longer be accused of disrespect or disloyalty in endeavouring to emancipate ourselves from the inconveniences due to our superstitious adherence to an effete and unworkable nomenclature. The present needs of our science demand, with a unanimous voice that partizanship can no longer silence, a distinct title for the rocks of the *Second Fauna*. The experiment of naming them in such a way as to recognize the claims of both Murchison and Sedgwick has been tried again and again with the same result. It has invariably ended in prolonging and greatly intensifying the original controversy. But one course remains to us. We must give it a new title, which, though it might have been originally suggested by either party, shall contain no element of future discussion.

So long as present systems of nomenclature survive, nothing can disturb the application of the title of Cambrian to the rocks of the *Primordial Series*, and that of Silurian to the strata of the *Third Fauna*. In these systems, as thus restricted, the most perversely ingenious partisan could scarcely find room for controversy. Within these limits the labours of their respective founders were comparatively perfect and complete, and the propriety and harmony of their original classifications, though slightly modified in detail by subsequent research, has never been impugned, either by friend or foe. It is vastly different, however, as we have seen, with the intermediate system. From the day it was recognized until now, it has been the object of incessant disputes. Its co-discoverers both committed the gravest of errors regarding either its proper limits, its relationships, or the sequence and fossils of its component formations. It has been the subject of almost as much passionate argument as the Wernerian theory itself; and the whole subject is a disgrace to modern science, and an obstacle to its progress that must be got rid of—whatever the sacrifice.

Time has already done justice to the value of the discoveries of both Murchison and Sedgwick, by assigning them each a system in which their labours were accurate and complete. We shall do their memories the greatest service by giving the system in which their work appears to our eyes—in the light of later discovery—to have been more or less inaccurate or deficient, a title which shall bear no personal reference to either.

Sedgwick, with his well-balanced and philosophic mind, named his system after the entire Principality in which his rocks were typically developed. His title of Cambrian is thus comprehensive enough to embrace the whole of the Lower Palæozoics. It not only calls up before the imagination the majestic mountains where they may be studied under their most typical aspect, but it reminds us that they formed the fortress-homes of the early Britons—those proud old savages, who, like the *Greywackés* upon which they trod, were the last to succumb to the irresistible march of conquest.

Murchison, on the other hand, with his military proclivities, and a keener instinct for locality, had already made choice of the term Silurian; associating the rocks of his system with that classic Cambrian tribe, the *Silures,* whose indomitable struggles for liberty had hallowed the very hills upon which he sought his types; and thus, in a measure, he may be said to have erected an everlasting monument to British valour and love of freedom.

But, as has been more than once pointed out elsewhere, the *Silures* were a nation inhabiting the *southern* parts of Wales, and Murchison distinctly availed himself of the privileges of genius in thus extending their rule into Shropshire and the regions to the *north.*

North Wales itself—at all events the whole of the great Bala district where Sedgwick first worked out the physical succession among the rocks of the intermediate or so-called *Upper Cambrian* or *Lower Silurian* system; and in all probability much of the Shelve and the Caradoc area, whence Murchison first published its distinctive fossils—lay within the territory of the *Ordovices;* a tribe as undaunted in its resistance to the Romans as the Silures. It was indeed the *last* of the old British tribes to yield to their invincible legions; and it is consequently quite as well worthy of scientific commemoration as the Silures themselves.

Camden thus refers to the Ordovices:[1] "Those countries of the *Silures* and *Dimetæ,* which we have last surveyed, were in after-times, when Wales came to be divided into three Principalities, called by the natives Deheubarth (or the Right-hand part), and in English, as we have already observed, South Wales. The other two Principalities (which they call Gwynedh and Powys, and we North Wales and Powisland) were inhabited by *the Ordovices,* called also *Ordevices,* and *Ordovicæ,* and in some authors, though corruptly, *Ordulucæ.* A courageous and puissant Nation these were, as being inhabitants of a mountainous country; and receiving vigour from native soil; and who continued, the longest of any, unconquered either by Romans or English. For they were not subdued by the Romans till the time of the Emperor Domitian; when Julius Agricola subdued almost the whole nation. Nor were they subjected by the English, before the reign of Edward the First. For a long time they enjoyed their liberty, confiding as well in their own strength and courage, as in the roughness and difficult situation of their country, which seems to be laid out by Nature for ambuscades and the prolongation of war. To determine the limits of these *Ordevices* is no hard task, but to give a true reason of the name seems very difficult. However, I have entertained a notion, that, seeing they were seated upon the two rivers of Devi, which springing

[1] Camden's *Britannia,* Dr. Gibson's Translation, second edition, p. 778.

not far asunder, take their course different ways, and that *Oardevi* (Read *Ar-dhyvi*—Transl.) in the British language signifies—Upon the rivers of Devi—they have been thence called *Ordevices.* To the *Ordovices* belonged those countries which are now called in English by new names—Montgomeryshire, Merionethshire, Caernarvonshire, Denbighshire, and Flintshire."

Here, then, have we the hint for the appropriate title for the central system of the Lower Palæozoics. It should be called the Ordovician System, after the name of this old British tribe.

Whatever arguments may be adduced in support of the term Silurian will apply equally well, or even with greater force, to this new title. Like the term Silurian, it is classic in origin, but at the same time thoroughly British. It is equally euphonious, and far more strictly significant of the geographical area where its strata are typically developed. Indeed, the employment of the one title almost of itself necessitates the adoption of the other; for only in this way is it possible to recognize the systematic equality of the two systems in their very designations—the one receiving its name from the ruling tribe in the *south* of Wales, the other from the dominant tribe in the *north.* If there is anything specially becoming in commemorating the warlike tribe of the Silures in the name of a geologic system, how strikingly appropriate is the title of *Ordovician* in erecting a similar scientific monument to the last and most valiant of the old Cambrian tribes.

On this arrangement the Lower Palæozoic Rocks of Britain stand as follows:—

(*c*) Silurian System:—Strata comprehended between the base of the *Old Red Sandstone* and that of the *Lower Llandovery.*

(*b*) Ordovician System:—Strata included between the base of the *Lower Llandovery* formation and that of the *Lower Arenig.*

(*a*) Cambrian System:—Strata included between the base of the *Lower Arenig* formation and that of the *Harlech Grits.*

That our attempt to cut in this way the knot which all the schools have already convinced both themselves and others of the utter impossibility of untying, will do much more than draw the attention of geologists in general to what we believe to be the more striking aspects of the question, can hardly be expected. It is almost certain that any suggestion that might have been made by either of the schools with the object of freeing this section of the science from the present dead-lock, would, as a matter of course, be opposed to the utmost by the others. How much worse is it when the hint is given from without. The great mass of the most influential of our living geologists have so long since given in their adhesion to one or other of the contending parties, that it is not improbable that our well-meant interference will be stigmatized by all as a most unwarranted and impertinent intrusion.

By those, however, who are weary of the interminable discussion, and who feel the necessity for some scheme of classification which, while it systematizes the known facts, holds the balance true with reference to the opposing claims of the two great pioneers in the study of the Lower Palæozoics, our suggestion may be tolerated now, and adopted later on, when the necessity for this course has become more strikingly apparent. To those who interest themselves in the attempted correlation of the Lower Palæozoic Rocks of the Northern Hemisphere, and who are continually hampered by the want of some clear and unmistakable generic terms expressive of the general parallelism among these widely-separated deposits, the ease and comfort of a classification which imitates Nature herself in placing the three grand members of the Lower Palæozoic Rocks upon an equal footing, is an advantage of which they are certain in time to avail themselves to the full. Those again, who feel how vain is the endeavour to parallel the special formations and minor stages of our British Lower Palæozoics with those of other areas, will hail with some approach to satisfaction the release of such convenient sub-generic terms as *Lower* and *Upper Cambrian*, and *Lower, Middle, and Upper Silurian*, with the list completed by the addition of *Lower* and *Upper Ordovician*;—terms all of easy and immediate application, and all expressive of epochs, which, so far as our present knowledge enables us to judge, embrace tolerably equal periods of geological time.

No earnest student of the history of discovery among the Lower Palæozoic Rocks, whose opinions are the natural outcome of his own careful generalization of presently known facts, and not the petrified remains of the views he so enthusiastically adopted a quarter of a century ago, can fail to perceive that the ideas of the extreme party which claims all the Lower Palæozoics for the Silurian are fated soon to become wholly extinct. The wave of backward opinion which led this party to revert in substance to the ideas of their predecessors was inevitable. We are now witnessing the as-inevitable return of the tide. Here and there this application of the term may linger on for a time, as in Bohemia, and possibly in Scandinavia, kept-alive by the very principle that must in the end prove fatal to it, when local conveniences become superseded by cosmopolitan necessities.

A single glance at the magnificent development of the Lower Palæozoics on the continent of North America is enough to convince every unbiassed investigator how much we have yet to learn regarding their British prototypes, and how ridiculously inadequate is our present estimate of their grand importance in the geological series. As this knowledge dawns upon us as the result of our discoveries in the future, some such classification as is here proposed will perforce be adopted by all; and the systematist will then be left free to work out his generalizations untrammelled by the defects of a cramped and unnatural nomenclature. Our British strata can in the end return but one answer to the most extended appeal. Every geologist will at last be driven to the same conclusion that Nature has distributed our Lower Palæozoic Rocks in three sub-equal systems, and that history, circumstance, and geologic convenience, have so arranged matters that the title here proposed for the central system is the only one possible.

25

PALEONTOLOGIC METHOD DIVISIONS OF THE SEDIMENTARY SERIES

A.A.C. DeLapparant

This excerpt was translated expressly for this Benchmark volume by Barbara Conkin, University of Louisville, from pages 704–712 of Traité de Géologie, *Savy, Paris, 1885, 1504p.*

IMPORTANCE OF THE PALEONTOLOGIC CHARACTER

Observation teaches us that the organic population of continents and oceans has undergone, since the beginning, numerous, but perfectly orderly, vicissitudes, by virtue of which each stage of the history of the globe has been characterized by certain special types. But every sedimentary bed encloses, in general, the more or less well-preserved remains of beings contemporaneous with its deposition. The examination of *fossils* thus offers a means of characterizing each episode with a precision that much surpasses that of the mineralogical argument. In effect, it is nearly impossible to say, in limiting oneself to a consideration of the composition of a bed, what the depth, temperature, or degree of salinity of the water in which it was formed could be or if the deposit had been in a gulf or on a largely open coast. On the contrary, the study of fossils, with the light that zoology and botany furnish us on the normal *habitat* of similar species, permits definition of the ensemble of physical conditions of the environment in which the deposit was formed.

Moreover, if it is a question of marine sediments (and those ought to occupy us above all because their extent is incomparably greater than that of terrestrial deposits), the principle of *continuity* becomes, thanks to fossils, of a much easier application. Without doubt, each kind of sediment carries its own population, and the beings of deep oozes are not the same as those of sandy beaches. However, the sea is no less continuous from one district to the other and, whether among floating animals, or among those of which the shells are not buried in their life position, it turns out that they frequent, indifferently, localities very dissimilar as to the nature of the deposits. The debris of these beings can thus become *characteristic fossils,* whose presence (supposing that it is a question of identical species and sufficient numbers) will establish the synchronism of deposits of different kinds, even where their lateral passage had not been seen.

This criterion can be applied moreover to the determination of the synchronism of deposits geographically distant provided that, in this case, one substitutes for the idea of identical species that of *representative species*—that is, being able to deviate from the chosen types in a feeble measure that the difference of geographic conditions will suffice to explain.

APPLICATION OF PALEONTOLOGY TO THE DETERMINATION OF RELATIVE AGE

It is not only on the synchronism of deposits that fossils provide the means of giving a verdict, however. One can also apply them with success to the determination of *relative age* in order to appreciate the value of gaps.

In effect, geologic maps show us that, if the sea covers today three-quarters of the surface of the globe, its domain has been, in the past, yet more extensive. The oceanic mass represents thus an environment whose existence has been *continuous* across the ages, and if, by exception, certains parts have been able to be isolated, as the interior seas are in our day, the greater portion of this mass has never ceased to form a whole, the divers sections of which would have communicated freely with one another.

However, observation also teaches us that external physical conditions have undergone, on the surface of the globe, a progressive modification such that this surface has had to pass little by little from tropical uniformity of the first ages to the variety of climates that prevail today according to latitude. At the same time, the extent of the continents always increased, while their form and their relief did not cease to diversify, in a way that leads to more and more contrasting differences between diverse littoral regions.

That granted, due to the harmony that always related the organic world to the circumstances of the ambiant environment, this orderly variation of physical conditions has had to make itself felt on the animal and vegetable population of continents and oceans. But on the solid earth, subject to even more varied vicissitudes, the faunas and floras, influenced by a thousand local circumstances, could offer sharp contrasts, whether at the same epoch in diverse parts of the globe, or in two consecutive epochs at the same place. It was otherwise in the maritime domain and especially in the pelagic regions. There the variations have rarely been abrupt, and if, in each successive fauna, one seeks with discernment the types it is suitable to compare, one can establish a series of which each term will be truly characteristic of a determined epoch.

We do not wish to contest that we can draw great enlightenment from the study of continental floras or that mammals had not often been employed with great success for the identification of tertiary lake beds.[1] Once again, however, the criterion the terrestrial beings furnish is of less general, and at the same time less sure, application because of what one can call their *impressionability,* which renders them fit for interpretation of phenomena the significance of which can be exclusively local. Also, to obtain homogeneous divisions, it will always be prudent to base them, as much as possible, on the comparison with animals of the high sea.

Thus, thanks to the continuity of the oceanic regime, paleontology furnishes the means of putting all the local series studied by stratigraphers in order. Moreover, it permits the resolution of most of the difficulties occasioned by faults and dislocations of the ground in teaching us to identify, by faunas, sections of one same assise now separated. Finally, always by virtue of the same principle of continuity, when two beds, certainly deposited in analogous conditions, present, although directly superposed one on the other, very great differences in their faunas, one will be legitimately justified to conclude that their epochs of deposit must be separated by an interval of time much longer than one would suppose from their concordant superposition.

IDENTIFICATION OF ASSISES

Ideas of synchronism and of equivalence

Thus, the paleontologic argument can be employed at the same time for *identification* of assises and for establishment of homogeneous *divisions*. In the first case, it is often indifferent whether the compared faunas are littoral or pelagic, provided that the organisms serving to establish synchronism are of the same kind. One must only distrust apparent similarities that the return of the same conditions often produces. Two not contemporaneous coralline limestones can differ much less one from the other, by their fossils, than will an arenaceous sediment and a clay bed of the same age. It is very necessary that the *sensibility* of all the organic types be the same, and this is the only way by which the difference of age is interpreted, by characters very little visible at first examination. One must apply this remark especially to coral species.

Concerning the establishment of divisions, one can, when one does not seek to embrace the entire surface of the globe, found them on the littoral types, in general much easier to collect than pelagic fossils. But then it is necessary to understand that the chronologic worth of the divisions thus obtained would be of a lower order. In effect, in case two nonpelagic assises, observed in two different regions but little separated, enclose faunas in fact similar, it does not follow that their deposition corresponds rigorously to the same instant of the history of the earth. Organic identity implies only identity in physical conditions. However, these conditions can cease to exist in one place, while they are carried wholesale into other more or less near places, causing a corresponding *migration* of animal or vegetable species so that two groups identified as identical do not always represent two *synchronous* formations but two *equivalent (homotaxial)* and perhaps *successive* assemblages, of which one, due to exterior conditions, has been able to prolong itself at the place it occupies much longer than the other.

Thus, there would be (Fig. 175) two distinct regions, 1 and 2, each presenting a similar system of three stages such that A would be equivalent to A′, B to B′, and C to C′. If one allots to these three stages height proportional not to their real thickness but to their respective duration, it can very well happen that their limits do not coincide in a vigorous manner. However, even when observation of continuity would succeed in establishing that the first beds of A were deposited at the

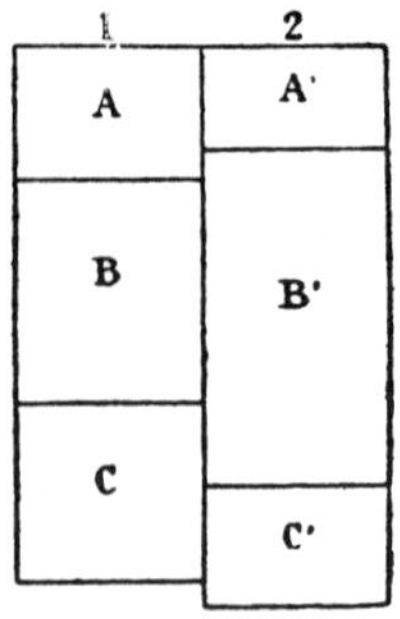

Figure 175. Diagram relating geologic equivalences

same time as the last beds of B′, it usually will be better, in the execution of geologic maps of regions 1 and 2, to place limits to the separation of local stages in order to have, in each region, divisions truly conforming to its history.

In effect, if it is necessary to state precisely the great stages of sedimentary phenomena, it is no less important to make a reconnaissance of the distinct regions between which the surface of the globe has been divided in the past. The reason for this is that each of these regions has its history, marked by events that have the value of *dates* but that can, in this quality, not apply in any way to other countries.

The same as the division of Middle Ages and of modern times, legitimately based, with respect to Europe, on the invention of printing and the discovery of America has not any sense when it is applied to the history of China or Japan, so a boundary of the first order, marked at one place by an accentuated discordance, as much stratigraphic as paleontologic, will be effaced, little by little as we have seen, in a certain direction in order to make one continuous series of concordant beds. One can thus say that the task of the geologist is double and that beside great divisions that he can hope to recognize in resorting to the study of pelagic faunas, he must not neglect the succession of *regional histories* in order to establish as best he can the more or less perfect concordance of these diverse histories.

PRINCIPLES OF THE DIVISION OF SEDIMENTARY FORMATIONS

Zones, Assises, Stages

According to what has been said, it seems rather easy to formulate the rules that ought to preside over the division of the sedimentary succession. At the present epoch, when geognostic knowledge is far from having acquired the generality and precision that would be suitable for proceeding by way of synthesis, it is prudent to go from the particular to the general and to lean above all on the regional successions, well established in a great number of places.

Each local stratigraphic succession comprises, in the last analysis, *beds, layers,* or *strata,* distinguished by the constancy of their mineralogic characters and, if there, by those of their fossil remains. These layers form *horizons,* of a rather limited extent, that one can designate by the name of the most typical species, among those which compose their *faunule* or their *florule.*

Several layers possessing certain common characters will be grouped in an *assise,* of which the fauna will be characteristic of a fixed *age.* The homogeneous subdivisions of assises succeeding one another in the manner of zones of agate or jasper are often designated, from a paleontologic point of view, under the name of *zones,* which are defined by the mention of the dominant fossil.

Each assise is truly the representation of what we have called a sedimentary *episode,* with its phases marked by the diverse zones. To define this episode completely, it would be necessary to know all the surface occupied by the assise and to trace in it the contours, not like one sees today but like they ought to have been before erosion removed a portion. In a word, it is the *restoration of ancient shores,* maritime or lacustrine, that, joined to the consideration of the thickness of the assise, can permit appreciation of the importance of this phase of the regional history.

In working thus on behalf of the diverse synchronous assises, one would have a series of *paleogeographic* maps, the comparison of which would permit one to know, at a glance, those epochs when the terrestrial geography has undergone the more important modifications, if not on all the surface of the globe at once, then at least in one region of notable extent. Then one could legitimately reunite the assises comprised between two instants of this kind and form of them homogeneous divisions to which one would apply the name *stages,* except to subdivide them into *substages,* in the districts where the intermediate vicissitudes would have been more pronounced. Each stage would correspond to an interval of time to which would be applied the name *epoch.*[2]

Unhappily, the restoration of ancient shores presents great difficulties, and there are again only a few places where it has been done with some certainty. Also, the paleogeographic criterion has need of being corroborated, often even to be replaced, by employing the paleontologic method. It is impossible indeed that important changes in the geography of a region would not introduce, in the physical conditions of neighboring oceans, some modifications capable of influence on their population. These will induce modifications in the species of the littoral zone and a considered study of these, above all where they are not missing ought to lead to establishment of divisions, completely concordant with those the stratigraphic method had permitted us to establish.

Control of the Divisions of Stages

However, stages thus based on littoral fossils will have in general a too pronounced local character, and if one does not arrange some means of control, it would be difficult, even at times impossible, to establish the concordance of *paleontologic scales* erected for two distinct regions. The elements of control will be furnished by the study of *pelagic fossils.* The beings of the high seas, and especially chambered cephalopods, among which the family *ammonites* offers, in all the extent of secondary time, a great variety of very orderly types, will serve to establish the importance of stages. The homogeneity of these last will be guaranteed by the condition that each of them could be characterized by *genera* or *families* that would be proper to it.

Thus, stratigraphy, in making evident the internal phenomena that have deranged the stable position of the crust of the earth and littoral or continental paleontology, in making the influence of these phenomena on the faunas and floras stand out, will have permitted the establishment of *boundaries*—that is, the *dates* of regional histories. In contrast, the study of pelagic faunas will make known the *groupings* by which it is suitable to reunite the local divisions in order to form groups a little more recognizable everywhere. These diverse modes of appreciation ought not, as often happens, be opposed one to the other. The role of the first two is essentially to define the phenomena of *discontinuity,* whereas the third has for its function making evident the different terms of a *continuous* succession. It will thus be illusory to claim to reject, in the name of observations made in one region of uninterrupted pelagic sedimentation, separations of stages justified by the study of littoral zones of the same date. All division based on a well-established fact has its raison d'être in the history of the globe, and it is only by right of control, to fix the degree of its importance, as well as its equivalence with other equally localized phenomena, that it is legitimate to bring in the *criterion* drawn from the study of faunas of the high seas.

Systems, Series, Groups

Just as great variations of regional geography, combined with the vicissitudes of organic families, have served to establish the stages, so these latter can be joined in divisions of a higher order, if one takes into consideration either geographic phenomena relative to some notable part of the earth or the appearance or disappearance of characteristic organic groups. We will give to these divisions the name *systems*[3], in endeavoring to compose them in such a way that each of them, corresponding to what we will call a *period,* can be recognized over nearly all the surface of the globe. The very extended limits of a given system will be, less yet than those of stages, susceptible of mathematic precision, the variations which define them not having been instantaneous. Moreover, one must remember that the limits are necessarily more or less variable with the regions—that is, the corresponding period did not have to begin or end everywhere at the same instant.

This observation is suitable *a fortiori* to geologic divisions that one can form by the union of several systems. In a normal manner, we will give to these divisions the name of *groups,* reserving that of *series* for intermediate assemblages, if there is grounds for establishing them, and each group will represent sediments that were deposited during a given *era*. To choose these groups, it will be fitting above all to be preoccupied with the great lines of the organic series, easier to define than geographic conditions of the ensemble, of which the trace is too often effaced.

DIVISION OF GEOLOGIC TIME INTO THREE GREAT ERAS

Applying these principles, we will distinguish, in the succession of sedimentary formations, three great groups: First is the *primary* or *paleozoic* group, characterized by the great distance that separates its organic types from those of the present time. Fish are therein, nearly until the end, the only well-developed representatives of the vertebrate branch, and among the mollusks, one sees predominate certain families of brachiopods such as *Spirifer* and *Productus.* The second group is the *secondary* or *mesozoic* group, wherein begins to appear the precursors of the present organic world; reptiles are preponderant, while in the seas the great family of *ammonites* rules. Third is the *tertiary* or *neozoic* (*caenozoic, kainozoic* of various authors), in which the fauna and flora offer so to speak only modern types and where the *mammals* are amply developed. As for the present era, it can bear the name of *modern* or *quaternary* era, and it is essentially characterized by the appearance of man on earth.

One can ask if the three groups that have just been enumerated have the same importance and if the first would not represent, as for length, something equivalent to the other two together. It is certain that the normal thickness of the primary deposits is enormous. Mr. Dana[4] evaluated it as nearly *double* that of the secondary and tertiary groups, and if this thickness should be considered as the measure of the interval of corresponding time, there would not be a proportion between the first group and the two following. At the same time, it is important to consider that the energy of sedimentary action has certainly not been constant on the surface of the globe and that time passed does not constitute, of itself, an element sufficient for the division of periods. What characterizes the last ages of the earth is the more and more accentuated *localization* and *specification* of physical and physiological conditions. From this point of view, the division, for a long time sacred, of three groups primary, secondary, and tertiary appears well enough based, and until the progress of science has demonstrated the opportunity of another mode of grouping, we will content ourselves with what, until now, has formed the basis of geologic thinking everywhere.

CHOICE OF SYSTEMS AND STAGES

Concerning *systems* and *stages,* one will strive to remain in accord with the general principles formulated here. Nevertheless, we will not venture to assert that divisions of the same order, adopted in this work, are always sufficiently well chosen to represent exactly equivalent units. Descriptive geology has yet much progress to make so that the classification of sedimentary series can be considered definitive. It has seemed to us, however, that there would be more danger in wishing to innovate prematurely than in seeking only to improve divisions that for many years appeared to have sufficed for the needs of the science.

We strive only to inspire ourselves, so far as possible, with a fact that seems to dominate all the history of the terrestrial globe: We wish to speak of this arrangement that has procured, from ancient date, by making emerge, since the first ages, a series of primitive cores in the temperate zone of the northern hemisphere, a division of the principal sedimentary regions into two distinct zones; the northern one remarkable by the littoral character of the deposits, with which the great variability of the faunas harmonizes; the other, mediterranean, where the sediments, chiefly pelagic, have harbored less varied faunas but where cephalopods form a very rich succession and allow one to notice a much more regular order. By the first zone we will order, above all, the divisions of stages, while the second in aiding us to assign them values of the same order, ought to furnish us the elements of more general classifications. In other words, northern and central Europe will intervene above all in the determination of *dates* or *boundaries* of geologic history, while the Mediterranean region will make known the *groupings* by which it is suitable to group the divisions thus established.

NOTES

[1]See, notably, Albert Gaudry, *Mammifères tertiaires,* 1878.

[2]The word *epoch,* proposed at the time of the Congress of Bologna, is perhaps not a very happy choice to designate a succession of ages because, from a grammatical point of view, this word must be applied to a definite *moment* of history. Nevertheless and for want of a richness of terms that our language does not comprise, we adhere to this designation.

[3]There has been an objection made to the word *system* by the Congress of Bologna, based on the fact that this expression is applied on the whole, in usual language, to an assemblage of propositions and of principles connected between them. However, on one hand, this word has the advantage of being able to be easily accepted by foreign geologists, and on the other hand, the sense given to it here is perfectly conformable to the traditions of the science. Thus, the *Dictionary of the Academy* defines *system,* in natural history, "a methodic and artificial distribution of beings, in order to facilitate their study," and it adds that this word signifies also "an assemblage of bodies, reunion of parts of a whole." It is moreover evident that it is always a *systematic* idea that presides in grouping several stages in one ensemble. Finally, this expression has made in science, by the publication of the book by Barrande, *Silurian System of Bohemia,* a rather brilliant appearance because, in spite of the popularity the word *terrain* enjoys in France, we have no repugnance to accept the vote of the geologic Congress on this point.

[4]*Manual of Geology,* 2nd edition, 1875, p. 112.

26

Reprinted from *Geol. Soc. London Quart. Jour.* **49:**480–483, 518 (1893)

ON THE BAJOCIAN OF THE SHERBOURNE DISTRICT: ITS RELATION TO SUBJACENT AND SUPERJACENT STRATA

S. S. Buckman

[*Editors' Note:* In the original, material precedes this excerpt.]

On Zonal Correlation.—The geological unit for the correlation of strata has hitherto been the 'zone.' Gradually, however, it has been felt that either the zones must be increased in number, or some modification adopted, if the true faunal sequence is to be expressed with that accuracy which is now necessary. Thus, in 1887, Hudleston, in his very able survey of the Inferior Oolite,[1] although he adopted the zonal divisions which I had briefly sketched out in 1881,[2] showed plainly that the strata required more subdivision; and this subdivision he accomplished by the term 'beds,' and by designatory letters. In dealing with the Toarcian I have shown the necessity, for the purposes of correlation, of adopting greater accuracy in division than the existing zones allowed; and for my subdivisions I used the term 'beds.' Quite recently, in communications to the Geological Society of France, Munier-Chalmas [3] and Haug [4] have shown the geological value of small divisions. The work of the former in Normandy is most valuable for its exactness; in comparison, our researches on this side of the Channel are still much behindhand.

To such subdivisions, however, H. B. Woodward [5] is opposed; but then he is not looking at the matter from a biological point of view. For correlation in certain cases fewer zones may be used; but any such correlation is deficient in the precision which the biological palæontologist demands, and fossils labelled on such a plan are useless to him. In this connexion it may be remarked that the more minute the correlation the better for the student: he can easily dispense with detail; but if such detail as he does need be not given in any paper, he searches in vain for information.

The term 'zone' is said to be used in geology in a zoological sense; but in zoology it is really used in a different sense. However, I take it that of late years the stratum (or strata) characterized by an assemblage of organic remains, more or less peculiar thereto, has been regarded as a zone; and, as H. B. Woodward practically implies, though his definition of a zone is faulty,[6] the presence of any of the species known to be sufficiently peculiar to a given horizon is considered to denote the zone in localities from which the index-species is absent. This may be called the geologist's zoological use of the term. Charles Moore [7] seems to have been the only one to use the term in a strictly palæontological sense: he confined it to the exact horizon of a particular fossil.

The term 'Hemera.'—It is for a palæontological purpose similar to Moore's use of 'zone' that I propose the term 'hemera' (ἡμέρα).[8] Its meaning is 'day,' or 'time;' and I wish to use it as the chronological indicator of the faunal sequence. Successive 'hemeræ'

[1] 'British Jurassic Gasteropoda,' Palæontograph. Soc. 1887, No. i. pt. 1.

[2] 'Ammonites from the Inferior Oolite,' Quart. Journ. Geol. Soc. vol. xxxvii. p. 588.

[3] Compte-rendu sommaire des Séances, No. 14, p. 164, 1892.

[4] *Ibid.* p. 174.

[5] 'On Geological Zones,' Proc. Geol. Assoc. vol. xii. (1892) p. 295.

[6] *Ibid.* p. 298.

[7] Proc. Somerset Archæol. Soc. vol. xiii. 1865–6.

[8] [Originally the word 'emar' (ἦμαρ—the Homeric form of ἡμέρα) was chosen; but in deference to the wish expressed by the Council of the Geological Society I have substituted the term 'hemera.'—July 5th, 1893.]

should mark the smallest consecutive divisions which the sequence of different species enables us to separate in the maximum developments of strata. In attenuated strata the deposits belonging to successive hemeræ may not be absolutely distinguishable, yet the presence of successive hemeræ may be recognized by their index-species, or some known contemporary; and reference to the maximum developments of strata will explain that the hemeræ were not contemporaneous but consecutive.

The term 'hemera' is intended to mark the acme of development of one or more species. It is designed as a chronological division, and will not therefore replace the term 'zone' or be a subdivision of it, for that term is strictly a stratigraphical one. Our present 'zones' give the false impression that all the species of a zone are necessarily contemporaneous; but the work of Munier-Chalmas in Normandy, and my own labours in other fields, show that this is an incorrect assumption. The term 'hemera' will therefore enable us to record our facts correctly; and its chief use will be in what I may call 'palæo-biology.'

Description of the Strata.—Anyone unacquainted with the Dorset-Somerset Inferior Oolite, and the richness of its deposits, would scarcely credit the difficulty experienced in these investigations from want of names for the Ammonites. Species which have been perfectly well known for years as indicators of certain horizons are altogether devoid of any specific name. They could not, therefore, be recorded with precision—only the generic name could be given.

In this connexion the great value of the new generic names becomes most apparent. By their aid it is possible to narrow the identity of new or unnamed species within reasonable limits, and to record the fact with brevity in a paper. With their help, the characteristics and chief features of unnamed species can be described in a way comprehensible to other workers in the same field, by the use of a single word. Lastly, by employing them in the case of known species, the remarkable changes in the Ammonite-faunæ of successive deposits can be illustrated; and this important fact it is impossible to show by any other means.

I consider the strata dealt with in this paper to have been deposited during at least twelve hemeræ; and the names of these hemeræ will appear in the Sections at the tops of the respective divisions of strata assigned to them. These hemeræ[1] are as follows:—

1. *Fuscæ* and 2. *Zigzag*. (*Oppelia fusca*, Quenst., sp., and '*Stephanoceras*' *zigzag*, d'Orb., sp.) Owing to the unfossiliferous nature of the strata deposited during these hemeræ, in this district, an exact apportionment of the beds to each one

[1] In connexion with the specific names which designate the hemeræ, I have considered it advisable to use the genitive inflexion as more truly expressing what is intended. Thus '*fuscæ* hemera,' which is short for '*Oppeliæ fuscæ* hemera' = the hemera of the tawny *Oppelia*, means the hemera of the tawny one (*Oppelia* understood); whereas '*fusca* hemera' might seem to suggest the 'tawny hemera.'

separately was not possible in the time at my disposal. In the neighbourhood of Broad Windsor the strata of these hemeræ may be studied to far more advantage.

3. *Truellii* (*Strigoceras Truellii*, d'Orb., sp., *sensu stricto*).
4. *Garantianæ* (*Parkinsonia Garantiana*, d'Orb., sp.).
5. *Niortensis* (*Parkinsonia niortensis*, d'Orb., sp.).
6. '*Humphriesiani*'[1] (*Stephanoceras Humphriesianum*, Sow. sp.).
7. *Sauzei* ('*Stephanoceras*' *Sauzei*, d'Orb., sp.).
8. *Witchelliæ*, sp.[2] There are numerous species of *Witchellia* in the strata of this hemera; but, as very few are named, it is desirable not to appropriate any particular specific name at present.
9. *Discitæ* (*Hyperlioceras discites*, Waag., sp.).
10. *Concavi* (*Lioceras concavum*, Sowerby, sp.).
11. *Bradfordensis* (*Lioceras bradfordense*, S. Buckm.).
12. *Murchisonæ* (*Ludwigia Murchisonæ*, Sow., sp.).

It may be noted as a matter of history that in my paper of 1881[3] the divisions—

1–4	were included as		*Parkinsoni*-zone,
5–8 (*pars*)	,,	,,	*Humphriesianum*-zone,
8[4] (*pars*)–10	,,	,,	*Sowerbyi*-zone,
11, 12	,,	,,	*Murchisonæ*-zone.

Subsequently the divisions 9–10 were called the *concavum*-zone; and Waagen's term '*Sowerbyi*-zone' was given to that which on palæontological grounds was assumed to be an horizon situated between 7 and 9. Practically speaking, the correctness of the surmise has been confirmed.

[*Editors' Note:* Material has been omitted at this point.]

Conclusion.

By way of conclusion I would make a few additional remarks concerning the proposed term 'hemera.' It must be particularly understood that it is used in a chronological sense as a subdivision of an 'age.' Here I would take occasion to state what seems to be a paradox—that species may occur together in the rocks, yet such occurrence is no proof that they were contemporaneous. If the species have not been proved to be successive in other places, then their joint occurrence in the same bed may be assumed to indicate that they were contemporaneous until such proof be forthcoming. It is, however, only negative evidence. If they have been proved to be successive species, then their occurrence together only shows that the deposit in which they are embedded accumulated very slowly. Had this been rightly understood we should have been spared the stress that has been laid upon a so-called intermixture of zonal species; though I must say that too often an investigation of such supposed intermixture proves it a fiction due to inaccurate palæontology or incorrect geology.

[*Editors' Note:* Material has been omitted at this point.]

[1] The confusion concerning this specific name, and the large number of species of *Stephanoceras*, make this appellation very unsuitable. It would be desirable to apply the name of a species belonging to a less prolific genus.

[2] This term has been suggested by French authors as a name for the strata of this horizon, from the prevalence of species of this genus in France; see later, p. 519.

[3] Quart. Journ. Geol. Soc. vol. xxxvii. p. 588.

[4] So far as the Ammonites from Sandford Lane were concerned; but the strata were covered up—I had not seen them.

27

Reprinted from *Geol. Soc. London Quart. Jour.* **54:**442–444 (1898)

On the GROUPING *of some* DIVISIONS *of so-called* 'JURASSIC' TIME. By S. S. BUCKMAN, Esq., F.G.S. (Read April 6th, 1898.)

CONTENTS.

		Page
	Introduction	442
I.	General Considerations	443
II.	The Ages of the 'Jurassic' Period	447
III.	The Divisions of a Portion of 'Jurassic' Time, with Table I	451
IV.	A Genealogy of some 'Jurassic' Ammonite-Genera, with Table II.	451
V.	Notes on certain Generic Names	452
VI.	List of Genera arranged under Families	459
VII.	Summary	462

INTRODUCTION.

A CONSIDERABLE portion of this paper was written before the appearance of a communication on Dundry Hill.[1] A reference to the ideas set forth in this paper will be found in that communication, in Table IV, facing p. 696.

[When this paper was first presented to the Society the Author had designated the Ages by the terms Bathonian, Bajocian, etc., proposing, in effect, to continue the usage carried out in the communication above mentioned, to employ these terms for chronological purposes only. They have been used formerly, as for instance by Renevier, for both stratigraphical and chronological divisions:— Bathonian Stage or Bathonian Age.

However, it was pointed out to the Author that this plan would lead to confusion; and he then proposed to use, as distinctly chronological terms for Ages, names taken from dominant ammonite-genera. There is an advantage in this plan, seeing that the chronological divisions are based entirely upon zoological phenomena. And there was an admitted disadvantage in retaining for chronological purposes names such as Bathonian, Bajocian, etc., taken from places where particular strata were developed.

But the change has a wider effect if a congruous series of terms is to be produced. Jurassic is equally open to the same objection as Bathonian, and it will require to be replaced by a term taken from some zoological phenomenon. The divisions of Jurassic—Eo- and Neojurassic—fall into the same category; they may be easily replaced by such terms as Arietidan and Stepheoceratidan Epochs, which appropriately mark the changes of the ammonite-fauna. Jurassic and Triassic it is not so easy to replace; so, for the present, they remain. They are strictly stratigraphical terms; but they are used in this paper with inverted commas for chronological purposes. Thus 'Jurassic' Period denotes really the period

[1] 'Dundry Hill,' by S. S. Buckman & E. Wilson, Quart. Journ. Geol. Soc. vol. lii (1896) pp. 669–720.

during which the strata of the Jurassic System were deposited. On the other hand Mesozoic is a chronological, and not applicable as a stratigraphical, term.

This note is inserted for the purpose of explaining the reason for the changes introduced since the paper was read before the Society. They are only changes of name; the principle of the paper remains the same.—June, 1898.]

I. General Considerations.

In the present communication I have two main purposes in view. The first is to demonstrate the application of the chronological system of nomenclature in regard to a certain portion of time; the second is to construct what may be called a geological calendar, in order that those who are collecting fossils may be able to date the species which they find. Particularly do I give this for the benefit of those who are concerned with the biological aspect of palæontology, to whom a knowledge of the exact succession of organic beings in time is an important feature, to enable them to study the evolution.

It may be noted, with one or two exceptions, to be dealt with later, that all the hemeral names are taken from the names of ammonites. For the divisions of Mesozoic time this plan would be adopted so far as possible; and it needs no defence, for Haug remarks with regard to the zonal work of Oppel, Quenstedt, Neumayr, and others:—'L'importance des ammonites pour l'établissement du parallélisme des couches était reconnue, et en même temps se trouvait posé le principe que les autres fossiles marins, tels que gastropodes, lamellibranches, brachiopodes, échinodermes, ne devaient intervenir qu'en seconde ligne dans les assimilations. Quant aux caractères lithologiques, ils étaient naturellement refoulés tout-à-fait à l'arrière-plan.'[1]

Thus the next step in the arrangement follows by logical necessity upon this one; for if the ammonites are recognized as the best indicators of the faunal sequence, and since the chronological subdivisions depend upon this sequence, then the further grouping of the chronological subdivisions must be controlled by the zoological affinities of the ammonites. For instance, the shortest geological time-division is a hemera: that is, the time during which a particular species—generally in Mesozoic chronology, of an ammonite—had dominant existence. A longer space of time contains so many hemeræ; it is at present designated by the very faulty title of 'an Age'; but it is obvious that, as the hemera depends on the ammonite-species, an 'Age' must depend on the duration of allied series of ammonite-species. In the present state of ammonite-nomenclature the duration of an Age is dependent upon the duration of an ammonite-family, or at any rate, if not the whole period of duration, at least that portion of time during which the family was of the greatest importance as a constituent of the ammonite-fauna. In other words, as a family has its periods of

[1] Article 'Jurassique' in 'La Grande Encyclopédie,' Paris, 1894, p. 325.

rise, of maturity, and of decline (or, in scientific language, its epacme, acme, and paracme), so the duration of the Age would be principally governed by the period of acme, and would be less concerned with the epacmastic and paracmastic periods of the same family; because during any one such period should be found the acme of another family.

Following on this again as a logical conclusion is the recognition that any time-division larger than an Age must depend upon the duration in time of allied ammonite-families. When it is found that a certain number of ammonite-families allied from a zoological and evolutionary point of view flourish and disappear, to be succeeded in a similar manner by another quite distinct group of ammonite-families similarly allied among themselves, then it is obvious that, so far as is practicable, the time-division which embraces so many Ages should correspond to the period of time marked by this zoological phenomenon, and that the end of this time-division and the beginning of the next should coincide, as far as practicable, with the time when the one family-group has declined so low as to become unimportant, while a succeeding family-group has risen to importance.

Such are the theoretical considerations which should guide the application of a chronological scale. Their practical application, however, may not fully conform to the theoretical premisses because, in the first place, our knowledge of the faunas is very incomplete, even in connexion with the most fully-explored regions; and secondly, the faunal successions (as known to us) do not follow with that faultless regularity which would be necessary for the complete fulfilment of the conditions laid down. All that is possible, therefore, is to so arrange the chronological scale that while it expresses what we do know, it shall be able to show good reason for the arrangement where the actual facts of faunal succession do not possess the regularity which would be desired to make the arrangement perfect and indisputable.

[*Editors' Note:* Material has been omitted at this point.]

28

Reprinted from *Geol. Mag.* **9**:554–557 (1902)

The Term 'Hemera.'

By S. S. Buckman, F.G.S.

NEARLY ten years ago I proposed the term 'hemera.'[1] The term is coming into use; but, so far as I am aware, in the majority of cases it is greatly misunderstood. From this cause arise objections as to its use, such objections being based on the misconception that the hemera is a subdivision of a zone. It is nothing of the kind.

The original description may be quoted. "It is for a palæontological purpose that I propose the term 'hemera.' Its meaning is 'day' or 'time'; and I wish to use it as the chronological indicator of the faunal sequence. Successive 'hemeræ' should mark the smallest

[1] "Bajocian of the Sherborne District": Quart. Journ. Geol. Soc., 1893, vol. xlix, p. 481.

consecutive divisions which the sequence of different species enables us to separate in the maximum developments of strata. . . . . The term 'hemera' is intended to mark the acme of development of one or more species. It is designed as a chronological division, and will not therefore replace the term 'zone' or be a subdivision of it, for that term is strictly a stratigraphical one."[1] "The proposed term 'hemera' is used in a chronological sense as a subdivision of an 'age.'"[2]

These statements should be definite enough. They show that while there were the stratigraphical terms 'zone,' 'stage,' and the chronological term 'age' for the time of a 'stage,' there was no chronological term to denote the time of a 'zone.' So 'hemera' was proposed to supply the deficiency. 'Hemera,' 'age' are the time terms; 'zone,' 'stage' denote the amount of work done in the way of deposition during these times; they are the stratigraphical terms. As I said, a hemera marks the acme of development of one or more species, as a chronological term. Just as a 'day' marks the phenomenon, sunrise, high noon, sunset, or, if one likes to put it so, from sunrise to sunrise again, so a 'hemera' was designed to mark the time from species-rise to species-rise—that is, from the time when one species or set of species becomes dominant to the time when another species or set of species does so. Such time intervals were to be termed 'hemeræ.' And so many 'hemeræ' make an 'age,' just as so many 'days' make a 'week.' However, as 'hemera,' 'age' must be of universal application and always have existed, 'zone,' stage' may be absent locally: either there may have been no strata deposited, or the strata may have been destroyed. At one locality may be seen the strata of the zones A, B, C, representing the work done during the hemeræ A, B, C. At another locality may be observed only the strata of zones A and C. One may rightly say that zone B is absent; but one cannot say that hemera B is absent. Hemera B has been proved as a time-period between hemeræ A and C, by the records at one locality of work accomplished. Therefore, absence of records at another locality cannot show non-existence of time. Time is not local. A man cannot say that there was no Monday in the week because the parish clock happened to be out of order. Yet remarks similar to this, made in regard to hemeræ, are what I have to protest against. "Absence of any hemeræ would not indicate a break in the series of strata" is a sample. A Dorset labourer once came to his master and complained, "Sir, your dawg a bin an yet my fittle." He indicated exactly the agent of destruction. He knew that the deposit (of his dinner) in a particular place had been duly made. He was aware, by facts *subjecta fidelibus oculis*, what agent had been concerned in its removal. And he said so; in effect he said, "There is a break in the series of my dinners due to destruction by a certain agent." But had he employed the language which is used about a hemera he would have said, not "My dinner is gone," but "One o'clock is absent"—"The absence of one o'clock indicates a break in

[1] Op. cit., pp. 481, 482. [2] p. 518.

the series of my dinners." But one o'clock is just what cannot be said to be absent—dinner or no dinner one o'clock must arrive.

So to return to the localities mentioned above: what has to be said with regard to the second is that during hemera B either there was no deposit or there was denudation of a deposit, according as the evidence may warrant.

Since hemera has nothing to do with zone properly speaking, so it has nothing to do with subzone. It has been supposed that hemera is a sort of finer division of a subzone. But it is not. Strictly speaking, there should be just as many zones as there are hemeræ; for the hemera is the time during which a certain piece of work, namely, the deposition of what is called 'the zone,' was done. But it may be said that zones coalesce. I confess to a deal of scepticism about this; for, generally, careful work shows that very attenuated as the deposits may be, yet there are the zones in order of superposition characterized by their distinct faunas. When there is coalescence it is more of the lion and the lamb kind, with the lamb inside: the earlier zone in pieces in the later one. Such is the case in the Radstock district in the Middle Lias, where the *raricostatus* beds occur as pebbles in the *armatus* zone. Here the value of the term 'hemera' is seen. One cannot talk of the *raricostatus* zone at this place, because there is really no such zone; but one can say that the strata deposited during the hemera *raricostati* were broken up and redeposited during the hemera *armati*.

Much of the trouble about zones and hemeræ has arisen from attempts to make the term 'zone' a kind of 'portmanteau word,' one into which several meanings were to be packed. Thus, a 'zone' has been spoken of as a 'zone of life,' as a 'band of deposit,' as a 'portion of geological time.' A 'zone' has even been defined as an "assemblage of organic remains"—which would make the contents of a museum a 'zone.' Now in scientific work we cannot have several meanings packed into one word; and so to take away the portion-of-time idea the term 'hemera' was proposed.

With the life-zone idea there is more trouble. In zoology the zone indicates the horizontal extension of species; it might be required and used in this sense in geology; but the life-zone in geology has reference to the vertical range. Why cannot we call this a *biozone*, using the term to signify the range of organisms in time as indicated by their entombment in the strata? Thus we might have the biozone of a species, of a genus, of a family, or of a larger group. Thus the biozone of the Trilobita would be, say, from Cambrian to Carboniferous; the biozone of Ammonites would be equal to Mesozoic time; the biozone of an Ammonite family, Arietidæ, would about equal the time of the Lower Lias; the biozone of an Ammonite genus, *Coroniceras*, would be through two or three hemeræ; the biozone of Ammonite species would be about equal to a hemera.

Then we are left with zone in its stratigraphical sense, best defined by Mr. J. E. Marr. But we have no business to use it for this, because of the zoological meaning which the term already has.

Now we cannot separate the stratigraphical from the palæontological phenomena; the two go together. Zones are not bands of rock merely, for the strata of a zone may be heterogeneous vertically or horizontally. The zones are really so many entombments of organisms, which entombments are shown to be successive by the way in which they occur in the strata. For these successive entombments as they are exhibited in the strata, why should we not use a term *faunizone*? With distinct terms for the different meanings we shall express ourselves much more clearly.

To return to hemera, I have been asked if it is advisable when faunas differ, and when the index species is absent, to use different names as hemeral designators. I say that it is not advisable. Temp. Edward VII may date an event in Australia as well as in England, even though Edward VII never visits Australia. So hemera *Murchisonæ* will date a deposit in a locality where *Murchisonæ* never resided.

Certainly in historical records each nation has dated its events by the reigns of its own monarchs. Yet it would, in the interests of a universal system, be quite correct, from a chronological point of view, to say that certain events happened in Spain temp. Queen Elizabeth (of England). It is the aim of science to be universal and independent of nationalities. It is the aim of geology to have a universal chronological system, not to apply different and therefore confusing names to contemporaneous periods of time merely because of local faunal differences. And when, as in the Jurassic rocks, the sequence of faunal phenomena from Würtemberg to England can be accurately described by one set of hemeral terms, we may reasonably hope that, as our knowledge increases, if we take the necessary trouble, a universal chronology may be possible. And if that can be done in the Mesozoic rocks how much more may we not expect it in the Palæozoic, seeing that, as Dr. H. Woodward tells us, "the ancient faunas of the earth were far more widespread, more simple, and more uniform than are our recent faunas." [1]

To sum up, a *zone* indicates the horizontal extension of species, and so is a geographical term; a *biozone* is the range of any organism or group of organisms in geological deposits, so it may be said to indicate vertical extension; *faunizones* are, to paraphrase Mr. Marr, "belts of strata, each of which is characterised by an assemblage of organic remains," with this provision, that faunizones are independent of lithic structure—the strata of a faunizone may vary horizontally or vertically, or the strata may not vary and yet may show several successive faunas. So faunizones are the successive faunal facies exhibited in strata. Lastly, *hemera* is a time term—the subdivision of an 'age'; it indicates the period of time from the rise of one dominant species to the rise of the next. It is the time term which corresponds to 'faunizone' as the stratigraphical term, just as 'age' corresponds to 'stage.'

[1] Anniversary Address of the President to the Geological Society, Q.J.G.S., vol. li, p. lxxxvii.

Part V

CHAMBERLIN (1898) TO GRABAU (1936)

Editors' Comments on Papers 29 Through 32

29 CHAMBERLIN
The Ulterior Basis of Time Divisions and the Classification of Geologic History

30 CHAMBERLIN
Diastrophism as the Ultimate Basis of Correlation

31 ULRICH
Excerpts from *Revisions of the Paleozoic Systems*

32 ARKELL
Excerpts from *The Classification of the English Jurassic Rocks, and the Partition of Jurassic Time*

In these final selections, and in Grabau's (1936) *Oscillation or Pulsation,* we find reasons to believe that precise correlation of strata and precise boundary determination are possible.

In Chamberlin's two well-known and discerning papers of 1898 and 1909 (Papers 29 and 30), he discussed causes of marine regressions and transgressions, and although he did not (and could not) fully understand the nature of the processes operating between his ulterior (ultimate) cause and the time divisions and correlations, he did see a reason for the existence of natural chronostratigraphic units.

Chamberlin (Paper 29) attempted to correlate regressions and transgressions of the sea with periodic movements in the ocean basins, which increased their capacity, with base leveling of the continents and extension of shallow marine plains during periods of relative quiescence. He believed that during times of uplift, carbon dioxide would be consumed in the weathering of rock, causing cooling of temperatures and glaciation (with further lowering of the sea level, to be followed by glacial melting and rise of sea level). In Paper 30, he again expressed the view that "each great diastrophic movement tended toward the rejuvenation of the continents and toward the firmer establishment of the great basins" (p. 687). Now, with evidence for sea-floor spreading and multiple glacial periods, Chamberlin's ultimate basis for dividing geologic time stands greatly enhanced because both processes surely effect eustatic sea level change.

The two concluding paragraphs in Chamberlin's second paper (Paper 30) concisely summarize his belief that stratigraphy and paleontology "go hand in hand, each sanctioning the other. *Diastrophism lies back of both and furnishes the conditions on which they depend*" (p. 692).

In Grabau's paper on some New York formations (1905, 258), he pointed to the greater emphasis being put on the biological history of the earth with consequent slighting of the physical or stratigraphic history. Not that he believed paleontology received too much attention—far from it. He pointed to the fact that historical geology rests on a paleontological foundation, on the life succession and not on breaks in the record, and therefore that standard sections must be in continuous series of marine sediments rather than in the incomplete sections where they often have been placed. Grabau remarked that "I am not here to speak of the imperfection of the geologic record, an imperfection which I think is more apparent than real, nor of the imperfection of our classification, which is more real than apparent" (p. 529). He believed the neglect of the physical side of earth history was due to its "clumsy and unrefined" nomenclature. One new term that Grabau suggested "until a better term is proposed," was *disconformable* for the "relation where two formations . . . conform in their bedding but comprise between them a time break of greater or less magnitude" (p. 534). Grabau first used the term *disconformity* in 1913 (p. 821) although he said then that he had proposed it in the 1905 paper. As described by Grabau, his disconformity would encompass both the paraconformity of Dunbar and Rodgers (1957) in the case of the greater time break and the paracontinuity of Conkin and Conkin (1973) in the case of the lesser break.

Ulrich (Paper 31) expressed his belief in relatively rapid transgressions over the low Paleozoic continent, with tilting and oscillation of the continental platform causing fluctuating sea levels and widely shifting strandlines. Because of the relatively rapid nature of these Paleozoic transgressions, faunas throughout the extent of a thin marine stratigraphic unit were, for practical purposes, essentially contemporaneous. Ulrich, like Chamberlin, believed that "faunal, diastrophic, and lithologic criteria are complementary, and each kind may and should be used to check the other two. None of the three classes of evidence is competent by itself to insure a refined and true classification of geologic time, but between them we hope to construct one" (p. 290).

Ulrich (Paper 31) truly stated, "There are limits beyond which correlation can not go," but he agreed that fossils "give a reliable clue as to the boundaries of the major divisions. Finally, the principles of diastrophism fix these with greater precision and supply data on which correlations of minor rank may be based" (p. 291). Ulrich presented diastrophic criteria for stratigraphic classification on pages 394–433. In this discussion we note, in light of today's understanding of sea-floor spreading, Ulrich's use of the terms *oceanic thrusts* (p. 398) and *suboceanic spreading* (p. 401). He pointed (pp. 401, 402) to a fundamental fact:

> The only thing that moves apparently in the same way and degree, and which, therefore, offers the most reliable criteria in determining the periodicity and contemporaneity of diastrophic events, is the level of the sea. . . . whatever the qualifications, there yet remains the fact that the strandline is contemporaneously and universally displaced. [pp. 401, 402]

Ulrich believed that correlation "will be done by correlating negative displacements of the strandline on corresponding parts of the several continents" (pp. 410, 411), displacements which "are always connected with some body deformation of the earth" (p. 404). He divided diastrophic movements into two generalized and not sharply differentiated groups:

> (1) those that are active and relatively impulsive in operation, and (2) those that are relatively gentle in operation and much more gradual in the production of results. The former are chiefly responsible for the wrinkles of the earth's crust, and their function is primarily orogenic. . . . The movements of the second group consist chiefly of gradual displacements occasioned by temporary accommodation to slowly accumulating stresses . . . connected with the early accumulation phases of the stresses. [pp. 407, 408]

These two types of movements would normally result in different kinds of unconformities, ranging from large scale angular unconformities and nonconformities in his first case, down to the smallest that is of stratigraphic significance, the paracontinuity, in his second instance, one of which he illustrated in Figure 17C (p. 450), which exists at the boundary between the Middle and Upper Ordovician in Missouri.

Ulrich (Paper 31) discussed structural criteria for stratigraphic classification on pages 448–467. We note that he believed that on the Cincinnati Arch and Nashville Dome, basal arenaceous deposits "are almost wanting [and] overlaps are but seldom indicated by quartzose initial deposits" (p. 457). We have found that with regard to thick basal sands this is true, but very thin basal sands, overlying paracontinuities, although somewhat difficult to detect, are consistently present between genetic stratigraphic units within the Devonian and at the Devonian-Mississippian boundary.

In a similar vein to the previously cited comment by Buckman on Lyell (Arkell, Paper 32, 27, 28), Ulrich remarked, "Let us, then, seek to interpret geologic history according to the facts as we find them and not solely according to criteria suiting conditions prevailing today" (p. 458). Ulrich expressed his belief that Paleozoic seas were "very shallow and frequently shifting bodies of water. . . . Evidence of sea withdrawal has been discovered in many parts of the stratigraphic sequence where none had previously been suspected" (pp. 461-462). Ulrich presented various physical criteria of stratigraphic unconformity on pages 462-467.

In discussing correlation by diastrophic movements, Ulrich (Paper 31, 532-535) pointed out that "stratigraphic correlation, whether we use physical or biological criteria, or both, is a slow, laborious process, each method requiring constant corroboration by other methods, with no certainty of results until we have closed the circle and proved the process" (p. 533). Ulrich knew that diastrophic criteria "by themselves are utterly valueless in correlating widely disconnected outcrops" (p. 533). Nevertheless, he realized that "the most important boundaries" have been frequently overlooked; these important subtle boundaries are the paracontinuities that separate "transition sequences".

In his discussion of principles of classification, Ulrich (Paper 31) stated with respect to location of a boundary that "the fossils give us the general position; the lithologic criteria checked by fossil evidence narrows the limit of possible error and the diastrophic criteria [but not angular unconformities, nonconformities, or paraconformities] finally establish the exact location [defined by a paracontinuity] of the boundary sought" (p. 575). He believed:

> Whatever the cause of faunal change, be it great or small, local or general, it probably was occasioned by some crustal deformation, and movements of the lithosphere necessarily were accompanied by corresponding displacements of the strandline. . . .It is therefore by the correlation of these displacements . . . that we finally establish the sequence of marine deposits and define the stratigraphic units." [p. 577]

Ulrich distinguished stratigraphic terms, which refer to rocks that are preserved and known, from the time terms, which "completely fill geologic time" (p. 580). He continued, "Accuracy in the delimitation of time division . . . depends solely on our ability to correlate exactly the stratigraphic breaks in neighboring and distant basins" (p. 581). First, he advised, we must determine if the hiatus is small or great and how much of it is represented elsewhere and if the break is seen at the base of the most complete section of that age. Only then can we aspire to a geochronologic classification. He warned (p. 585) against acceptance of so-called transition beds, which in many cases include a break (paracontinuity in our terminology) or overlie one.

In discussing the principles and giving examples of introductory facies of faunas and deposits (pp. 590–593), Ulrich (Paper 31) stated that clastic beds that overlie breaks and truly grade upward should be included as the initial deposit of the overlying unit and that first appearance of species and faunas may mark the beginning of a new age, because the "new thing shows that something has happened" (p. 593).

As Buckman, Ulrich, and others have noted, there are no gaps in the time continuum even though there are many gaps in the geologic record. Arkell (Paper 32), following a discussion of stratigraphical and chronological terms, stated:

> One day, perhaps, it may become possible to attach our present chronology to an absolute time-scale, measured in fixed, definite units. Either we may gain some conception of the average duration in years of ammonite species, or we may discover the frequency of the pulse of earth-movements controlling the cycles of sedimentation (and so, indirectly, the life-succession). The realization of this ideal seems as far off as ever, but meanwhile it is something to perfect our shadow-chronology and have it ready, awaiting the time when the key will be supplied to interpret its relative conceptions in absolute terms. [p. vii]

While a detailed absolute time scale (geochronologic scale) may someday be possible, it is not achievable by classic stratigraphy alone, not even with the aid of paleontology; rather, the capability, and goal, of stratigraphy is correlation, and the essence of correlation is the recognition of discrete chronostratigraphic units that, once identified, are the unit blocks of the worldwide composite chronostratigraphic column.

At present, very few methods (perhaps only magnetic reversals, Oxygen-isotope ratios in carbonate rocks, Carbon-14, and pyroclastic intervals) are as applicable to precise chronostratigraphic determination (and thus to ordering of the chronostratigraphic record) as the paleontologic method based on organic evolution. In any case, the four methods cited (and any possible others) finally must be related to the organic evolution-based method by which the geologic column has been compiled and has reached the degree of perfection it has attained and to absolute time-lines (geochronologic boundaries) if more perfection of the chronostratigraphic scale is to be achieved. Only studies based on minute scrutiny of the paleontologic-stratigraphic context are presently universally applicable to the task of more refinement of the chronostratigraphic column. In fact, the limits of error in radiometric dating vary from 0.1 to 2 percent (Hedberg, 1976); thus, we have an average limit of 3 to 5 million years for a point in the Paleozoic and 1.6 to 3.2 million years in the Mesozoic. The fact is that we cannot delimit any moment in time finer than an average stage (assuming an average series is divided into three stages) in the Paleozoic, or possibly a substage in the Mesozoic, by means of radiometric dating. Oppel (1858) recognized thirty-three zones in the Jurassic, and this number has since been tripled, which would allot an average of about 570,000 years per zone, far greater accuracy for the location of a moment in time than that provided by radiometric dating, and compared to which the time needed for dispersal of a species is negligible. Thus, a given zone is everywhere essentially of the same age.

Arkell (Paper 32) restated Quenstedt's belief expressed seventy-five years earlier that:

> When geologists of every nation have studied, described, and redescribed the succession of rocks, and above all of the fossil faunas, in their own countries, bringing their correlations up to date with the advances of the sole international medium, paleontology, only then will the time be ripe for standardizing stratigraphical nomenclature. [p. 2]

Arkell advised that we "get to know our successions minutely, as the very first step in stratigraphical geology" (p. 36).

We have selected several passages from Arkell's discussion of Buckman's polyhemeral chronology (Paper 32) in which he (p. 29) considered the time element in the distribution of species over large areas, (p. 31) criticized the use of the acme of a species, and (pp. 34, 35) suggested the advantages of old zones or faunizones.

Our final quotation, from Grabau's (1936) *Oscillation or Pulsation,* gives us reason to believe that natural divisions of the geologic time scale can be drawn if we will study the rocks as we find them, applying physical and biological criteria to the solution of the problem. Grabau stated:

> But whether we consider that oscillation—that is epeirogenic movements—or orogenesis is of primary tectonic significance, the fact remains that both imply movements of the land masses.
>
> I believe, however, that the conviction is growing among stratigraphers that the great oceanic transgressions and retreats, which involve not only one section but all continents, are due primarily to rise and fall of the sea level, the continental movements being secondary or contributory factors, which guide the transgressions and locally accelerate or retard both transgressions and regressions. As our study progresses, the fact becomes more and more evident that transgression and regression are primarily 'hologeodic'—that is, universal so far as the earth is concerned—and that, moreover, they proceed in a rhythmic manner, regression following transgression and being in turn succeeded by renewed transgression. If it can be shown that such transgressions and regressions are essentially simultaneous in all

> parts of the earth, then they can be explained only by a periodic rise and fall of the sea level. It is this rhythmic pulsation, which can be shown to have taken place, that I would express by the term 'pulsation theory,' and though I have an ever-growing conviction of the correctness of this theory, I am content to propose it as a working hypothesis and to invite a rigid test by my fellow stratigraphers. [p. 540]

In this faith, Grabau (p. 552) apocalyptically stated:

> For the ultimate cause of the rise and fall of sea level I offer no explanation. If Joly's contention that it is produced by radioactive processes can be substantiated, it may serve as an adequate explanation; if not, and if the fact of pulsation is firmly established by future research, geophysicists will no doubt furnish us other explanations of its cause.

REFERENCES

Conkin, J. E., and B. M. Conkin, 1973, The Paracontinuity and the Determination of the Devonian-Mississippian Boundary in the Type Lower Mississippian Area of North America, *Univ. Louisville Studies Paleont. Stratigr.*, no. 1, 36p.

Dunbar, C. O., and J. Rodgers, 1957, *Principles of Stratigraphy,* Wiley, New York.

Grabau, A. W., 1905, Physical Characters and History of Some New York Formations, *Science* (n.s.) **22**(565):528–535.

Grabau, A. W., 1913, *Principles of Stratigraphy,* A. G. Seiler, New York.

Grabau, A. W., 1936, Oscillation or Pulsation, *16th Internat. Geol. Congr. Wash.,* Rept. 1, pp. 539–553.

Hedberg, H. D., 1976, *International Stratigraphic Guide,* Wiley, New York, 200p.

Oppel, A., 1858, The Jurassic System of England, France, and Southwestern Germany, *Wuerttemb. Naturwiss. Jahresh.,* Stuttgart, pp. 148–262.

29

Reprinted from *Jour. Geology* **6**(5):449–462 (1898)

THE ULTERIOR BASIS OF TIME DIVISIONS AND THE CLASSIFICATION OF GEOLOGIC HISTORY.

T. C. Chamberlin

IT was intimated in the introduction to the symposium on the classification and nomenclature of geologic time divisions published in the last number of this magazine that the ulterior basis of classification and nomenclature must be dependent on the existence or absence of natural divisions resulting from simultaneous phases of action of world-wide extent. If there have been such universal phases and if they can be detected, they must ultimately be accepted not only as the true basis of division, classification and nomenclature, but their exposition must constitute the major work of research and of instruction. The most vital problem before the general geologist today is the question whether the earth's history is naturally divided into periodic phases of world-wide prevalence, or whether it is but an aggregation of local events dependent upon local conditions uncontrolled by overmastering agencies of universal dominance.

That there were no universal breaks in sedimentation or in the fundamental continuity of life is not only admitted but affirmed without hesitation. The old doctrine of physical cataclysms attended by universal destruction of life has passed beyond serious consideration. And so, in the judgment of the writer, have all doctrines which attribute profound effects on the life of the globe or the progress of sedimentation to the violence of physical disturbances of any kind. That sedimentation has been in constant progress somewhere since the inauguration of the pre-Cambrian lands and seas, and that life has been likewise continuous and self-derivative, may be accepted as fundamental postulates. If, therefore, we seek for *absolute* divisions we doubtless seek in vain. But this does not dismiss the question whether this continuity of physical and vital action proceeded by heterogeneous impulses or by correlated pulsations. If the latter, then the history of the earth, when deciphered, will assume a rhythmical periodicity susceptible of natural classification and of significant and rational nomenclature; if the former, the contradictory phases of local actions will inhibit all but the most general unity and render classification and nomenclature either arbitrary or provincial.

I venture to urge three general grounds upon which I entertain the former view. These grounds, if valid, hold out the hope that the history of the earth will be found not only susceptible to natural division, but capable of its truest exposition only through the recognition of its inherent periods.

I. The first of these grounds is the presumption that great earth movements affect all quarters of the globe. Minor stresses may find relief in local readjustment, but profound stresses cannot be relieved, it is assumed, without generating appreciable stresses in other portions of the globe and leading to general readjustment. In a globe, all of whose parts owe their positions to the stress and tension of other parts, every rearrangement that rises in magnitude above the limits of local support extends its influence to the whole. Any massive earth movement must change the gravitative stresses of all parts of the globe unless the movement be divided into contrary phases so adjusted as to be compensatory, the possibility of which in the strict sense may be questioned. The recognized causes of profound movements such as secular refrigeration, change of speed of rotation, progressive molecular rearrangement, and like agencies, are comprehensive in their action and accumulate general stresses whose natural issue is coöperation in a common movement of relief.

The validity of this presumption of general coöperative movements will perhaps not be so much questioned as the mode of their execution. There are those who believe that a downward movement in one region is correlated with an upward movement in some other region. The correlated movements have, therefore, opposite phases and if the distribution of these is not controlled by some unifying agency the general terrestrial effects are heterogeneous, or, if not that, at least uncertain. This view is the natural sequence of the doctrine of a thin, floating crust warping to satisfy its own changes of density and tension, and wrinkling to adapt itself to a shrinking nucleus. Accepting the truth that lies under this view, but rising to a broader generalization, there are those who entertain the conception that the depressions of the earth's surface habitually became more depressed with every readjustment to smaller dimensions (local exceptions aside), while the protuberances became more protuberant. In other words, the oceanic basins became progressively deeper and more capacious, while the continents became higher (degradation aside). In this assumption of habitual downward movements of the ocean bottoms and of correlative upward movements of the continents, there lies, if it be true, a basis for the natural division of geologic events, these movements being in themselves and in their immediate consequences the basis of such division. The full establishment or overthrow of this assumption must await the extension of critical research to at least the major part of the earth, and it is not the purpose of this paper to seriously attempt its advocacy by the citation of the evidence already gathered in its support. Incidentally it will be touched upon in the discussion following.

II. The second ground of belief in a fundamental periodicity in terrestrial progress is founded on the conviction that the major movements of the earth's surface have consisted of the sinking of the ocean bottoms and the withdrawal of additional waters into the basins whose capacities were thereby increased. This belief is not quite identical with the assumption last made, for it does not necessarily involve the simultaneous action of the different ocean beds, although the conclusions about to be urged would be strengthened if such common action could be demonstrated. But quite apart from this, it is believed the following argument rests upon rather firm observational and inductive grounds except in the matter of two fundamental postulates which are almost universally assumed by geologists. These are as follows: (1) It is assumed that the earth was at first a nearly perfect spheroid, the surface being essentially plane. (2) It is assumed that the great movements of the earth's crust have consisted fundamentally of shrinkage. Probably no serious geologist maintains that the earth has enlarged its average diameter during geological history by expansive action, whatever he may hold respecting local expansion.

These two propositions being accepted, it follows that the radial shrinkage of the ocean bottoms has surpassed the radial shrinkage of the continental platforms to the average amount of some 10,000 or 12,000 feet. This excess of radial shrinkage is to be multiplied by four to measure the excess of volumetric shrinkage, since the area of the ocean bottoms is about four times the area of the continental platforms. The master factor, therefore, in the surface movements of the earth has been the sinking of the ocean bottoms and the formation of the great oceanic basins. Most geologists will probably agree that the continental platforms have also sunk, in the sense that they have shortened their radial distances from the center of the earth. Opinion seems to be divided, however, on the question whether there have been actual epeirogenic uplifts or not. Probably most geologists would regard the rising of the Thibetan plateau in late Tertiary times as involving an actual increase of radial distance. Probably very few geologists would question the absolute elevation of the crests of the loftier mountain ranges. However, the question of absolute as distinguished from relative sinking does not seriously affect the question in hand. If the earth has absolutely grown smaller by some notable amount the average ideal ocean has, as a consequence, grown deeper (if its volume has remained constant), for its circumferential expanse has been reduced. Aside from this, so far as I can see, our argument holds as firmly for relative as for absolute sinking.

From the greater depths to which the ocean bottoms have sunk, the presumption follows that in every great crustal readjustment the major factor consisted of the descent of the ocean bottom or some part of it. Logically, as here stated, this is only a presumption which might be set aside by the assumption of a single or a few great depressions, while the other movements might be upward or indifferent, but this will appear less tenable in the light of further considerations.

Not only has there been increase in depth, but increase in capacity also. From a capacity essentially zero at the outset the basins have developed a capacity sufficient to hold nearly all the water of the globe. In the aggregate, therefore, the capacities of the ocean basins, as well as their depths, have been increased by crustal readjustment, and the presumption is that this has usually been the case in individual readjustments, although this does not rigorously follow. It will be sustained, however, by further considerations. The crustal readjustments here referred to are those resulting from internal causes. External readjustments work to precisely opposite ends, the degradation of the land and the filling of the basins. This opposing action strengthens the presumption that the internal causes have habitually increased the capacity of the basins, for they have grown more and more capacious in spite of this constantly opposing action. This constant filling in affords a presumption of frequently repeated increases of capacity; otherwise the land should have disappeared.

Proceeding upon the presumption that internal readjustments habitually increased the capacity of the ocean basins, it is important to note in detail the consequences that follow. These are involved in the functions of the circumcontinental terrace, and will be more easily followed after an explicit statement of these functions. These are more or less fully apprehended by all acute students of continental evolution, but like the correlative functions of baseleveling previous to the explicit exposition of Powell, they have not come into that large service as working principles of which they are susceptible.

Every continent which stands in a given position with reference to the sea for any prolonged period develops a submarine terrace about its borders. This is formed from the débris of the land deposited beneath the edge of the sea. In its initial stages it is nothing more than the familiar shore terrace; but as it develops it becomes a broad submerged platform with a steep face dropping away to the abysmal depths of the ocean. The submerged platform has its outer limit at the depth at which detritus can be effectively moved off shore by the agitation of the surface waters. This, though varying with conditions, may be roughly taken to be one hundred fathoms. The breadth attained by the upper surface of the terrace is conditioned upon the length of time the continent remains in a fixed attitude and the activity of land wash. Simultaneously the sea cliffs are moving inland, and the valleys are developing base plains which are the correlatives of the terrace plain which is growing seaward, as illustrated in Fig. 1.

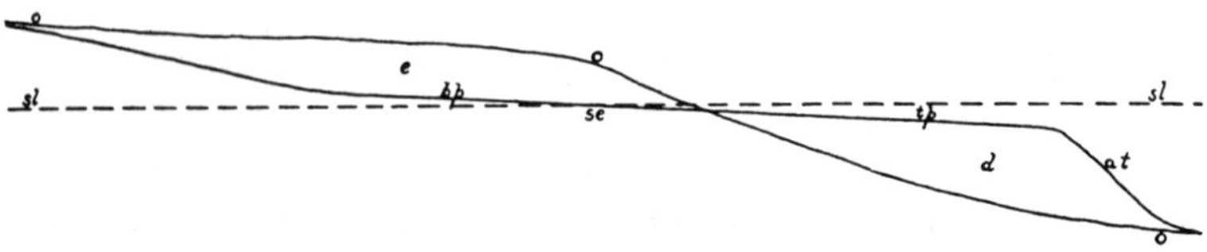

FIG. 1.—*o-o* Original surface. *s l* Sea level. *e* Land carried away by erosion. *d* Detritus built into circumcontinental terrace. *t p* Terrace plain. *a t* Abysmal terrace face. *b p* Base-plain developing landward.

The extreme limit of development is attained when the continent has been baseleveled and no farther detritus is furnished for the extension of the terrace. The baselevel of the continent then becomes essentially continuous with the submerged terrace surface, and the whole constitutes the perfected continental platform, as shown in Fig. 2.

The development of the circumcontinental terrace and of the perfected continental platform is subject to intercurrent disturbances from local and from general sources. There appear to be two systematic sources of slight but very critical modification that require special consideration.

FIG. 2.—*b p* Continental base-plain. *s l* Sea level. *t p* Submerged terrace plain. *a t* Abysmal terrace face.

1. The transfer of débris from the land to the sea displaces an equivalent amount of water, and raises the sea level proportionately, and causes an advance upon the land. The effects are volumetrically small compared with the great body of the ocean, but a slight rise in the surface as the baselevel stages of the continent are attained is peculiarly effective. This coöperates with the cutting back of the sea cliff, and, combined, they become effective in advancing the edge of the sea upon the border of the land.

2. There are both theoretical and observational grounds for the belief that in the process of periodic readjustment of the earth to its internal stresses, portions of the crust are thrust up to heights notably above the plane of isostatic equilibrium, and that these portions gradually settle back toward equilibrium by virtue of the slow fluency or quasi-fluency of the rocks. Recent pendulum studies by Putnam and Gilbert seem to indicate that the portion of our continent most notably lifted in late Tertiary times still stands appreciably above isostatic equilibrium, and there is little doubt that the same is true of other continents, as is, indeed, indicated by partial pendulum data. There is, however, a large mass of concurrent data which shows an aggregate subsidence of the continent since late Tertiary times, data which have been industriously marshaled in the interests of an epeirogenic explanation of the glacial period. This leads to the impression that in late Tertiary times, when the upward move-

ment reached its maximum effects, the land stood very notably above isostatic equilibrium, and that it has been settling back, but has not even now reached isostatic equilibrium. While the generalization cannot be rigorously established, there seem to be sufficient data to warrant entertaining tentatively the doctrine that in periods following crustal upheavals which pass beyond the plane of equilibrium the lifted portions slowly settle back toward equilibrium. If so, this retrocession would coöperate with the filling of the basins in causing an advance of the sea upon the land. At the same time the conditions for the seaward growth of the terrace plain may still continue and the plain be thus simultaneously extended on both borders.

As already noted, the evolution of this peri-coastal plain is subject to interruptions and local modifications to an extent comparable to the interferences in the development of a base-plain, and perhaps to a greater degree, but I think it has like claims to acceptance as an effective general process.

Now the development of such submerged terraces around the several continents for any given period is accurately correlated by the sea level. They are all built immediately beneath its border at a common level. The continental baselevels are correlated by the same controlling horizon. So, necessarily, the final continental platforms are likewise reduced to the same common natural datum plane.

If, therefore, it be admitted that there are periods of general quiescence, it follows that there are periods of simultaneous platform-making just below and just above the sea level on all continents. And this is accompanied by an inevitable tendency of the sea to advance upon the land. Now this submerged sea shelf is the special zone of sedimentation, and hence it is the peculiar locus of registration of geologic events. It is at the same time the peculiar habitat of shallow-water marine life, and this is the life which specially enters into the geologic record. We know almost nothing of the ancient abysmal life and relatively little of the land life. Both the physical and the biological record, which are our chief dependence in reading the earth's history, are therefore made upon the surface of the peripheral terrace and of its inland extension, and hence this becomes preëminently a critical geologic zone.

To follow out the sequence of a typical cycle, let it be supposed that a circumcontinental submarine platform of ample dimensions has been developed, and that it is peopled by a fauna comparable to its extent and resources. It has been suggested that a typical crust movement has for its major feature the depression of the sea bottom and an increase in the capacity of the basin. Let such a movement succeed. The effect of this, whether it involves one ocean basin or all, must be the withdrawal into itself of water from the submerged platforms of all the continents alike, since the oceans are connected. If the basin movement has sufficient magnitude to draw down the sea surface to the terrace edge, the shallow water zone becomes narrowed to a mere strip on the rapidly shelving abysmal face of the terrace, as illustrated in Fig. 3.

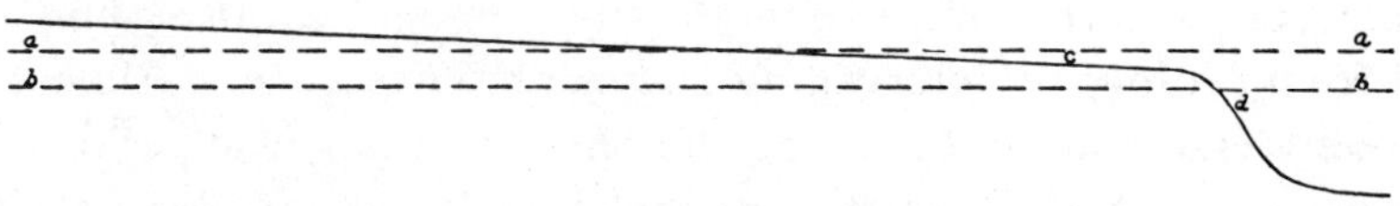

FIG. 3.—*a* Former sea level. *b* Succeeding sea level. *c* Former ample shallow-water tract. *d* Succeeding constricted shallow-water tract.

The ample fauna of the previous broader tract is thus forced into the constricted zone and brought under the direst stress of competition and scant room. The destruction of the larger part is inevitable, and the residue is forced to undergo repressive evolution to meet the severe conditions of the new environment. As this is common to all continents, it constitutes a comprehensive evolution of the severely competitive phase. There would, to be sure, be such exceptions as the local variations from the typical configuration of the continental border afford. These might be very considerable. Portions of the continents may have been previously carried down to moderate depths in the sinking of the ocean basins and may become shallow water ground by the lowering of the sea level. But even then there would remain a community of dominant action that would give a decisive aspect to the progress of life and to the sedimentation on all continents alike.

If now a long period of quiescence follows, a new universal terrace will begin to form and will extend its marine plain seaward and its baselevel inland until at length an ample zone for the evolution of a new shallow water fauna is provided. If to the cutting of the sea edge and the filling in of the sea basin there be added the settling of the continent, the sea may make a wide incursion upon the low parts of the land, as it did in Cretaceous times, and unusual facilities be thus afforded for that form of life-evolution which follows rich and genial conditions.

Thus on the one hand the sinking sea bottom induces that form of evolution in which stress is the dominant factor, and on the other, quiescence induces that form of evolution in which new ground and rich opportunities constitute the dominant condition. Both of these follow simply and inevitably from the sinking of parts that have been already predominant in sinking, and from prolonged intervening stages of quiescence. Almost the only essential postulate of the one evolution is a periodic increase of sea basin capacity; of the other, periodic quiescence. No profound catastrophe is involved; rather on the contrary it is inhibited by the conditions postulated.

Both the evolution of restrictive environment and the evolution of expansive environment, in the opinion of the writer, are effective in the change of faunas, though their respective results may be as different as their modes. In the rhythmical action postulated there is an alternating application of these opposed evolutionary processes with the natural result of an effect of the maximum order; for the evolutionary effects of restrictive conditions are believed to reach their greatest magnitude when they follow conditions of expansion, and, reciprocally, expansive conditions realize their greatest results when they follow conditions of restriction.

Such a succession of shallow sea incursions and withdrawals

reciprocating with crustal movements and quiescence seem to me to be well indicated as the master features of geologic progress from the beginning of the Palæozoic era to the present time. To these features I look for the primary terms of a natural and permanent system of classification and nomenclature.

III. The third agency which affords some promise of becoming a means for strict correlation of transoceanic events and for the division of these events into their natural epochs is an assumed fluctuation in the constitution of the atmosphere. Too little has yet been learned by direct induction respecting the nature of the successive atmospheres of the geologic periods to render this a firm ground for conclusions, but I venture to invite attention to the doctrine enunciated some time ago[1] that the exposure of the crystalline areas to the action of the air necessarily led to changes in the constitution of the atmosphere, especially in the critical element of carbon dioxide. The principle was urged that the greater the exposure of the decomposable crystalline rocks in area and in elevation, by leading to wider contact and deeper penetration of the atmosphere and atmospheric waters, the more rapid must have been the decomposition of the crystalline rocks and the consequent consumption of carbonic acid in the carbonation of the alkalis and alkaline earths, which is the most important part of the decomposing process. This greater exposure obviously followed the crustal readjustments, for at these times the land was largest and highest. It then not only exposed the greatest surface to atmospheric contact, but the atmospheric waters penetrated deepest because of the hydrostatic pressure arising from great differences of water level. At times of approximate degradation to baselevel and of sea-border encroachment the area of action was reduced and the power of penetration of the atmospheric waters became slight because of the low elevation and consequent slight differential pressure. In a word, the consumption of the carbonic acid proceeded rapidly at times of broad and

[1] A Group of Hypotheses Bearing on Climatic Changes. JOUR. GEOL., Vol. V, No. 7, October-November 1897.

high elevation of the land, and slowly at times of low altitude, grand averages being always understood.

If the atmosphere were once excessively burdened with carbonic acid and its later history has been merely a progressive depletion, these stages of rapid consumption only introduced specially rapid reduction of the superabundant supply, and the effects on tangible geological processes may have been quite beyond detection. But if, on the other hand, the atmosphere was limited in amount at the beginning and has been gradually supplied as well as gradually consumed throughout the ages and has been susceptible to serious change, an unusually rapid consumption of the carbon dioxide at the stages of land elevation would result in appreciable depletion of the atmosphere unless the supply were correspondingly increased. On the other hand, at those stages in which the continents were reduced well toward sea level and the land areas were diminished by the incursion of the ocean, the consumption of the carbonic acid would be checked, and if the supply were not correspondingly reduced, reënrichment in carbonic acid would follow. Under this hypothesis, the history of the atmosphere involved alternate enrichment and depletion.

The carbon dioxide is critical because of that peculiar thermal capacity by virtue of which it retains the heat of the sun to a relatively extraordinary degree, a capacity which is shared by water vapor, but which is possessed in very low degree by oxygen and nitrogen. The amount of aqueous vapor, however, is dependent upon temperature, while the carbon dioxide is stable and active at all terrestrial temperatures. Whenever, therefore, there is a notable percentage of carbon dioxide in the atmosphere, it performs a most important function in conserving the heat of the sun and raising the temperature of the lower atmosphere and of the earth's surface. By this rise it increases the aqueous vapor in the atmosphere, which in turn aids the carbon dioxide in retaining the heat of the sun, the two acting conjointly. On the other hand, when the carbon dioxide is reduced to a small factor, the heat of the sun is less effectually retained at the surface of the earth, the water vapor enters less into the atmosphere and low temperature and aridity are the consequences.

If these considerations are valid, the history of the earth has been marked by periods of relative cold and aridity resulting from stages of rapid rock disintegration, alternating with periods of warmth and moisture correlated with periods of limited rock disintegration and of carbonic acid accumulation. These stages are genetically connected with periods of continental elevation and rapid subaërial degradation, on the one hand, and with slight degradation and sea incursion, on the other. It will be observed that continental elevation as a purely topographical condition contributes to cold and aridity, while continental degradation correlated with oceanic extension contributes to equalization of temperature and to warmth. We have, therefore, the conjoint action of topographic agencies with atmospheric constitution in producing alternations of cold and aridity with warmth and moisture. The aridity is thought to express itself in salt and gypsum deposits and in the red sediments with which these are habitually associated; the cold, in glaciation; the warmth and moisture, in the polar extension of tropical life.

Now these atmospheric influences are strictly simultaneous for all parts of the globe, latitudinal effects, of course, being neglected, for the diffusion of the atmosphere is such as to render its constitution practically uniform for all parts of the globe. In so far, therefore, as atmospheric conditions of a constitutional nature affect the progress of terrestrial phenomena, they affect them universally, and if these influences are pronounced and can be identified they furnish an additional basis for the strict correlation of transoceanic action and for the division of geological history into its natural epochs.

In summation, therefore, I rest in a somewhat confident hope that under continued study adequate natural bases for the more important divisions of geologic time and for a stable and fitting nomenclature will be found (1) in simultaneous internal readjustments alternating with intervals of relative quiescence, (2) in the periodic development and emergence of circumcontinental terraces and their critical effects on the evolution of life, and (3) in the successive depletions and enrichments of the atmosphere. For subdivisions of lower order the migration of faunas and the special features of continental development will furnish appropriate bases, and below these again, the local phases of sedimentation and faunal adaptation will afford the provincial terms of a natural classification. If this hope be well grounded, the arbitrary divisions that now vex our system may be largely eliminated.

30

Reprinted from *Jour. Geology* **17**(8):685–693 (1909)

DIASTROPHISM AS THE ULTIMATE BASIS OF CORRELATION

THOMAS CHROWDER CHAMBERLIN
The University of Chicago

There are many and diverse views relative to the nature and the causes of diastrophic movements. To keep as largely as may be on common ground, most of these divergencies of view may be set aside as immaterial to our present purpose. We may all agree that the fundamental factors of the case are a lithosphere with a deformable surface, a liquid, covering part of this surface and determining erosion and sedimentation, and a gaseous envelope. We may easily agree that the outer part of the lithosphere is solid and has a sufficient measure of rigidity to maintain the surface inequalities. I do not see that we need to agree as to the causes of deformation. In some sense, I do not see that we need even to agree as to just what the absolute movements were, i. e., I do not see that it is material for us here to know whether the deformative movements were shrinkages, or expansions, or lateral shifts, provided we agree as to the general nature of their effects on the agencies at work on the surface of the lithosphere. We do not need to entertain the same conception of the nature of the earth's interior, if we are at one as to the working conditions which have prevailed on its surface.

No doubt we can easily agree on the present great working factors: (1) abysmal basins occupying about two-thirds of the earth's surface, bordered by terrace faces rising at angles of 2° to 5° for say 12,000 feet to a quite definite terrace-angle about 100 fathoms below the sea-level; (2) continental platforms whose upper faces slope gently up from this angle to the coast-line and thence ascend into the various reliefs of the land. If we thus agree that the upper face of the continental platform is bounded by the edge of the continental shelf, and that this edge is equally the boundary of the abysmal basins, whether the waters overlap the edge or not, we may also agree that the edge of the oceanic waters, whether they agree with the edge of the abysmal basins or not, form the chief line of demarkation between the great erosions and the great depositions the world over. It is not the only line of such demarkation, to be sure, for degradation gives place to aggradation at many other local horizons, but in this discussion let us agree to deal only with factors of the larger order and to neglect incidentals; let us deal with body deformation, rather than local or provincial warpings. We all recognize further that the sea-level is not only a dividing plane between two great divisions of physical agencies, but between two great biological divisions.

To this list of agreements, there are two other propositions which we cannot add quite so unhesitatingly, because we need to weigh them well, and if we cannot all agree respecting them, we must agree to differ, for they are fundamental to the further discussion. These relate to the effects of body deformation on the relations of land and sea.

If deformation were confined to the abysmal bottoms and were compensatory, no effect would be felt on the relations of land and sea. If deformation were confined wholly to the interior of the continents, it would be similarly ineffectual. Deformations so limited are, however, likely to be only provincial, and fall outside our discussion.

There remain two conceptions of general or body deformation between which choice must be made. In the one, the deformations are supposed to be indifferent to their predecessors, and to disregard the configurations produced by previous deformations. Their successive effects upon continental outlines and basin capacities are thus heterogeneous and the combined results irregular and uncertain. It is not clear to me how they can be made a very trustworthy basis of systematic correlation. The submergent phase of one continent or fraction of a continent may, in this case, be contemporaneous with the emergent phase of another continent or fraction of a continent, and the progress of events on one continent is as likely to be contrasted with those of another continent as to fall in with them co-ordinately.

According to the other view, deformations are inheritances, one of which follows another in due dynamical kinship. The succession is therefore homogeneous and the results co-ordinate. If, for example, the first depression of the abysmal basins was due to the superior specific gravity of the basin-bottoms, this specific gravity remained and participated in the next deformation. If the continental masses, at the outset of continental formation, were relatively low in specific gravity, this low specific gravity was handed down to later periods and helped to renew deformation of the same phases in the same regions. Under this view, ocean basins and continental elevations tended toward self-perpetuation. It is not assumed that this prevented shell crumplings, provincial warpings, or block movements of diverse phases within the continental or the abysmal areas, for these might obviously be necessary effects of the general deformative movements, or at least inevitable incidents connected with the dynamics lying back of them.

A choice between these two conceptions is imperative to this discussion, as they lie at the parting of the ways in the interpretation of the larger events of geologic history. I accept the second view with much confidence. It should be more fully qualified respecting the incidental accompaniments just mentioned, but time does not permit.

According to this view, each great diastrophic movement tended toward the rejuvenation of the continents and toward the firmer establishment of the great basins. The distinction between continent and basin must not, however, be interpreted on the superficial ground of the water-line, for the water-line merely shows that the basin is over-full, just full, or under-full, as the case may be. The average water-line undoubtedly helps to give a definite terrace border to the abysmal basin, but the water-line freely abandons this and often is far from coinciding with it.

The base-leveling processes have shown that they are able to lower the continents approximately to the sea-level in a fraction of geologic time. The continents would therefore have long since disappeared, if they had not been rejuvenated by renewed relative elevation or the withdrawal of the sea. I am able to find no evidence of lost continents. There are submerged margins, and matter has been carried continent-ward from denuded borders. There are some submerged dependencies and inter-continental connections. There are also some rather deeply submerged ridges that probably connected the present continents at remote stages in their history. In the earlier eras, when the differentiation of platforms and basins was less advanced, ridges which have since been submerged are perhaps recognizable. In the interpretation of the earlier periods, these should probably be restored as continental connections. In the earliest known ages, these may have been rather numerous and their combined area considerable, but these seem to me to be only qualifying features which, by the natural place in evolution which they fill, support, rather than weaken, the general conception of a systematic succession of deformations in which the offspring of each is the parent of the next, and in which both continents and ocean basins were progressively segregated and unified.

I trust that many of you will agree that, in general, the relatively upward movements of diastrophism have been located continuously in the continents, and the broad downward movements continuously in the ocean basins, and that, setting aside incidental features, the dominant effect of the successive diastrophic movements has been to restore the capacity of the ocean basins and to rejuvenate the continents. This conclusion seems to me to be strongly supported by the general course of geologic history, wherein sea-transgressions and sea-withdrawals have constituted master features. Perhaps our firmest ground for this conviction is found in the present relations of the continents and the sea basins. If heterogeneity had dominated continental action in the great Tertiary diastrophisms, the results should stand clearly forth today. Some continents should show recent general emergence, while others should show simultaneous general submergence. The dominant processes today should be those of depressional progress, on the one hand, and those of ascensional progress, on the other. As a matter of fact, all the continents are strikingly alike in their general physiographic attitude toward the sea. They are all surrounded by a border-belt, overflowed by the sea to the nearly uniform depth of 100 fathoms. These submerged tracts are all crossed by channels, implying a recent emergent state. None of the continents is covered widely by recent marine deposits, and yet all show some measure of these. Wide recent transgressions in one part do not stand in contrast with great elevations in another. Even beyond what theory might lead us to expect, when we duly recognize the warpings incidental to all adjustments, the recent relations of the continents to the seas conform to one type. The 10,000,000 square miles of continental margin, now submerged, is distributed around the borders of all the continents with a fair degree of equability. May we not, therefore, agree that in the world-wide phases of diastrophic movements, the basins have been additionally depressed and the continents repeatedly rejuvenated.

It is important that we should agree, or agree to disagree, on one further point. Have diastrophic movements been in progress constantly, or at intervals only, with quiescent periods between? Are they perpetual or periodic? The latter view prevails, I think, among American geologists. This view has acquired especial claims since base-leveling has come to play so large a part in our science, for it is clear that the doctrine of base-leveling is specifically inconsistent with the doctrine of perpetual deformation, for the very conditions prerequisite to the accomplishment of base-leveling involve a high degree of stability through a long period. The great base-levelings, and the great sea-transgressions, which I think are little more than alternative expressions for the same thing, have, as their fundamental assumption, a sufficient stability of the surface to permit base-leveling to accomplish its ends. Shall we not therefore agree that there has been periodicity in the world-warping deformations? Let this not be held with such exclusiveness that we fail to recognize duly the effects of the adjustment of minor stresses, at other times. These may be preliminary or after-effects of the larger movements, or they may be due to local stresses more or less independent of the general body-stresses. These quite certainly have been present, and have produced intercurrent departures from the strict tenor of the great systematic movements.

If there is need for additional argument on periodicity, it may be found in theoretical considerations, but these we have tried to avoid in the main. Whatever we may regard as the fundamental agencies that give rise to those stresses in the earth which are precedent to deformations, we may easily all agree that the earth opposes some resistance to deformation. There is certainly some rigidity in the body of the earth. According to the fundamental laws of rigidity, the deforming stresses must reach a certain magnitude before a movement can start. Now, if we recall that every such deformative movement, affecting a free surface, in its very nature, throws the resisting crust into an attitude of relative weakness, it follows that, with such progressive easing, the movement will go on until the stress is accommodated and a state of equilibrium essentially restored, after which another period of accumulation is prerequisite to another movement.

If we are agreed on the periodicity of great deformations, it clearly follows that in a quiescent state the base-leveling of the land means contemporaneous filling of the sea basins by transferred matter, and hence a slowly advancing sea-edge which is thus brought into active function as a base-leveling agent. This water movement is essentially contemporaneous the world over, and is thus a basis for correlation. *The base-leveling process implies a homologous series of deposits the world over.* At first these represent the conditions immediately following continental rejuvenation. Later they are succeeded by the deposits representing the modified conditions to which the first stage gives rise, and so on through the series up to the climacteric ones when base-leveling has reached its greatest development. After this a declining series follows. The deposits of the more advanced stages of base-leveling are, as now well recognized by most American geologists, markedly different in physical constitution and physiographic aspect from those of the earliest stages of continental rejuvenation. The criteria for discrimination between these earlier and the later members of the series are indeed of the collective rather than the individual type; they have character as distinctive assemblages of criteria rather than as single or isolated criteria, but they are perhaps all the safer for this composite character.

Correlation by base-levels is one of the triumphs of American geology; correlation by its complement, transgressive deposits on a base-level, may easily be added, and perhaps on quite as firm or even firmer physical grounds. If we add the biological element the case is immeasurably strengthened, for correlation by cosmopolitan faunas, the very best of faunas for the purpose, is added to the physical correlation. Migration at the climax of base-leveling and sea-transgression is freer and more prompt than at other times. Correlation to the foot, as by an unconformity, may not be practicable, but the precision of correlation by unconformities has more apparent than real value, for the different parts of the same unconformity vary much in time. All distant correlations involve some measure of inexactness, and the more frankly it is made obvious, the less its liability to mislead.

Correlation by general diastrophic movements takes cognizance of four stages: (1) the stages of climacteric base-leveling and sea-transgression, (2) the stages of retreat which are the first stages of diastrophic movement after the quiescent period, (3) the stages of climacteric diastrophism and of greatest sea-retreat, and (4) the stages of early quiescence, progressive degradation, and sea-advance.

(1) The characteristics of the climacteric stage of base-leveling and sea-transgression need little further characterization here, for the function of base-levels is known to all American geologists and the function of great sea-transgression to every stratigrapher and paleontologist. We have in base-leveling conjoined with sea-transgression, just that combination of agencies which is competent to develop the broad epicontinental seas of nearly uniform depth requisite for an expansional evolution of shallow-water life. At the same time, it furnishes broad pathways around and across the continental surfaces for wide migrations and the comminglings that lead to cosmopolitan faunas of the shallow-water type.

(2) The stages of initial diastrophism and sea-retreat find their criteria in the deposits that spring from an increased erosion of the deep soil-mantles accumulated in the base-level period, in the effects of increasing turbidity, in the lessening areas suitable for the shallow-water life, and in the limitation of migration.

(3) The climacteric stages of diastrophism are marked by the stress of restrictional evolution among the shallow-water species; by increased clastic deposition in land basins, on low slopes, and on sea

borders, by great land extension, but often, perhaps dominantly, by diversity of land surface and by liability to climatic severities and diversification. Areally, land life is favored, but it is hampered by the climatic and topographic diversities, and these may prove graver obstacles to migration and intermingling than even the tongues of sea that previously traversed the land surface. Correlation by glaciation in these stages is likely to prove a valuable adjunct, but we must first test our criterion, for we are not as yet quite sure that contemporaneity of glaciation is inferred on reliable grounds. The shallow-water life of the diastrophic stages is driven into narrow border tracts and into local embayments, and is thus forced into special adaptations and into narrowly provincial aspects.

(4) The early stages of quiescence and of base-leveling, with advancing seas, are peculiarly fruitful in biological criteria, for they are marked by re-expansions of the narrowly provincial shallow-water faunas of the previous stages. The progressive development of these provincial faunas and their successive unions with the faunas of neighboring provinces, as these come to coalesce by means of the progressive sea-advances, form one of the most fascinating chapters in life evolution, and give some of the most delicate of criteria for correlation.

This rough outline is quite too meager duly to set forth the criteria of correlation connected with the stages of general diastrophism. It rather suggests them than sets them forth.

It remains to consider the precedence among themselves of the three factors, diastrophism, deposition, and life development.

We are accustomed to look to the life record as our chief means of correlation. Its very high utility is quite beyond discussion. Thoughtful students, however, recognize that the paleontological record is based, in an essential way, on stratigraphy and that it is corrected and authenticated by the precise place the life is found to occupy in the stratigraphical succession. Stratigraphy and paleontology thus go hand in hand, each sanctioning the other. *Diastrophism lies back of both and furnishes the conditions on which they depend.* The relationship is not reciprocal in any radical sense. The life does not, in any appreciable way, affect diastrophism. Deposition has been thought to be related to mountain-folding.

Erosion in one area and deposition in another has been assigned as an initial agency in deformation. While some influence of this kind may be conceded, I think it is rather a localizing influence than a fundamental one. If wrinkling must take place from other causes, quite possibly previous erosion here and deposition yonder may localize the wrinkling. But that is quite apart from fundamentally causing the wrinkling. Reasons are growing yearly in cogency why we should regard the earth as essentially a solid spheroid and not a liquid globe with a thin sensitive crust. I think we must soon come to see that the great deformations are deep-seated body adjustments, actuated by energies, and involving masses, compared to which the elements of denudation and deposition are essentially trivial. Denudation and deposition seem to me clearly incompetent to perpetuate their own cycles. It seems clear that diastrophism is fundamental to deposition, and is a condition prerequisite to epicontinental and circum-continental stratigraphy.

Diastrophism thus seems to me fundamental both to stratigraphic development and life development. Diastrophic action seems to be the forerunner of both these standard means of correlation. It therefore seems to be the ultimate basis of correlation. The criteria of this correlation include at once its own specific criteria, the criteria of stratigraphy as dependent on diastrophism, and the criteria of paleontology as modified by the direct and indirect effects of diastrophism.

31

Reprinted from *Geol. Soc. Am. Bull.* **22:**289-292, 394-411, 448-459, 461-464, 532-535, 574-577, 579-585, 590-593, 595-597 (1911)

REVISION OF THE PALEOZOIC SYSTEMS[1]

BY E. O. ULRICH

(*Presented in abstract before the Society December 30, 1908, and December 28, 1910*)

[*Editors' Note:* The contents list has been omitted.]

[1] **Published by permission of the Director of the U. S. Geological Survey, who, in view of the fact that this paper presents a new classification, at variance with adopted usage, disclaims responsibility for the classification, correlations, names, and redefinitions.**

Manuscript received by the Secretary of the Geological Society February 25, 1911.

PART I. INTRODUCTORY CHAPTERS

Present Instability in stratigraphic Classification

CAUSES

The obvious unrest of the present day in matters pertaining to geologic chronology should be ascribed, I think, chiefly to the fact that stratigraphic knowledge has progressed far beyond the stage of prevailing schemes of classification. These schemes do not fit the known conditions. Not only were they founded on a mere fraction of the mass of data now available, but also the criteria used were frequently misinterpreted. Besides, the pioneers in this branch of geology had no conception of the complicated and variable conditions for which we now have to account; hence their practice has resulted in an inconsistent and incongruous arrangement of stratigraphic units.

But the weakest feature in these classifications is that their original framers availed themselves almost exclusively of a single kind of evidence, namely, that afforded by the fossil faunas and floras. If the fossils of geographically separated formations agreed in general aspect and composition, and especially if the percentage of the identified or supposed vicarious species was large, then the stratigraphic units were correlated. The possibility of locally surviving species or faunas, or of local variations in development, was not considered.

But these early practices were merely the extreme swing of the pendulum. The great success attending the use of fossils in correlation when compared with the results accomplished under the wholly incompetent methods previously employed not only placed stratigraphic geology on a firm basis, but induced a blind and unquestioning confidence in paleontologic evidence. The word "blind" is used advisedly, because an unprejudiced analysis of correlations by fossils in the fifty years preceding the past decade shows beyond denial they were made in practical disregard of factors vitally concerned in the production of local facies of faunas.

Under the circumstances it is not surprising that the efficiency of fossil evidence began to be questioned; that good philosophers began to discuss "shifting of faunas," to use expressions like this, "it required a long time for the *Olenellus* fauna to encircle the globe"; or for another to say "the very fact that geographically separated faunas are the same proves they are not contemporaneous." Presently others, chafing under the yoke of the "Paleontological autocrat," were suggesting classes of evidence and modes of correlation that might entirely supersede that based on fossils. Dual nomenclatures are proposed, new principles of correlation are suggested, and a symposium of correlation is arranged. It is the unrest preceding a revolution. What shall we do to direct it toward the promotion of our science?

DUAL NOMENCLATURES

As a result of my investigations I am opposed to a dual nomenclature, or, indeed, to any scheme tending to divorce paleontology from stratigraphy. The study of fossils apart from their stratigraphic relations is pure biology. With geologists the study of fossils should be a means to a definite end, and that end the elucidation of the geologic history of our earth. Geologic time should be measured by stratigraphic units, not by the indefinite duration of successive but overlapping dominant types of life. The time scale was not built up of life units, but of stratigraphic units. Indeed, the only competent means of estimating geologic time is based on ratios of deposition and on other considerations that are equally distinct from biologic investigations. The geologic time scale, therefore, is an essential feature of geology and not of biology.

The lithologic scale, like any faunal scale, is necessarily a local affair. In other words, both scales are founded on factors that are more or less strictly local in their typical development and subject to great variation when pursued beyond these areas. Neither, therefore, is capable of supplying data for a positive standard. So far as I know, there is but one possible standard by which geologic events and sedimentary rocks may be classified, and that is by *time* units—in other words, the chronologic standard.

MUTUAL CONCESSION

It is evident that the remedy for the present difficulties lies not in divorce, but in mutual concession and adaptation to prevailing conditions. Faunal, diastrophic, and lithologic criteria are complementary, and each kind may and should be used to check the other two. None of the three classes of evidence is competent by itself to insure a refined and true classification of geologic time, but between them we may hope to construct one. Let us reconstruct, or rather amend, our stratigraphic columns so that they shall indicate all well defined lithologic boundaries as well as the important faunal changes and the broader crustal movements that as a rule caused both.

Further, let us draw the various boundaries as nearly as possible in accord with the composite evidence of all classes of criteria. I should not object to considering even such non-geological matters as exigencies of mapping. Often a lithologic boundary could be drawn just as well or better at a line indicated by diastrophism or by a sharp faunal change as at the plane selected by the geologist without regard to either of these criteria. There is need for closer cooperation, and, more important still, for a spirit of concession and compromise in matters of detail and method between the lithological stratigrapher and the stratigraphical paleontologist. Neither should be asked to sacrifice important principles.

POSSIBILITIES OF CORRELATION BY FOSSILS

There are limits beyond which correlation can not go. Within a single stratigraphic province, as, for instance, the Mohawkian area, extending from Cincinnati southwestward to northwestern Alabama and southeastward to the southwestern corner of Virginia, or, in the Silurian province, embracing southern Indiana on the north and west middle Tennessee on the south, satisfactory correlation is possible to the smallest units that it is practical to map. And in such cases no other evidence than that afforded by the fossils is essential. But when it comes to correlating minor stratigraphic units of distinct basins, then the fossils need the help of other criteria. The fossils supply the first step in the process. They give us a reliable clue as to the boundaries of the major divisions. Finally, the principles of diastrophism fix these with greater precision and supply data on which correlations of minor rank may be based. But with no fossils at all it would be simply impossible to attain satisfactory results. Had not a Walcott found remains of *Olenellus* in the quartzitic sandstones on the east side of the Appalachian Valley, many areas now known to be lower Cambrian would still be classed as Medina; and had not a Ruedemann discovered certain Eurypterids in the Shawangunk sandstone, we might never have been able to prove that this formation also is not of Medina age. We might cite local successions of sandstone, shale, and limestone differing in no physical respect from similar cycles of sedimentation elsewhere that could never have been distinguished except for the infallible clue given by a few or, may be, but a single fossil.

On the other hand, the evidence of the fossils—often used by the paleontologist with such confidence and finality against the conclusions of the structural geologist that the rather disrespectful appellation of "autocrat" may sometimes appear justified—is not an obvious quality of the fossils themselves. The sequence of faunas as known today is the result of long continued and laborious research in areas of undisturbed stratigraphic succession. Faunal zone after faunal zone was located and their respective relations to certain lithologic units noted. Thus piece by piece the column has been built up. But the very foundation of the training of the modern paleontologist, who is not merely a biologist, makes him a good stratigrapher. Besides, there is no phase of diastrophism or of biology that he does not use or is not willing to use in the improvement of his determinations. His training is broader and he must work harder than any other class of geologist. Yet the paleontologist has often been disconcerted when the unquestionable succession of beds showed he had gotten things upside down or on the wrong plane. And he knows better than any one else that his information on the mutation and succession of faunas and floras is far from complete. Still it is a fact that fossils in experienced hands afford by far the most competent and reliable evidence now available in stratigraphic correlation. Likewise in the elucidation of the physical as well as the organic history of the earth fossils play a most important part. Finally, it is true that without their aid stratigraphy could never have become a science. The biologic part of paleontology may be considered separately from geologic processes, but stratigraphy can not exist apart from paleontology.

[*Editors' Note:* Material has been omitted at this point.]

PART II. CRITERIA AND PRINCIPLES OF STRATIGRAPHIC CLASSIFICATION

Diastrophic Criteria

BODY DEFORMATION CAUSING VERTICAL MOVEMENT OF LAND AND SHIFTING OF CONTINENTAL SEAS

General discussion and permanence of oceanic basins.—An adequate discussion of the criteria and principles of diastrophic correlation would fill a volume—and doubtless the effort would be worth while. But on account of lack of space summary treatment, confined as much as possible to the practical aspects of the main purpose of this work, must suffice for the present. It is to be said, however, that much of the matter to be presented under the heads "gradational and lithological criteria" (pages 467 to 480) and "correlation by lithological similarity" (pages 519 to 532), also in several chapters of Part I, is really a discussion and application of diastrophic criteria. Even the organic criteria are, to a large extent, expressions of diastrophic activity.

For purposes of stratigraphic correlation and classification the primary cause or causes of diastrophic movements—especially those resulting in considerable horizontal shortening—are not of vital importance. We are here concerned chiefly with the fact that such movements have taken place and with the determination of the general nature of their effects on agencies at work on the surface of the earth. It will be agreed by most of us that such deformation had a fundamental bearing on the migration of the strandline and on the relative elevation or depression of parts of the surface, hence on the extent of marine deposition and the character of the deposits. It also affected the evolution and distribution of faunas.

Chamberlin's views.—Two widely different conceptions of body deformation prevail among modern geologists. The important features of these divergent views have been briefly stated by Chamberlin in a paper entitled "Diastrophism as the ultimate basis of correlation."[33] As I concur in general with the principles advocated therein by this authority, a few sentences quoted from this paper will serve present purposes perhaps better than any I might write. He says: "There remain two conceptions of general or body deformation between which choice must be made. In the one, the deformations are supposed to be indifferent to their predecessors and to disregard the configurations produced by previous deformations. Their successive effects upon continental outlines and basin capacities are thus heterogeneous and the combined results irregular and uncertain. . . . The submergent phase of one continent or fraction of a continent may, in this case, be contemporaneous with the emergent phase of another continent or fraction of a continent, and the progress of events on one continent is as likely to be contrasted with those of another continent as to fall in with them coordinately." This conception, entertained chiefly by certain European geologists, is rejected by Chamberlin on what, especially as regards the indifference of deformations to their predecessors, seem to me valid grounds. The last of the quoted sentences seems less objectionable, since, as explained on pages 405 to 409, the conditions mentioned are probable also under the second conception.

"According to the other view, deformations are inheritances, one of which follows another in due dynamical kinship. The succession is therefore homogeneous and the results coordinate. If, for example, the first depression of the abysmal basins was due to the superior specific gravity of the basin-bottoms, this specific gravity remained and participated in the next deformation. If the continental masses, at the outset of continental formation, were relatively low in specific gravity, this low specific gravity was handed down to later periods and helped to renew deformation of the same phases in the same regions. Under this view, ocean basins and continental elevations tended toward self-perpetuation. It is not assumed that this prevented shell crumplings, provincial warpings, or block movements of diverse phases within the continental or the abysmal areas, for these might obviously be necessary effects of the general deformative movements, or at least inevitable incidents connected with the dynamics lying back of them. . . . Each great diastrophic movement tended toward the rejuvenation of the continents and toward the firmer establishment of the great basins. . . . The baseleveling processes have

[33] T. C. Chamberlin: Journal of Geology, vol. xvii, 1909, pp. 685-693.

shown that they are able to lower the continents approximately to the sea-level in a fraction of geologic time. The continents would therefore have long since disappeared, if they had not been rejuvenated by renewed relative elevation or the withdrawal of the sea. . . . There are some submerged dependencies and intercontinental connections. . . . In the earlier eras, when the differentiation of platforms and basins was less advanced, ridges which have since been submerged are perhaps recognizable. . . . In the earliest known ages, these may have been rather numerous and their combined area considerable, but these seem to me to be only qualifying features which, by the natural place in evolution which they fill, support, rather than weaken, the general conception of a symtematic succession of deformations in which the offspring of each is parent of the next, and in which both continents and ocean basins were progressively segregated and unified."

The latter of these two conceptions is held rather generally by American geologists. The results of my own investigations tend to show that it holds true not only for the continents and oceanic basins on the whole but also for relatively small portions of each continent. Once started, the areas of relative uplift have always continued to be areas of stratigraphic overlap or of land, while the relatively depressed areas remained as a rule areas liable to subsequent submergence. Though the surface of the lithosphere was affected by innumerable minor differential vertical and horizontal movements and the strandline is forever changing, the continents and the oceanic basins yet permanently retained essentially similar geographic relations to each other. The truth of this conception, especially in so far as it concerns continental areas of elevation and depression, is abundantly attested by detailed studies of the formations in and around areas of Eopaleozoic rocks. Some evidence of this kind may be gleaned from the introductory chapters, and more from following pages describing vertical movements of median continental areas and corresponding oscillations of continental seas, while considerable additional information respecting localities of repeated overlaps is in hand and ready for publication through other channels.

Willis's views on permanence of oceanic basins.—Chamberlin (op. cit., page 688) in saying that "there are some submerged dependencies and intercontinental connections," takes a broad and liberal stand respecting the probable former existence of considerable landmasses in areas now covered by the great oceans. Willis,[34] however, though he does not com-

[34] Bailey Willis: "Principles of Paleogeography." Science, vol. xxxi, No. 790, 1910, pp. 243-246.

mit himself positively, still seems to question the possibility of a Gondwana land, and of transatlantic lands, one connecting South America and Africa and another North America with northern Europe. That such dependencies and intercontinental connections did exist at various times in geologic history—their exact location is another matter—seems incontrovertibly established by the geographic distribution of certain fossil faunas and floras. These east-west connecting lands were emerged at times when the horizontal movements of the shell in a north-south direction dominated those acting in the opposite direction. The latter won in the long run, the connecting lands being finally dragged down in the aggregate growth of the oceanic geosynclines to depths from which it seems unlikely that they can ever again emerge.

I must take issue with Willis also when he says (Science, February 18, 1910, page 243) "in their genesis ocean basins and geosynclines may have been similar; but in their dimensions, histories and structural relations they are *radically* different;" and again to his succeeding statement (op. cit., page 244) that the "distinction between geosynclines and ocean basins is fundamental" because the former "occupy positions among the positive continental elements" the latter "separate and surround continents . . . and no ocean basin has been compressed, crumpled and raised, after the manner of the Appalachians or Alps." As I see it, the differences are merely relative—matters chiefly of magnitude—and therefore neither radical nor fundamental. Indeed, in conceding similarity in genesis, Willis himself tacitly admits this. Had he compared a whole continent and not merely one of its compound synclines with the oceanic basins, he might perhaps justly have called one the antithesis of the other, the one being characterized by dominance of positive tendencies, the other by negative. Both, however, include areas that are synclinal in structure, being relatively depressed, with respect to adjacent areas, and each of the oceanic basins comprises several pronounced geosynclines and weaker geanticlines just as the continents are made up of several strong geanticlines and relatively subordinate geosynclines. That compression and resulting folding and crumpling was developed to a less degree in the irregular floor of the oceans than on the continents is probably true, but to say that no ocean basin was compressed "after the manner of the Appalachians" seems unreasonable. That there is no sharp boundary of fundamental structural significance between those parts of the earth-crust that lie beneath the sea and those which form the land is shown, as I believe, conclusively by the fact that compressed and folded areas, both new and old, still high or reduced, strike from the land into the sea. And the contours of the ocean bottoms surely indicate great diversity in the way of broad and deep valleys and undissected ridges. These submarine ridges naturally receive less deposit than the valleys, and they are belts of relative weakness for much the same reasons that have made the similar ridges of the lands less resistant than the heavily loaded downwarps. Why, then, should they not have been at times compressed and raised in the "manner" of the Appalachians? Further, as each geosyncline is essentially a compound syncline, so each again is included in a still broader structure that might be called a compound geosyncline. Under this view the differences in dimensions and structure are less formidable. So far as I can see, the whole difference, then, is limited to variations in degree of development and to the circumstance that on the one hand we are dealing with land expressions of the originating process and on the other with its development under water.

Periodicity of movements—General discussion.—The statement previously made that "the strandline is forever changing" does not imply that diastrophic movements have been in constant progress in any given era. Adjustment of minor and local stresses, which may be preliminary to or follow stronger movements, and the slowly transmitted diminishing effects of oceanic thrusts, doubtless occurred in one or another place almost constantly. But there were other, more consequential deformations, that must have taken place periodically. These were movements in a horizontal direction, accompanied by warping, folding, and overthrusting, whose major effects, following the Proterozoic, were confined to the marginal areas of the continents. Though probably slow and long continued in their progress, the times of their prevalence would seem to have been rhythmically periodic. That these movements recurred at intervals separated by times of relative quiescence is in accord with theoretic considerations and proved by the fact that baseleveling, which involves as a prerequisite a considerable stability of surface, can be shown to have occurred during the later submergent stages of most of the periods. Considering the exceeding shallowness of the continental seas, especially of those most concerned in the inquiry, the great extent and rapid advance of these submergences would have been impossible except under conditions of approximate baselevel.

That the movements were rhythmic in time of occurrence and kind is strongly suggested by the approximate coordinateness of the successive systems as measured by their respective limestone values (see page 381). The average thickness of accessible deposits (reduced to a limestone basis) for the eight Eopaleozoic and Neopaleozoic systems is not far from

5,000 feet. Those exceeding this amount (Cambrian and Devonian) are thought to comprise more of the sedimentary record in accessible situations than usual, while those of which the known amount falls considerably beneath the average suggest longer intersystemic and intrasystemic intervals of inaccessible depositional records. This suggested explanation of the difference in the known aggregate thickness of the respective systems is based on the wide extent of the Warsaw, or first Tennessean invasion, and the similarly wide extent of the Pottsville transgression, implying in both cases a preceding long period of baseleveling, and it finds support in the well marked differentiation sustained by the southern Atlantic fauna in the time between the Keokuk and the Warsaw and again between the last of the Chester and the first of the Pottsville. In both instances the change is greater than between the latest Devonian and the early Waverlyan facies of the same fauna. Great caution, however, is required in making such faunal comparisons, they being truly significant only when the successive facies are derived from the same oceanic basin.

In my opinion a rhythmic relationship connects nearly all diastrophic movements. For a few the meter is very long, for others shorter, and for still others much shorter. The last may be arranged into cycles and these again into grand cycles, the whole arrangement probably corresponding in units to the divisions of an ideal classification of stratified rocks and, so far as these go, of geologic time. The determination of the relative importance of diastrophic phenomena is of course exceedingly difficult. At present their classification is fraught with uncertainties, whose elimination offers a never-ending task to present and future generations of geologists. Although prevailing interpretations vary greatly and progress is much impeded by inherited erroneous opinions, we may yet claim that enough has been established to give a reasonable generalized conception of the order of geologic events. And as to the broader relations of the minor occurrences, these, too, in many instances, are beginning to be understood and used in correlation. As I shall endeavor to show in succeeding parts of the present work and on future occasions, diastrophism affords a true basis for intercontinental correlation of not only the grander cycles but also of their subordinate stages. As yet I am unwilling to acknowledge any limit to the possibilities of stratigraphic correlation. The principle of rhythmic periodicity being recognized, it seems to me merely a matter of time and close comparative study of sedimentary records and faunal associations to determine the time relations of interruptions in sedimentation in any one section to similar interruptions in another.

Willis's views on periodicity of diastrophism.—Willis[35] recently published a brief statement of his views on the periodicity of geologic events and processes. He recognizes the principle, but qualifies it by grouping the phenomena of diastrophism according to several distinct dynamic regions, each of which he supposes to have had its individual history, "expressed in cycles of movement and quiescence peculiar to itself." "The cycles of one region have been, however, to some extent parallel, though not coterminous, with the cycles of other regions, and thus major cycles of world-wide conditions are constituted by coincidence of regional conditions." This qualification is based on the belief that the diastrophic history of the lands adjacent to the north Atlantic differs in essential particulars from that of the lands bordering the Arctic and Pacific oceans. He says that the "lands about the Arctic Ocean did not share in the Atlantic movements of Silurian, Devonian, and late Paleozoic epochs;" that "in the Atlantic provinces the Paleozoic era closed with marked diastrophism, while comparative tranquillity reigned around the Pacific," and that "the Pacific provinces were greatly disturbed in the middle Mesozoic when quiet had supervened about the Atlantic."

On the face of the accessible sedimentary record in the respective regions these allegations may seem well founded. But even the visible part of the evidence is not altogether in favor of the quoted statements. Neither are we justified in claiming, because abundant clastic deposits and other indications of active diastrophism are preserved in one area and little or nothing of the kind is found in another at approximately the same age of the geologic time scale, that the two areas did not share in the same deformative movements. The share in the one case may have been small, in the other large. In fact, no two areas, however near, participated in exactly the same way or to exactly the same degree in the movements of any period. The more conspicuous phenomena induced by body deformation of the earth, such as mountain making, are merely local effects of the movement as a whole. While the positive areas were elevated, immediately adjacent negative areas were relatively depressed; and more distant regions may have shared in the movement only to the extent of either gentle subsidence or perhaps slight emergence. Whether it built mountains in one place and caused but gentle warpings in another, or whether the mountains bordering one oceanic basin are contemporaneous with a similarly located slight elevation in another, or even a subsidence in a third, they are all only different manifestations of the same period of diastrophism.

[35] Bailey Willis: "Principles of Paleogeography." Science, vol. xxxi, No. 790, Feb. 18, 1910, pp. 246-249.

That the lands about the Arctic Ocean have not been folded like those bordering the Atlantic and Pacific oceans and those flanking the east-west Gulf of Mexico and Mediterranean depressions, of course implies decided differences in diastrophic history. Evidently, tensional stresses, with crustal flow toward the equator, predominated in the circumpolar regions, while compression, resulting in part from equatorial flow of lands and seas, and in part from suboceanic spreading acting from the water side and continental creep from the other, prevailed as a rule in the folded tracts. But these differences in history are not prohibitive to exact correlation of geologic events transpiring in the several regions. It is highly improbable that mountain making took place at any one time on all the continents; and it is certainly not true that the folding observed in any mountainous tract dates all from a single period of diastrophism. To postulate a period of major diastrophic activity, we need but to establish that strong folding actually occurred at such time in certain regions, and this proof should be sufficient warrant for the beginning of a new geologic period. What occurred elsewhere at about the same time becomes merely a matter of interesting detail—doubtless important in so far as it shows the extent and relative taxonomic importance of the deformative movements. However, this detailed information is very desirable and to a large extent essential to properly define and discriminate the respective periods in regions of varying dynamic history.

And is the evidence wholly satisfactory on which is based the postulate of marked diastrophism in the Atlantic provinces at the close of the Paleozoic and comparative tranquillity around the Pacific at the same time? Or for the supposition that quiet had supervened about the Atlantic in the middle Mesozoic when the Pacific provinces were greatly disturbed? Since the areas in which decided folding occurred are largely confined to the marginal parts of the continents, may it not be possible that tracts bearing the physical record of important diastrophic movements are now submerged or buried beneath more recent undisturbed sediments? Finally, is the evidence on which it is inferred that the folding of the Appalachian Valley was accomplished and marks the close of the Paleozoic quite incontrovertible? As will appear from later statements (see pages 468 to 477), I think not.

Displacement of strandline chiefly relied on in proving periodicity of deformative movements.—The only thing that moves apparently in the same way and degree, and which, therefore, offers the most reliable criteria in determining the periodicity and contemporaneity of diastrophic events, is the level of the sea. Even this statement must be qualified, since, under the perhaps well grounded conception of a periodic equatorial heaping of the waters, the level of the several oceans may rise or fall in the same measure only at the same latitude, and then only when the proportionate bulk and altitude of adjacent land-masses remains unchanged. But, whatever the qualifications, there yet remains the fact that the strandline is contemporaneously and universally displaced. Obviously the exact correlation of these displacements at times of great and diversely manifested diastrophism, as at the close of periods and eras, is more difficult than are the more eustatic migrations of the strandline which occurred at intervals within the durance of the periods. Doubtless the times of excessive crustal deformation are everywhere indicated by some marked change or break in sedimentation; but the break is likely to be a long one and the determination of its time relations in two or more areas correspondingly indefinite. During parts of the interval represented by a well defined intersystemic hiatus deposition may have been going on elsewhere; and it is sometimes very difficult to decide whether these intervening deposits represent later stages of the old period or earlier beds of the new or, as often happens, of both. In the last event several perhaps equally distinct boundaries may occur within the intervening mass, and as all the criteria relied on in distinguishing the two systems in the first case are more or less weakened through transition, it may indeed become a matter of no small difficulty to decide which of the boundaries is the most important. The Siluro-Devonian boundary in southeastern North America and in central Europe affords a good illustration.

The displacements of lesser rank, such as define series and groups, are more readily amenable to intercontinental correlation. It is not that the deposits of a given epoch can be more easily recognized in the several continents than those of the much larger period unit, but that the average time value of the breaks is less and the competent accessible data more numerous, thus enabling us to reduce the limit of error in determining the equivalence of intrasystemic boundaries to less formidable dimensions than when we seek to correlate the terminal deposits of a system. For instance, it will be less difficult to establish the exact time relations of the subdivisions of our Niagaran series to the Silurian zones recognized in England and Sweden by diastrophic criteria—using that term in its broadest sense—than it will be to prove that the Manlius of New York is not represented in the upper part of the Wenlock, or that the Ludlow group—yes, even the Dudley limestone of England—does not include beds of the age of the Helderbergian in America.

It is not my intention to imply that the systemic boundaries can not be drawn in the several continents so as to insure against the possibility of referring contemporaneous beds to the top of, say, the Silurian in Europe, and to the base of the Devonian system in America. On the contrary, I believe that accuracy in correlation, whether narrow or intercontinental in scope, depends solely on the uniform application of the criteria and principles adopted, and that if our practice is thoroughly consistent we shall finally succeed in discovering physical boundaries that will separate the systems so that none will include beds of ages elsewhere referred to either the preceding or the succeeding period. This may sound unduly optimistic, but judging from my own experience in the still young science of stratigraphy it is not so. The greatest difficulties lay not in the correlation of sections personally studied; they arose rather from the continual expansion of the stratigraphic column and the frequent readjustments and changes in classification thereby necessitated. In short, they have been taxonomic rather than objective; and withal never so great as to discourage the hope that where I have failed others may succeed.

Results of recent methods of correlation.—The feature of advancing knowledge of stratigraphic correlation that more than any other is responsible for the expansions of the sedimentary record briefly outlined on pages 379 to 381 lies in the gradual conviction that instead of stable, slowly growing continents covered with great, long enduring and deep interior seas, the surface of the continents was exceedingly unstable and subject to frequent oscillation and more or less local warping, the seas were correspondingly inconstant, shallow, relatively small, and frequently withdrawn entirely, only to reappear in forms simulating preceding geographic patterns. The effect of this revised conception seems so far-reaching that if proved true it must greatly modify—perhaps revolutionize—stratigraphic geology. Possibly this statement is stronger than is warranted by the evidence. But the facts bearing it out are not only abundant and of varied kinds, but they are repeated over and over again so much that I can not help being convinced of the truth of the principle of oscillating seas and lands. In fact, this principle is at the basis of whatever new philosophy is established or suggested in this paper.

Not so many years ago it was commonly accepted that extensive migrations of the shoreline resulted principally from *en masse* elevation and subsidence of the continents. Latterly, and more particularly since the publication of Suess's "Antlitz der Erde," the transgressions of the sea are being ascribed chiefly to changes in or beneath the hydrosphere itself without vertical movement in the land. That the latter view is more nearly right than the former can not be denied. Neither can it be denied that most of the great sea withdrawals, also many of the minor withdrawals, were immediately and chiefly occasioned by deepening of the permanent oceanic basins or by some other condition inducing negative fluctuation of the water level independent of land movement. But these changes within the hydropsphere were perhaps only seldom or never solely responsible for the migrations of the strandline. If the deepening of the oceanic basins was a broad synclinal effect of the shrinkage of the lithosphere, then we may for equally good reasons assume a contemporary anticlinal effect on the continents that must have contributed in a small or larger degree to the conditions causing general sea withdrawal. After all, then, displacements of the strandline are always connected with some body deformation of the earth.

As to the alternating withdrawals and transgressions of the sea, these seem to have occurred, at least in Paleozoic times, much more frequently than even Suess conceived. The diagrams on pages 346 and 347, illustrating invasions of seas, give a generalized idea of these migrations of the shore; but the presentation is confessedly inadequate in the matter of number and detail. To what extent land deformation contributed to the production of these broad migrations perhaps can not be definitely determined. Sea filling, continental attraction, deepening of oceanic basins, and the equatorial heaping of waters incident to accelerated rotation of the earth about its axis at times of pronounced shrinkage, the last suggested by Suess and more recently by Schuchert (Bull. Geol. Soc. America, vol. 20, p. 505) as possibly important, are all forces or processes that may have operated concomitantly or separately in producing displacements of the strandline. The forces being in part compensatory, the advance or retreat of the waters, as the case may be, is the resultant of their variant activities. While the competence of these various factors to effect general submergence or emergence is admitted, it is denied that this fact necessarily excludes the operation of deformative movements within any part of the land-masses themselves.

That the marginal areas of the continents were at times elevated and folded is, of course, accepted by all—even by Suess and his followers, who speak of the continents as having the character of "horsts" and of the ocean basins as being permanently "sunken areas." This authority, however, tends to the belief or conviction that the median areas of the continents are essentially stable, a conviction shared and recently expressed by Schuchert, who holds "that the continent (North America) is a horst, that the great medial region remained unmoved, while the margins were often folded and elevated. The seas periodically flowed over this medial land—in fact, were elevated over it—owing to the detrital materials unloaded into the oceanic areas, thus filling them and causing them to spill over on to the lands."[36]

I cannot subscribe to this view. On the contrary, though accepting the idea of permanent oceans and continents, it seems to me that the crust of the lithosphere, not unlike the more mobile hydrosphere, was subject to periodic movement away from the poles; that the surface of the lands was exceedingly unstable and that this instability pertained, though in an inferior degree, to the median areas as well as to those along the borders of the continents. Schuchert's paleogeographic maps, indeed, offer convincing proof of such instability. The general nature and the reciprocal features of these "small movements" are excellently stated by Chamberlin and Salisbury. (Geology, vol. 1, 1906, pp. 540.)

DIFFERENTIAL VERTICAL MOVEMENTS OF THE LITHOSPHERE

Tilting of land areas.—The varying distribution of marine deposits of successive ages naturally suggests differential upward and downward movement of the lands as the immediate cause of such variation. If the submergences had been occasioned solely by rise of the waters, whatever the cause of the overflow, the successive submergences of the relatively featureless lands would have been always similar in geographic pattern and different only in lateral extent. In fact, a general similarity or repetition of old patterns is recognizable, but there is also exceeding diversity of expression, and the significant feature is that this diversity is often greatest when directly succeeding stages are compared. One stage may be very different from the next, but the third or fourth again may be very much like the first. Only oscillatory movements of the land surfaces could produce such results. The area affected by such movements may be very large, as, for instance, during the middle Ordovician and middle Silurian, when nearly half of the continent of North America was involved. During this period the Gulf waters seem at certain times to have been completely withdrawn from the southern part of the

[36] Bull. Geol. Soc. Am., vol. 20, p. 438.

It is to be remarked that this statement is inconsistent with others of this writer in the same paper. Thus, on pp. 501 and 502, he speaks of the medial areas being buckled and warped and of variable local loading and compensatory isostatic vertical movements. Suess also noted conditions that are not in harmony with his general conclusion. Thus, for instance, he says (Antlitz der Erde, English edition, pt. 3, p. 545): "A close examination of the stratified series often leads us to suspect the existence of numerous smaller oscillations which are hard to reconcile with eustatic processes."

continent, the middle and northern parts at such times being tilted so that the boreal sea extended southward beyond Chicago and occasionally as far as northern Tennessee. Possibly the invasions incident to such broad tilting were slightly accentuated, according to the theory of Schuchert (Bull. Geol. Soc. America, vol 20, page 505), by the return to the poles of waters previously piled up at the equator during periods of active earth shrinkage. However, assuming essential permanence in location of the poles, it is inconceivable that such returns could by themselves produce the condition of advance in the north and retreat in the southern States, the latter, according to theory, being in latitudes of negligible change in water level. But granting that for some unknown cause they were competent, then the exclusion of the Gulf waters would have continued to the next period of great shrinkage—a prohibition that we have no reason to believe obtained. Hence, even these extensive reversals of sea invasions are regarded as unquestionably indicative of differential vertical movements. (See figure 8, page 407.)

These north-south tiltings of the continent are relatively modest phases of diastrophic movements of a high order of magnitude that are expressed in a more striking and indubitable manner by marked orogenic phenomena along the southern borders of the northern tier of continents. The crustal folds which have resulted from this series of movements have a general east and west strike and are especially well developed in areas adjacent to the Mediterranean and Indian seas in the eastern hemisphere, and to the Gulf of Mexico in the western. The available data suggest a tendency of the outer part of the crust to move away from the poles, with sinking in the polar regions and corresponding heaping of the crust equatorially. This process long continued might be expected to result finally in great depressions of the lithosphere about the poles and in nearly compensating elevations within the median east-west zones of orogenic activity. For two reasons chiefly the present conditions fall short of the expectations. First, the elevated areas have been subject to consequent acceleration of erosion, the disparity in relief being thereby reduced; and, second, local loading of the areas in which mountain-building was in progress (the loading being occasioned by folding and consequent heaping of the superficial crust) unsettled the isostatic equilibrium, and thus aided erosion in bringing the high parts back to former levels. Sedimentary loading of adjacent submerged areas assisted by tending to drag the high lands down with them. Finally, with sinking going on, say, in lands bordering the Gulf of Mexico, deep-seated, northward flowage of rock matter doubtless took place, presumably resulting in refilling of the polar depressions from below and in gentle, upward movement of the

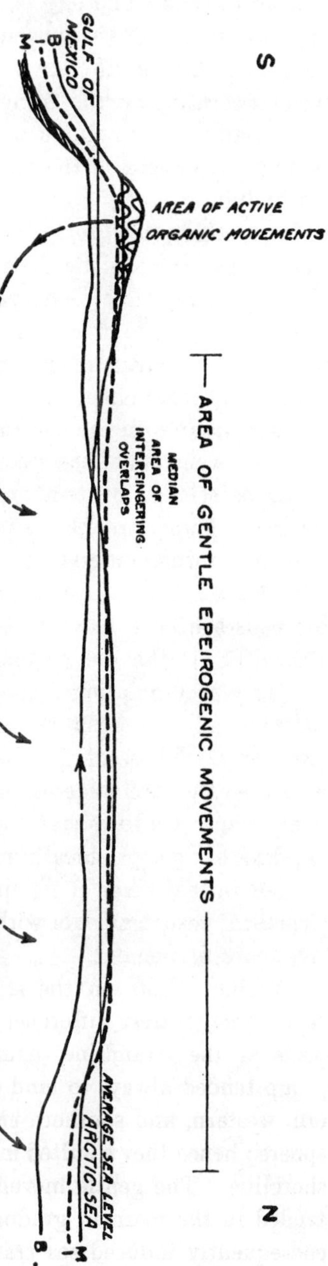

FIGURE 8.—*Schematic Sections of North America from the Arctic Sea to the Mouth of Mississippi River*

The diagram illustrates tilting of continent in Ordovician and Silurian times and isostatic adjustment following southerly thrusting of the earth's crust. M = southerly tilt when the Gulf of Mexico invaded. B = northerly tilt when the Arctic Sea transgressed. I (dashed line) = intermediate stage in which the whole continent is emerged. In the B stage the area of active orogenic movements is represented at its maximum elevation. This was followed by subsidence and erosion to the submerged plane shown beneath it in profile M. Lower dashed line and arrows represent assumed deep-seated northward flowage induced by overloading of southern areas.

central and northern parts of the continent. Something like this is thought to have occurred repeatedly, and probably many times, during past geological ages, and the theory would serve very well in explaining the alternating north and south tilting of the surface of North America that is so clearly indicated by the distribution of Ordovician and Silurian faunas and sediments described on pages 367 to 371.

Two groups of movements—The movements defined. — Diastrophic movements are divisible into two generalized and not sharply differentiated groups, (1) those that are active and relatively impulsive in operation, and (2) those that are relatively gentle in operation and much more gradual in the production of results. The former are chiefly responsible for the wrinkles of the earth's crust, and their function is primarily orogenic. On the continents these relatively active movements originated and repeatedly rebuilt the mountain chains of the present day; in the oceanic basins they occasioned the great undulations of their bottoms. Though the same in ulterior causes and effects, there is still some difference in the progress and final results of the movements in the continental and oceanic parts of the crust. These differences are due to a secondarily contributing causal factor operating in the former but absent in the latter. On the continents, namely, the reduction of the inequalities of the surface is produced as much by erosion of the mountains as by filling of the hollows, whereas the undulations of the oceanic bottoms are lessened, if at all, chiefly and perhaps solely, by greater accumulation of deposit in the depressions. The doubt expressed by the words "if at all" is based on the suggestion that subsidence, due to depositional loading and to inherent differences in specific gravity, may have maintained topographical relations pending the periodic revivals of diastrophic activities when the inequalities must have been accentuated. Accordingly the aggregate vertical displacement is much greater on the continents than within the oceanic basins.

The movements of the second group consist chiefly of gradual displacements occasioned by temporary accommodation to slowly accumulating stresses. These stresses are produced by suboceanic spreading on the one side and continental creep on the other, by depositional loading, and by all other causes tending to disturb isostatic equilibrium. Ulteriorly, then, the causes of the movements are essentially the same in both groups. They differ in that the movements of the second group are connected with the early accumulation phases of the stresses, in that they are relatively continuous in operation and comparatively local in their contributions to orogenesis. Those of the first group, on the other hand, are associated with the saturated stages of accumulation of stresses, are likely to be relatively impulsive in operation, and their results in the way of mountain-building are geographically more extensive. Their dynamic effects differ further in that those of the first group are of a higher order, and that the structural readjustments within the lithosphere caused by them are as a rule more permanent.

Varying effects on the strandline.—From the standpoint of stratigraphy the greatest difference lies in their respective effects on the same parts of the strandline, namely, the impulsive movements of the first group tended always to land elevation and sea withdrawal from the eastern, western, and southern sides of the continents of the northern hemisphere; hence they resulted in negative displacements of these parts of the shoreline. The gentle movements of the second group, on the contrary, tended in the main to gradually lower these parts of the continents and consequently induced sea transgression and partial submergence of their flanks. Coming to the northern side of more particularly the North American continent, it is found that the displacement of the strandline is quite different from that noted on the other sides. Here, as explained and illustrated on page 407, the movements of the first group caused gentle though very widespread subsidence. In other words, while the results of the movements are strictly orogenic on the east, south, and west borders of the continent, on the northern side they resulted in sinking and gentle warping of wide land areas, and thus are distinctly epeirogenic in character. When the subsidence proceeded to the submergent stage, as in the late Black River, the earliest Trenton, and in one or more of the Richmondian ages,[37] the northern sea transgressed very broadly and with extraordinary rapidity over the great median flats of the continent. Proof of the relative rapidity of these transgressions is seen in the small thickness of the deposits of seas invading from the north, this being out of all proportion to the great extent of territory covered by them. The seas invading from the east, south, and west as a rule not only laid down much greater thicknesses of sediment, but the inland extent of the deposits was very much less, and the rate of sea advance, as indicated by the relative abruptness of the inland overlaps of the successive beds, was much slower.

From the foregoing it will be observed that the great north-south differential movements referred to under the term "continental tilting" are partly assignable to both groups. The northward tilt occasioned what may be called the submergent phase of the total result of a movement of the first group that manifested itself more typically in the south by crustal folding and land building. Both phases suggest impulsive and on the whole rapid development. The reversal of the tilt, occasioned by elevation of the northern and median parts of the continent and subsidence in the southern, was a much slower process and therefore belongs to the second group.

All other tilting of continental areas, especially the "minor tilting" of such prevailingly *positive* inland domes as the Cincinnati, Nashville, and Adirondack uplifts, which oscillated in east-west directions more commonly than north-southwardly, is similarly attributable to movements of both groups. However, in these cases the reversal or decided modification of the tilt of the domes is probably always more or less impulsive in origin and consummation; hence due to a movement of the first group. Apparently, then, only the gradual subsidences and consequent submergences of one or another of their sides may be attributed to movements of the second group.

---

[37] It should be admitted here that it is only when such northern invasions were accomplished that the differential character of these movements is demonstrable on competent stratigraphic and faunal evidence.

A reasonable deduction from these considerations is that, with the exception of the noted invasion by polar seas, all submergences are directly ascribable to movements of the second group, and that all negative displacements of the strandline, excepting again the retreat of polar waters, that are not purely local in development, are relatively impulsive and therefore due to or connected with earth movements of the first group.

Discrimination of general earth movements.—The details of the most active movements which resulted in great land growth are often very difficult to decipher. In these cases the competent sedimentary and organic criteria are, if not wholly, at least mostly, confined to parts of interior basins and to extracontinental areas now inaccessibly buried. We have, therefore, no adequate means of discriminating between the grander and the less important movements that took place during a highly emergent phase of the continents. Indeed, the land record is as a rule drawn in broad and none too definite lines. When it indicates diastrophic movements, the record shows both the major and the minor movements in a composite picture in which the individual occurrences can not be satisfactorily discriminated. Moreover, since the movements exerted themselves chiefly in areas and lands offering the least resistance, and as these non-resistant areas grew continually weaker and were thus perpetuated, it has become extremely difficult to say how much of the composite effect really belongs to the period to which it is commonly assigned and how much to preceding and succeeding periods. Appalachian folding certainly was not all accomplished at one time or during a single period. It began in a pre-Cambrian era and probably is going on in the present (see pages 436 to 439).

The grandest of the movements referred to in the preceding paragraph are thought to have occurred about the close of the eras. As they left deeply impressed records and their periods were long, the times of their occurrences are recognized the world over. Other movements of major import are indicated in the intersystemic intervals. These also are recorded in a recognizable manner—chiefly by evidence of great sea withdrawal—on all of the continents. Then there were movements, defining series, groups, and the more important formations, that resulted in similar though not always general evacuations of the marine continental basins. These, too, have left intercontinental physical records that will be correlated more or less definitely when the known facts are revised and mere matching of fossils shall have given way to comprehensive modern methods of determining stratigraphic equivalence. It will be done by correlating negative displacements of the strandline on corresponding parts of the several continents rather than faunas and floras.

Regarding the slow movements that induced positive displacements of the strandline, except on the polar sides of the continents, their records are to a large extent clearly preserved in the accessible deposits of the continental seas. Detailed stratigraphic studies in the Appalachian region and on the flanks of the Adirondack, Ozark, Cincinnati, Nashville, and Wisconsin domes have brought out much interesting information respecting their operation.

These studies show that shifting of the strandline, which, from the viewpoint of the systematic stratigrapher, is the most important of the criteria of diastrophism, is recorded, organically, (1) by abrupt changes in the aspect of faunas, and (2) by the sudden appearance or reappearance of species and genera in deposits of continental seas (see examples noted on pages 298, 302, and 514); physically it is recorded (1) by more or less obvious breaks and hiatuses in the stratigraphic column indicating sea withdrawals, (2) by changes in the character of deposits, especially when this involves abrupt transition from biochemical and organic sedimentation to clastics, or a change from land to marine deposition, and (3) by overlaps of marine sediments. The relative elevation and subsidence of the continents, each taken as a whole or considering only parts of each, also the varying climatic and other superficial conditions, are recorded by local or more or less general changes in character and volume of marine and land deposits.

The rate of transgression of stratigraphic overlaps is dependent on the relative relief of the land in course of submergence. It is most rapid and widest in extent when baseleveling of the lands, whether due solely to subaerial erosion or assisted by general subsidence and continental tilting, has been most nearly accomplished; and it is partly for the same reasons that the introductory submergence of a period or epoch sometimes spreads farther and more uniformly than its succeeding stages (see page 338). The rate and extent of the submergence, especially of the eastern, western, and southern parts of northern continents, therefore affords a means of estimating the duration of the emergent intervals. The character and volume of clastic sediments following a stratigraphic hiatus and the rate of change from clastic to non-clastic limestone deposition may also be used to the same end.

[*Editors' Note:* Material has been omitted at this point.]

Structural Criteria

General Discussion

The structural phenomena which have a more or less direct and correspondingly exact bearing on stratigraphic correlation and taxonomy are those indicating either positive or negative displacements of the strandline. Positive displacements—that is, advance of shorelines—are indicated (1) by progressive overlap of sediments,[49] and (2) by peculiarities in kind and arrangement of the initial deposits. Negative displacements or sea withdrawals are indicated (1) by evidence of shallowing, (2) by imperfections of the sedimentary record, beds or formations of ages elsewhere represented by deposits being absent, (3) by obviously unconformable relations of adjoining beds, and (4) by less clearly manifested evidence of discontinued marine sedimentation and emerged or land conditions. The general aspects of these phenomena, also many examples, are more or less fully discussed in preceding parts of this work.

[49] As used in this work the term "progressive overlap" refers only to the progressively spreading deposits of an advancing sea. "Regressive overlap," on the other hand, is applied only to the areally diminishing deposits left by a retreating sea.

The following remarks, therefore, will be confined so far as practicable to the criteria themselves.

Less definite in their bearing, and of use only in broader correlations, are those phenomena which indicate periods of rejuvenated orogenic activity. Having devoted considerable space in preceding chapters to the discussion of the earth movements which occasioned them, this phase of the subject may be passed without further comment. Moreover, since these periodic activities are chiefly indicated by consequent quickening of erosional processes, and as matter pertaining also to this aspect of the subject is presented in the introductory chapters comprising Part I, it is thought advisable to defer additional consideration of such criteria to a more appropriate succeeding chapter on degradational and lithological criteria.

PROGRESSIVE OVERLAP OF MARINE DEPOSITS

Rate of progress.—The criteria of progressive overlap are primarily such stratigraphic phenomena as indicate landward spreading of successive zones of deposition in an advancing sea. Obviously the rate of progress depends on the relative steepness or gentleness of the slope of the land in course of submergence. If steep the successive stages of the overlap as registered by the landward progression of each layer beyond the next preceding is correspondingly slow and small in horizontal dimensions; if very gentle the younger layer spreads with proportionate rapidity and in extreme cases may extend great distances beyond the limits of the next older stratum. Obviously again in the case of long and relatively narrow troughs the longitudinal progress of the overlap is proportionately much more rapid than the transverse advance. Figures A and C of the accompanying sketches illustrate relatively rapid transgression of overlapping formations, while in D and E, more especially the latter, the overlaps progress much more slowly. The Boone at the top of D is a widely transgressing formation; hence relatively rapid in overlap.

Most of the Cambrian, Ozarkian, Silurian, and early Devonian formations in the several depositional troughs of the Appalachian Valley constitute excellent examples of rapid longitudinal and slow transverse advance of overlapping deposits. Some of these formations, notably a reddish middle Cambrian shale, the Copper Ridge chert, and the typical Clinton, extend northwardly from Alabama and Tennessee to or beyond southern Pennsylvania, while their present distribution in an east and west direction is commonly confined to strips less than 75 miles in width. The Arctic invasions of the relatively flat median areas of the

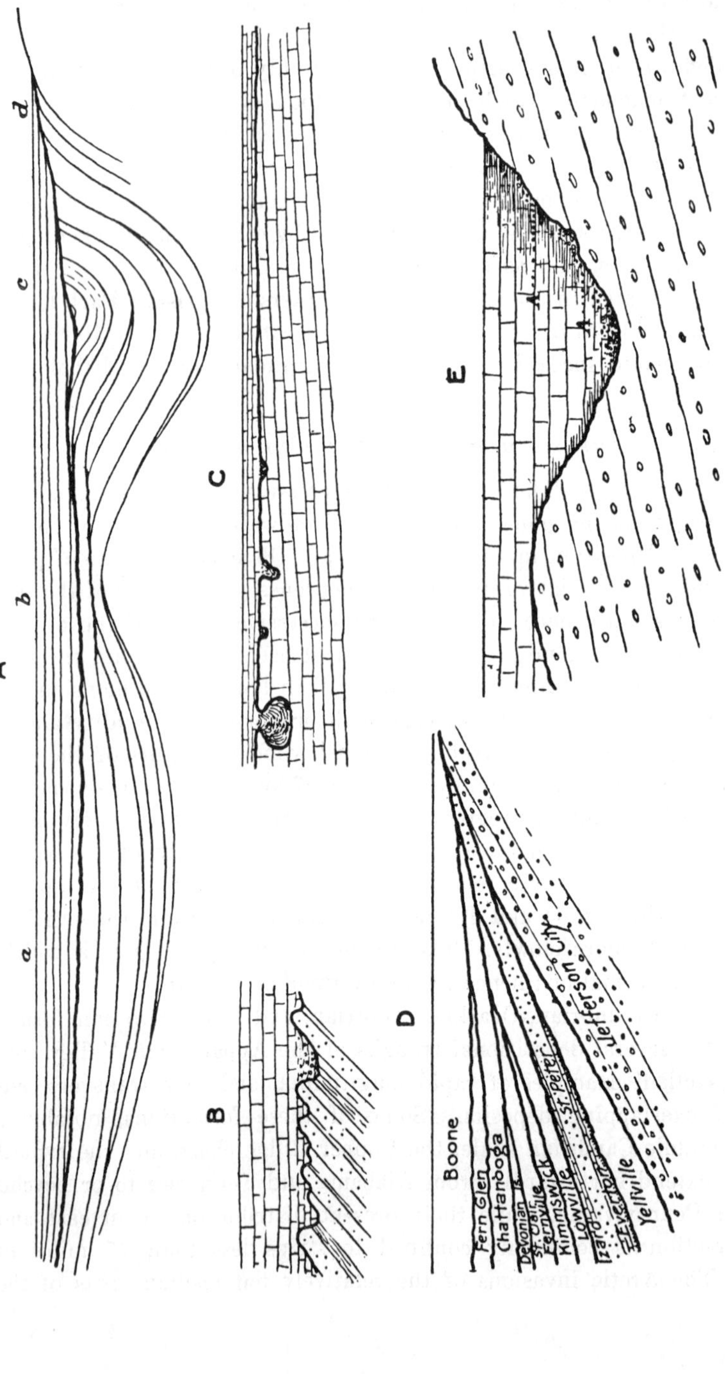

FIGURE 17.—*Diagrams illustrating Overlaps and Unconformities*

A shows a formation transgressing from a broad shallow basin across the eroded edges of beds in a deeper syncline, as, for instance, from the Allegheny basin eastward over an Appalachian Valley trough. The unconformity at the base of this overlapping formation varies greatly from place to place in the angle of divergence of beds on its opposite sides and in the time value of the hiatus. At *c* the structural discordance is much greater than at *a*, *b*, and *d*, though at each of the latter points the hiatus is greater than at *c*.

B shows an "angular" unconformity with a limestone resting on the eroded edges of a tilted sandstone formation. The contact is irregular because of inequality of resistance offered to erosion by the underlying formation, and the depressions may or may not contain accumulations of clastic material. Often they are swept entirely clear of such matter by the advancing sea. An unconformity like this is seen at Rondout, New York, where late Silurian limestone overlaps tilted late Ordovician sandstone.

C shows an unconformity between two limestones, as between the Kimmswick and Fernvale limestones south of Saint Louis, Missouri. In these cases, as shown in the right half of figure, the unconformity is often barely distinguishable from an ordinary bedding plane. At other places, however, the surface of the lower formation is more uneven and occasionally exhibits solution cavities or channels, which are filled with sweepings of the old surface, or even by ordinary marine deposits not seen in adjacent parts of the section.

D illustrates stratigraphic conditions observed on the southern slope of Ozarkia, and shows how the major unconformity on the right, where the Boone rests on the Jefferson City, wedges apart so as to permit the intercalation of numerous other overlapping formations that are intermediate in age and similarly unconformable in their relations to each other. Obviously the time value of the hiatus separating the intercalated wedges is greatly inferior to that of the combined hiatus found toward the summit of the island; but in physical manifestation and structural discordance one may be as clearly defined as another.

E shows a cross-section of the Foster Mountain embayment, north of Birmingham, Alabama. This was excavated in the western slope of the Rome Barrier and subsequently occupied by a small arm of the Ordovician sea coming in from the west. This occurrence is important in proving that considerable relief and consequent surface erosion took place occasionally in the Appalachian Valley as far back as pre-Ordovician times, and in showing (1) that the heavy chert conglomerate was dropped close to the shore, and (2) that the argillaceous near-shore deposit may give way to pure limestone deposition—in this case it does as a rule—in less than half a mile from the shore. Occasionally small chert pebbles were carried much farther out.

North American continent, as noted and explained on pages 407 and 409, effected in perhaps every case extraordinarily rapid extensions of sedimentary planes. Overlap structure in these cases is demonstrable only on the flanks of island domes and peninsular projections included in the general area submerged at such times by northern waters.

General prevalence of overlap structure.—With relatively few exceptions, lithologic or stratigraphic units, whether constituting the whole or merely subordinate divisions of formational units, are bounded below by progressive overlap structure. The exceptions consist almost solely of instances of limestone or shale formations which grade laterally into more clastic, near-shore, delta or fluviatile, deposits. Notable instances of such transitions have been shown to occur in the Devonian of New York by Clarke, Williams, and others who have contributed to our knowledge of the stratigraphy in this State. These cases are typical of a class in which land debris is accumulated in a bay or along a shore, finer sediments being laid down farther out. The Selma chalk band in the southern States which, according to unpublished investigations by L. W. Stephenson, passes at each end into less calcareous deposits, seems to illustrate a different class, namely, deposition around a low blunt peninsula influenced by lateral drainage, so that relatively clastic deposits formed on its sides and calcareous sediments off the rounded extremity.

Though lateral transitions in lithic character have hitherto been supposed to occur very commonly, a critical review of the cases has shown that true transitions of this kind are by no means common. As a rule the presumed lithically changing stratigraphic zones proved to comprise deposits that were not only laid down in physically distinct troughs or basins, but also that their respective beds are not contemporaneous. Faunal studies first suggested and have finally established the proposition of small, disconnected, oscillating continental seas; but, after all, the most convincing and therefore necessary evidence was furnished by stratigraphic overlaps. In the Appalachian Valley, for instance, most of the Eopaleozoic formations were found to be confined to a single or to two adjacent troughs. Other formations spread more widely and are found in three or more basins. In the broader interior basins the stratigraphic record as exposed on the flanks of low uplifts is punctuated at frequent intervals by evidence proving oft-repeated advance and retreat of marine waters.

Owing to overthrusting of the middle and eastern troughs in the Appalachian Valley, the overthrusting being incidental to the excessive folding of the valley rocks (see figure 16, section 6, page 440, and figure 9, page 414), the eastern edges of their respective formations are as a rule inaccessibly buried. The raised western edges, too, have in most cases been removed by erosion, but several satisfactory remnants are still to be found. Actual, indeed great thinning by westward overlap is shown by the Holston limestone and in a smaller degree by the Ottosee formation in the Ordovician belts next east and west of Loudon and Concord in Tennessee. Similar east-west thinning and final extinction is clearly shown by members of the Chambersburg limestone in southern Pennsylvania. (See description in Part I and figure 3.)

That overlap extinction of stratigraphic units on the sides of the original Appalachian troughs actually occurred is reasonably inferred from the fact that certain formations and faunas are confined to certain troughs. Thus, as shown on the map of Ordovician outcrops in the southern part of the valley (see map opposite page 412), limestone deposits that may be referred to with propriety as the Trenton and Stones River groups are confined to the Clinton and more western bands (see page 293). The intervening Black River group likewise occurs in the western troughs, but differs in spreading farther eastward, being found in considerable development in the Pearisburg trough, the Lowville part, indeed, extending as far eastward as the Knoxville trough. On the other hand, the sediments and faunas of the Lenoir, Tellico, and Ottosee formations are confined to the Knoxville and Athens troughs, while the Athens shale is seen only in the Athens trough. The Holston, finally, is absent in the Athens trough, best developed in the Knoxville trough, and locally overlaps westward into the Clinton and Newman troughs (see also table facing page 544).

Interfingering overlaps.—Exact correlations of such geographically limited formations are always difficult. The faunas often serve only in a broad general way, because they are frequently derived from distinct oceanic basins in which organic evolution progressed along lines peculiar to each. The most valuable aid in determining their age relations is occasioned by differential surface oscillation which permitted interfingering of overlaps from different directions, as, for instance, by alternations of westward transgressions of Atlantic waters into troughs more commonly occupied by invasions from the Gulf of Mexico and *vice versa.* The principle is fully illustrated by examples in a later section (see pages 538 to 568).

The broad differential movements of the continent and the minor oscillation of the interior structural domes described in preceding chapters under the term "tilting" have resulted in similar interfingering overlaps.

These have not only established that such movements actually took place, but their evidence has proved of the highest value in unravelling perplexing problems of correlation. Without such evidence the attempt to prove that the Carter division of the Stones River group does not outcrop on the east side of the Nashville dome, and that beds on that side which seemed to occupy the same stratigraphic position and were in fact correlated with the Carter by Safford are really Lowville, hence younger, might well have failed to convince. But the task lost most of its difficulties when a thin wedge of Lowville was found resting unconformably on the Carter near Nashville; and the correlation became easily convincing when typical magnesian Carter limestone was found at Highbridge, Kentucky, beneath a full Lowville section.

For similar reasons, although strongly suggested by differences in faunal contents, it would have been difficult to prove the alternating transgressions of Arctic, Pacific, and Gulf of Mexico waters, described on pages 367 to 371, if the thin wedges of northern formations had not been found in Kentucky and Tennessee and if it had remained unknown that the Kimmswick limestone passes under southward extensions of lower members of the Galena group in northeastern Missouri. Or, taking another case, to prove that the Maquoketa shale, which closely simulates the Utica in appearance and fauna, is really of middle Richmond age, if we did not know from unequivocal evidence in eastern Missouri and west central Tennessee that it overlies two lower Richmond formations—the Fernvale, and beneath this the Arnheim. Evidence of this nature is simply incontrovertible. In short, no other evidence is so trustworthy in disproving supposed synchrony of deposits which have been correlated chiefly or solely on fossil evidence, or because of apparent likeness in stratigraphic position, as that of stratigraphic overlaps. Nor is any so valuable in narrowing the limit of possible error in correlating disconnected lithologic units that happen to be included between two more widely transgressing formations.

PECULIARITIES OF INITIAL DEPOSITS AT BASE OF OVERLAPPING FORMATIONS

Ordinarily we expect to find a layer consisting wholly or in part of ill-assorted clastic material at and rising with the bottom of an overlapping formation. Sometimes it is a sandstone the grains of which are likely to be rounded; or it is a conglomerate the pebbles of which may have been derived from the immediately underlying rock or from many near to distant sources. Again the basal layer is a mudrock, in one case composed mainly of clay, in another calcareous enough to be called a limestone. As a rule, the material is imperfectly stratified, and often, especially between two limestone formations or between a shale and a limestone, it suggests agglomeration of loose surface matter into mud balls. When the transgression is over an old land area it is often red; occasionally it contains enough of iron to constitute a low-grade ore. At other times yellowish and green patches may be noted, while more rarely there is a suggestion of fossil black soil. In most cases this initial deposit consists mainly of wave-transported regolith, and its principal service is to pave the way for succeeding regular deposition by filling the decomposition and solution hollows in the old land surface now in course of submergence. Obviously, then, the deposit may vary greatly in composition, and it is always irregular in thickness and distribution. At one point it pinches out to a few inches or nothing; at another near by it may thicken to 50 feet or more. Its composition depends largely, and sometimes entirely, upon the character of the rocks exposed in adjacent contributing lands. Sometimes, however, prevailing winds and other climatic conditions that more commonly affect drainage and deposition may introduce relatively foreign constituents in considerable quantities.

The Chattanooga shale formation in Tennessee and in Arkansas and Missouri offers many interesting and instructive illustrations of overlap phenomena, and especially such as concern initial deposits. The features of the Tennessee occurrences have been described in sufficient detail by Hayes and Ulrich.[50] Those observed in Arkansas and Missouri are even more varied. In northern Arkansas the Sylamore sandstone, evidently a beach deposit, presents the prevailing aspect of a sandy basal deposit. This sandstone resembles the much older St. Peter sandstone, and in fact consists chiefly of the more or less cleanly washed soil of areas in which bodies of the St. Peter were then exposed. Though usually less than 2 feet in thickness, the Sylamore expands to much greater thicknesses in places where it fills depressions in the preceding land surface. Very interesting accumulations are found, too, that seem to represent fillings of late Devonian caverns in the underlying Ozarkian dolomites. Even more remarkable are the vertical, subtubular solution channels seen in the quarries near Clarksville, Missouri. These extended completely through a Niagaran dolomite before the Chattanooga submergence reached this point and filled the tubes with wave-swept black land debris.

When little or no coarse material, either native or foreign in derivation, is included in the initial deposits, the lowest bed or beds of the overlapping formation often resemble the underlying rock in composition.

[50] Geol. Atlas U. S., Columbia folio, No. 95.

If the latter is a shale, the first bed of the succeeding formation is likely to be an amorphous or obscurely stratified clay shale. If it is a limestone, then the new bed commonly is a highly calcareous, thin, and irregularly distributed mudrock. The resemblance is greatest when the sea transgresses over a sandstone surface. In such cases, as, for instance, in Arkansas and Missouri, where the Sylamore or some even younger sandstone is occasionally found in contact with the St. Peter or some other sandstone, it may be very difficult to draw the line between the old and the reworked new deposit.

Theoretically the littoral deposits of a sea should be of coarser grain than those laid down at depths beyond the reach of wave action. Strictly speaking, the facts are in essential accord with the theory, but viewed from the side of practical stratigraphy the difference is rather generally of small consequence and often quite negligible. Although a large amount of data respecting near-shore deposition in Paleozoic continental seas is in hand, the space here available permits only the briefest reference to a few. However, fuller references to these and other cases of obscurely indicated littoral deposits will be found in this and preceding chapters.

LITTORAL CONDITIONS RECORDED BY INITIAL DEPOSITS

Accepting as an established fact that nearly all diastrophically determined formations have an overlapping structure— the exceptions being formations based solely on lithologic differences—it follows that the basal layers of each constitute a continuous local record of littoral conditions during the time consumed in the respective transgressions of the sea. In each exposure of such basal layers we see, therefore, a representation of the littoral deposits of the sea at a particular time and place. Comparative studies of stratigraphic units carried on under this conception show that in the case of limestone formations the littoral deposits are but seldom, and at that only locally, conspicuously more arenaceous than the deeper water sediments. The relatively common exceptions are limestone formations, like the Boone in northern Arkansas, which transgressed old land surfaces, and in such situations begin with a few inches of sandstone. Despite the prevailing contrary conception, unquestionable major exceptions are few indeed. The best that comes to mind at this moment is the Joachim, a magnesian limestone in Arkansas, Missouri, and western Illinois that is regarded as having been deposited in the wake of the northwardly advancing St. Peter sea.

In a great majority of the limestone cases there is no sign of quartz sand at the base; and when such material is present it is usually confined to a layer less than a foot in thickness. Further, as suggested above, this layer owes its existence chiefly to the fact that the overlap extends to or across areas exposing older sandstone formations. For this reason a thin basal sandstone is a rather common feature of formations developed on the flanks of Ozarkia, whose broad dome of Eopaleozoic rocks comprises several important sandstone formations. On the Cincinnati and Nashville domes, however, where old arenaceous deposits are almost wanting, overlaps are but seldom indicated by quartzose initial deposits. The principal, not to say the only, exception is the Chattanooga shale, and this probably derived its basal sandstone from Ozarkia.

As a rule, the percentage of argillaceous matter in limestone formations increases toward the shore, but even this change is generally not very conspicuous. Indeed, it is with considerable surprise that we noted the small thickness of impure reddish limestone that separates in every way typical Pamelia limestone in New York and pure Lowville limestone in the vicinity of Kingston, Ontario, from the pre-Cambrian floor. Surprise was felt again when, after a long search near Theresa, New York, the contact between the base of the Pamelia (upper Stones River) and the top of the Theresa (earliest Canadian) finally was located, and nothing found, save an inch or two of fine conglomerate, to indicate that the hiatus is of sufficient magnitude to represent deposition of 5,000 feet of dolomite and limestone in central Pennsylvania. Also, again in New York, when we found no conglomerate at all, but only a thin "passage bed" at the base of an incomplete overlapping Lowville section to mark the even greater hiatus at Middleville, where it rests on Saratogan (Little Falls) dolomite, and at Little Falls, where the early Canadian Tribes Hill limestone is beneath. And all this on the flanks of old lands that at times have furnished large quantities of detrital material. After such observations it seems quite proper to find new limestone formations beginning with sediment no less pure than that above them.

STRATIGRAPHIC HIATUSES

Frequent absence of clastic material at base of overlapping formations not extraordinary.—That in many good exposures scarcely a trace is to be seen of the imperfectly stratified initial deposits referred to in preceding paragraphs will seem to many a strong argument against the assumption in such cases of emergent conditions. In other words, they may claim that in the absence of subaerially produced clastics, and perhaps of other clear evidence of land decay and erosion, other possible causes of stratigraphic hiatuses should be given preference. Willis[51] ad-

[51] Bailey Willis: Science, N. S., vol. 31, No. 790, Feb. 18, 1910, p. 249.

vances the suggestion that marine current scour is responsible for many of the local breaks in sedimentation, saying that "marine waters may not only deposit sediment, but may also prevent deposition, or even remove a deposit previously made." However, as shown on pages 362 to 375, marine currents can not possibly account for more than a very small percentage of the imperfections in the sedimentary column whose existence is proved beyond any reasonable doubt by faunal and physical evidence. So far as I can see, no other conclusion is possible under the circumstance than that land surfaces frequently existed in Paleozoic time which, on resubmergence, left no depositional evidence of subaerial decay and erosion.

But, after all, is the absence of such evidence to be regarded as extraordinary? I think not; and most certainly I can not concede that the condition is impossible, nor even improbable. Considering the kind of lands that must have prevailed in regions where such obscurely indicated emergences, the Eopaleozoic more particularly, are most common, the sedimentary record in the adjacent continental seas is as complete as it could be made under the circumstances. Let us, then, seek to interpret geologic history according to the facts as we find them and not solely according to criteria suiting conditions prevailing today.

The mere absence of distinctive shore deposits surely does not outweigh all the other evidence on which small low lands and shallow seas is inferred. This negative evidence goes no farther than to suggest that the lands were too low and the seas too shallow to favor the development and accumulation of such material. But there are other lines of reasoning that may be brought to bear on this problem.

That the Paleozoic interior lands were as a rule very low may be accepted as an established fact. Erosional processes consequently affected them only to a very small, not to say negligible, extent. There must have been some surface decay, but unless we assume that the lands were generally clothed with abundant plant growth—a most improbable condition prior to the Devonian—even this can not have been very great. Soil must have formed slowly and, with few or no vascular plants to facilitate its downward extension, it must also have required a long time to attain any considerable depth. Doubtless, too, its particles were very finely divided. Finally, study of Paleozoic unconformities shows that surface inequalities that are evidently due to irregularities in superficial rock decay became common only after the Silurian.

With slow submergence of such a land the advancing sea must have taken nearly the whole of the residual mantle into suspension. Some of it may have been swept into whatever hollows there were, but most of the finer particles were probably carried seaward and later deposited with the normal sediments of the time. If proper clastic material was available, then it was arranged to form a beach. If not, then whatever littoral fauna was present attached itself to or bored into the bare rocky floor, and was finally buried beneath normal marine sediment in which land detritus constituted but an insignificant part.

Time values of stratigraphic hiatuses. — The chief difficulty about overlaps and unconformities lies in the determination of the time value of the hiatus. None of the physical criteria are competent to give a reliable clue to its durance at any given point. A break corresponding to several geological periods may be no more clearly marked than the relatively brief interruption of sedimentation between two small formations or between diastrophically distinguished members of a single formation. In fact, in equally good exposures it is more difficult to detect the contact between lower Pennsylvanian and Ozarkian sandstones in the vicinity of Bolivar, Missouri, than to find the boundary between the Jefferson City and the Yellville, in Arkansas, or the sharp line between early and later limestones of the Black River group at many places in the Mohawk Valley in New York. It makes little difference whether the compared sequences are lithologically similar or dissimilar; the large hiatuses are, as a rule, no more clearly defined than the small breaks.

Faunal evidence alone gives an immediate clue to the time value of the hiatus. Wanting this, a fair idea of its relative importance may be gained by tracing the break into areas where the sequence is more completely developed. The stratigraphic method is illustrated by figure D on page 450, which shows how various formations are locally intercalated in the hiatus between the Boone and the Ozarkian Jefferson City dolomite in north Arkansas.

[*Editors' Note:* Material has been omitted at this point.]

PHYSICAL EVIDENCE OF SEA WITHDRAWAL

It is generally believed by geologists that in past geologic ages marine waters pulsated back and forth over areas now comprised in the continental land masses. The more important of these transgressions were legitimately inferred long ago on good stratigraphic and paleontologic grounds, and they rightly formed the basis on which geological time was subdivided. Progress in our science since its early and middle stages has been largely confined to the accumulation of data and to occasional attempts at refinement in the way of closer discrimination and readjustment of the long established major boundaries. More recently the local components of the several systems were subjected to critical study. On comparison it was found that great diversity of expression and development exists. The fossils were not exactly alike, the lithology was different, and the stratigraphic boundaries indicated in one area seemed not to harmonize with those in another. Evidently something was wrong.

The fault lies in holding too long to the idea of great continental seas and in overestimating their durance. Under this primitive and long treasured conception local absence of deposits of certain ages had to be ascribed to removal by erosion; but as the imperfections in the stratigraphic sequence continued to multiply, while the discovery of adequate amounts of erosion material, especially in the interior areas of flat-lying rocks, lagged far behind, and, more important yet, since detailed investigations of overlapping formations proved over and over again that their upper beds not only spread farthest, but were often preserved despite long exposure to subaerial conditions, some modification of the prevailing conception became imperative. The belief in broad, deep, and long enduring continental seas—seas that began early in the Cambrian and continued spreading wider and wider until well toward the close of the Ordovician—must be abandoned. In its place we should conceive of smaller, very shallow, and frequently shifting bodies of water, of seas that filled a given basin in one age and were withdrawn in the next, that returned again and again in familiar patterns, though perchance from different quarters, in succeeding geological ages. In short, seas that migrated in and out of the structural basins—sometimes extending far across the continents and at other times limited to much smaller areas—

as many times as the number of the diastrophically distinguished "ages" into which the geological time scale is divided.

This revised conception, however revolutionary it may seem at first, is after all merely a logical refinement of the old, an application of the same ineradicable principles to the minor as well as the major divisions of the geological column.

Evidence of sea withdrawal has been discovered in many parts of the stratigraphic sequence where none had previously been suspected. This is not because geologists then were either less diligent in their field work or less keen observers than are those of the present generation, but because they had bound their minds and perceptions by certain beliefs, and because they knew and employed only those criteria that under these conceptions alone seemed competent. They recognized shallowing of the seas and shores by the presence of ripple-marked and sun-cracked beds, by trails of creeping things, by tracks, impressions of rain drops, conglomerates, and other phenomena commonly seen along the present shores of the sea; and they knew that tilted marine deposits, the edges of which had been planed down and subsequently covered with approximately horizontal beds, signified previous elevation and subaerial erosion. Presently unconformities between horizontal beds were recognized, and, at first tentatively, then more confidently, interpreted as general elevation into land and resubmergence without conspicuous folding. And so it has gone on and on, perhaps unconsciously, to the present time, when some geologists at least, postulate land areas at almost all such times and places as are indicated by imperfections in the sequence of marine sedimentation. Conviction in this matter of frequent alternation of emerged and submerged conditions seems, indeed, to have progressed so far that the burden of proof now lies with those who deny the proposition rather than with those who affirm it.

PHYSICAL CRITERIA OF STRATIGRAPHIC UNCONFORMITY

Of the various physical phenomena indicating emergence, stratigraphic unconformities doubtless are the most common and most reliable. But it is only when the contact line is boldly irregular, or when the discordance in bedding is great enough to be easily measured, that the unconformity is immediately distinguishable from the average bedding plane. These strongly marked cases are relatively few as compared to the many that are far from conspicuous and which often require corroborative evidence before their true nature can be established. The characteristics of clearly defined unconformities are so well known that further description seems unnecessary. We will, therefore, proceed at once to the discussion of the more obscurely manifested instances.

Every sharp faunal or lithologic break suggests a stratigraphic hiatus and, hence, an unconformity and preceding sea withdrawal.[52] If the zone in which the suggested unconformity must, if present, occur can be narrowed down to a few feet by continuity of faunal evidence carried downward from above and upward from below the final detection of the plane of unconformity is usually accomplished quickly and satisfactorily. But when the fossils are scarce, or of kinds that are not very definite in their age relations, then the problem is more difficult. In such cases all sharp lithological changes in the undetermined interval should be investigated for possible clues. Assuming unfavorable distribution of conclusive fossil evidence, those unconformities which occur between two lithologically similar formations are, for obvious reasons, the most difficult to find. Still, many unquestionable unconformities between two limestone formations have been satisfactorily located. It is to be admitted, however, that in most of these cases abundant fossil evidence assisted materially. Instances of directly superposed unconformable shale formations without intervention of a bed of sandstone or conglomerate are more rare. When they occur the determination of the hiatus is almost impossible except by proving gradual intercalation of distinct lithologic units. The Woodford-Caney contact in Oklahoma, which shale formations are locally separated by the Sycamore limestone, is a good example.

When two well marked faunal zones, or two persistent beds easily recognized by lithic peculiarities, are traced from place to place and found to diverge or converge, as the case may be, the fact almost invariably indicates the presence of an unconformity in the intervening measures. The upper formation obviously must be limited below by overlap structure, and the first step in locating the unconformity is to find the line along which the additional beds are intercalated. All of these intercalated beds may be referable to the base of the overlapping superior formation, but when the lower formation has been truncated by erosion one or more may belong beneath the unconformity. Again,

[52] It is to be understood that the lithological aspect of fossils rather than their biological relations are referred to here. Viewed as lithologic features of the rocks, their significance is distinctly stratigraphic and useful in the same way that phosphatic pebbles, glauconite, and other mineral substances that are distinguished by peculiarities in crystallization, color, hardness, and composition are employed in the identification of certain beds. The stratigrapher can and often does use fossils in this limited way without pretense to paleontological training. Of course, with such training fossils assume the additional significance that enables the paleontologist to estimate the relations of isolated stratigraphic occurrences to the generalized standard time scale.

as shown in figure D, page 450, the intercalated beds may themselves bear unconformable relations to both the superposed and the underlying stratigraphic units.

The principal lithologic criteria employed in locating the line of unconformity were mentioned in discussing the peculiarities of initial deposits in relation to overlaps (see page 454). Here we need to emphasize only those bearing more directly on sea withdrawal. The most common of these, and from that standpoint the most important, is the red or rusty color that pertains so frequently, though perhaps but locally, to deposits immediately following an unconformity. While practically all red deposits are either directly or remotely related in their origin to subaerial decomposition processes, and are thus connected with land conditions, it does not by any means follow that all deposits of this color are indicative of stratigraphic unconformities. On the contrary, the larger accumulations of red sediments are not directly associated with such structural phenomena. Often they begin considerable distances above the recognized unconformity which, moreover, may show no red deposits at all, even when a basal conglomerate is present. An instance of this kind was observed in the Wichita Mountains of western Oklahoma, where presumably Permo-Carboniferous "red beds" overlapped on early Ordovician limestone. It is the thin beds of this color, say from an inch or two to not over two feet in thickness, that separate two limestones, or a calcareous shale from an overlying limestone, that are especially significant in this connection. This layer is but rarely well stratified, and usually it is softer than the rocks on either side. In weathering it breaks down quickly, leaving a conspicuous, hollowed-out space. Very commonly, too, it is marked by a line of springs. In some cases it seems probable that the rusty color is not original, but is to be attributed to percolation of ferruginous waters.

[*Editors' Note:* Material has been omitted at this point.]

CORRELATION BY DIASTROPHIC MOVEMENTS

General discussion of principles.—Diastrophism, the initial cause of physical and organic change, is urged by Chamberlin as the ultimate basis of correlation. The essential features of his conclusion are ably presented by this writer in a recent short paper on the subject[69], and also by Willis[70] who emphasizes certain aspects not definitely stated by the former. Assuming the principle of periodicity of great deformations, both conclude that the baseleveling of the land in the relatively quiescent stages between the periods of active diastrophism "means contemporaneous filling of the sea basins by transferred matter," and that the resulting advance of the sea "is essentially contemporaneous the world over" (Chamberlin op. cit., page 690). Willis thinks "the eras of inactivity, the baselevel eras for the whole world," were very long, and because of "their very great duration, from which their essential contemporaneity results," they are unfitted "for any except the broadest outline of classification." Further on he adds that these baselevel eras "afforded the best conditions for correlation by the other criteria." Regarding periods of active diastrophic movement he says, they "have been shorter than the eras of baseleveling, and consequently define time divisions, which are more nearly commensurate with those of current geologic standards. They are, however, still long, and their value is in broad fundamental classification." That Willis believes only very crude, generalized, correlations are possible by means of continental and orogenic movements is evident from the following statement: "Disturbance and quiet, erosion and continuous deposition, unconformity and conformity, have developed simultaneously in immediately adjacent districts many times" (Willis, op. cit., pages 255, 256). Finally, according to this author, only the occasional world-wide ebbs, resulting from subsidences beneath the oceans, would seem to constitute anything like an exact measure of contemporaneity.

Chamberlin and Willis have presented the theoretic aspects of the principles of diastrophic correlation in sufficient detail, but, so far as I know, no practical application of the proposition, except in the broadest and most generalized terms, has been published. Indeed, such application is no easy matter, the problems being exceedingly complex and quite insolvable except with the aid of the most refined paleontological methods. Chamberlin probably recognizes this fact when he says "if we add the biological element the case is immeasurably strengthened;"

[69] T. C. Chamberlin: Diastrophism the ultimate basis of correlation. Journal Geology, vol. 17, 1909, pp. 685-693.

[70] Bailey Willis: Principles of paleogeography. Science, Feb. 18, 1910.

but he seems to underrate and misapprehend the difficulties when he adds "for correlation by cosmopolitan faunas, the very best of faunas for the purpose, is added to the physical correlation." Aside from the pelagic faunas, and using the term in not too broad a sense, is there, or has there ever been, a truly cosmopolitan fauna? For reasons given on page 366, the forms comprising the so-called cosmopolitan faunas, instead of being contemporaneous, are more probably either later or earlier stages of slowly modifying species, and, consequently, of little value in exact correlation.

Except those afforded by pelagic and free swimming species, there are no highways, no "short-cuts" to world-wide correlations. On the contrary, stratigraphic correlation, whether we use physical or biological criteria, or both, is a slow, laborious process, each method requiring constant corroboration by the other methods, with no certainty of results until we have closed the circle and proved the process. As we proceed from a given locality, the organic and the lithologic evidence first depended upon grows ever weaker until one or both pass out entirely. Fortunately, the continuance of diastrophic lines frequently enables us to mend the broken thread; and herein diastrophism proves itself not only the theoretic basis of correlation, but its criteria show themselves to be eminently practical.

Though fully endorsing Chamberlin and Willis's claims for diastrophism as the ultimate basis of correlation, and though affirming the exceeding value of diastrophic criteria when abundantly checked and corroborated by paleontological and lithological evidence, I must yet frankly admit that by themselves these criteria are utterly valueless in correlating widely disconnected outcrops. Diastrophic criteria are so variable in local expression, and their successive manifestations so similar in general aspect and range of variability that positive and unerring discrimination seems impossible. We may determine the position of certain stratigraphic boundaries, as between bodies of limestone and shale or sandstone, or we may note and trace some persistent zone of overlap and possible conglomerate, but what these lines mean in the time scale may, as a rule, be determined only by fossils. Moreover, it has frequently happened that the most important boundaries were entirely overlooked when the geologist took into account solely the obvious changes in character of deposits. That the supposed lithological unit included one or more great discontinuities of sedimentation, or that the formation spanned two or more systems and thus made no provision for the discrimination of important, elsewhere perhaps more clearly defined, geological events, was either unknown or not considered so long as the beds seemed conformable and were of the same general lithologic type. With the aid of fossils, however, not only important breaks in deposition came to light, but also constant and easily recognizable differences in petrological characters were found that by themselves enabled the field geologist to distinguish the several parts of the previously supposed unit. Moreover, the fossils rarely failed to substantiate the persistence of the more striking lithological boundaries.

Willis's evident pessimism respecting correlations seems grounded in the belief that the continents, though permanent and essentially stable, were locally affected by periodic but otherwise unrelated movements, and that the continental seas were long enduring, often broad and deep interoceanic waterways, the sediments and faunas of which varied greatly from place to place according to nearness or remoteness from shores and to varying local efficiency of currents. My conception is very different. I see mostly small, shallow, often disconnected basins, occupied at times by Atlantic, and at other times by Gulf of Mexico or by Arctic waters. These were filled and emptied many times, and on each occasion the size and form of the basins differed more or less. As to the continents, I, too, believe they were permanent in general features and that they were affected by many deformative movements, but these movements were always related to a definite, rhythmically progressive plan. Under Willis's view detailed correlations are as impossible, as he believes, because it affords no means of determining the time relations of relatively local physical and organic phenomena to the general scheme of geologic events; under mine definite and often very detailed correlations can usually be made because it is based primarily on the displacements of the strandline which, of all geologic processes, were the most widespread and most nearly simultaneous in their operation. Under his view, again, the local imperfections of the marine stratigraphic column were largely occasioned by fortuitous conditions of uncertain character and relations and not, as under mine, by the periodic retreat and consequent absence of marine waters. Finally, in that it insists on land barriers between continental seas and on the integrity of the invasions from the several oceanic basins, my view, in contradistinction to Mr. Willis', offers sound and efficient reasons for the regional variations of the faunal aggregates.

Summary of diastrophic criteria.—Concisely stated, the criteria of diastrophism embrace all physical, and to a certain extent all organic phenomena implying horizontal and vertical movements of the crust of the lithosphere. These deformative movements were not always in operation at the same time, or entirely similar in degree of activity and in character of effects produced, on different continents. They varied also very greatly in intensity and character on different parts of the same continent. Often one part was in course of submergence while another remained emerged or was being elevated. The broader of such diverse movements are indicated more especially as affecting the continents in a north-south direction, but tilting, usually more limited in scope, occurred frequently also in other directions. The alternating Arctic and Gulf of Mexico invasions which, as partly described in preceding chapters of this work, occurred during the middle Ordovician, Silurian, and Devonian, are good examples of the broader differential movements.

Again, in certain parts of the continents, as for instance in the lands bounding the Arctic sea and in the broad median flats of North America, movements in either horizontal or vertical directions were relatively slight. Indeed the former were so broadly disseminated that their local manifestations are scarcely determinable. The vertical movements of these regions were similarly wide in their operations, but in this case, though the maximum vertical displacement was perhaps never very great, their ultimate effects on the size and pattern of the continental seas were manifestly of great and patent consequence.

Though all parts of the crust were affected, nearly the whole of the conspicuous orogenic results of the movements are confined to tracts adjacent to the borders of the continents; and in these the north and south trending orogenic wrinkles increase in development from the poles toward the equator. On the whole, further, the results of the east-west movements are greater on the west borders than on the east. There were also north-south movements of equal and possibly greater magnitude. These resulted in excessive folding and mountain building. The resultant ranges have a generally east and west trend, and were best developed in belts bordering the Gulf of Mexico in the western hemisphere and the Mediterranean and Indian Ocean in the eastern. These north-south movements differed from the east-west, in that while they caused land elevation in the belts mentioned, concomitant subsidence occurred in the Polar regions.

[*Editors' Note:* Material has been omitted at this point.]

PART III—STRATIGRAPHIC TAXONOMY

Principles of Classification

GENERAL DISCUSSION

In the study of geology, the first essential, of course, is some competent and consistent method of classifying geologic events. Hitherto the paleontological method—that is, the advent and disappearance of fossil faunas and floras—has seemed to serve the purpose. In later years, however, accumulated observations have tended to show that fossil evidence is not so uniformly reliable as had been supposed; that not only single species, but associations of species sometimes occurred at lower horizons, at other times in younger beds, than the one to which they were credited in the standard section. At first it was thought that the interval between these varying occurrences expressed the time consumed in the gradual dispersal or migration of the organisms. Later when three or four appearances of the same fauna were found to occur at long intervals above each other in the same section, paleontologists began to speak of "shifting of faunas." This shifting was supposed to indicate involuntary migration from place to place within broad long-enduring continental seas in response to local change in environment. These observations naturally suggested distinct paleontologic and lithologic classifications and dual nomenclatures, which while checking each other in a general way were necessarily independent in detail.

All such suggestions are believed to be detrimental to the progress of stratigraphic science. There is only one method of classifying geologic history and that is the chronologic; and in the determination of the sequence of events and in drawing the time boundaries every criterion, be it physical or organic, should be employed. In correlation each set of criteria is found to have its limitations, and none is alone sufficient for the purpose of the taxonomist. Nor can we justly say that the paleontologic criteria are more important than the lithologic or diastrophic. The organic criteria doubtless are the most useful and in many cases are by themselves determinative where the physical evidence by itself may be quite incompetent. But as each method of correlation commonly requires the corroboration of the other criteria before a detailed final determination of the geologic history of a given area becomes possible, and as each fills its own place and carries out indispensable functions, they are all equally important. As a rule the several methods are logically supplementary. In studying a section the fossils give us the general position; the lithologic criteria checked by fossil evidence narrows the limit of possible error and the diastrophic criteria finally establish the exact location of the boundary sought.

Regarding such ideas as "shifting" and migration of faunas and floras, providing their conception contemplates time consumed in migration of sufficient duration to be definitely expressed in the geologic time scale, or if they are promulgated with a view to show the futility of endeavors to establish the contemporaneity of geographically separated occurrences, such views, it seems to me, are practically groundless. They do not sufficiently take into account the immensity of geological time. I tried to illustrate this with the case of a living shell, *Littorina littorea* (page 295), which has extended its range in the past fifty years so fast that at the same rate it might encircle the globe many times before the smallest stratigraphic unit now thought correlatable over wide areas could be laid down. The groundlessness of the fear that shifting of faunas may seriously affect the accuracy of correlation by fossils is all the more apparent if my belief that marine faunas were modified and developed only to a very limited extent in the continental seas is accepted. (See pages 495 to 501.) In my opinion, the evidence as a whole points very strongly to the conclusion that faunas passed from the oceanic basins into and about in the continental seas so rapidly that their appearance in New York and central Kentucky or Tennessee is to all intents and purposes of stratigraphic correlation practically simultaneous.

The presence or absence of a fauna in a given continental sea at a certain time is wholly dependent on whether or not its migration is opposed by physical obstacles. When conditions for expansion of range were favorable, then the fauna instantly took advantage of them; if they were not propitious, then the organisms either succumbed or they retreated to regions where existence was still possible. In the shallow continental seas life conditions could never have continued unbrokenly favorable through long ages; hence the faunas which inhabited them required continual replenishment. And it is to be remembered that geological time was always long enough for the consummation of every such purpose. Also that the fossil record of these small breaks in the life history of the beds is rarely measurable stratigraphically.

The modification of species and faunas that originated and developed in the same oceanic basin was very slow. This is shown by comparison of successive invasions from a given oceanic basin. Even when we can prove by stratigraphic evidence that long ages intervened the change is often deceptively small. The Hamilton, the Spergen, and the Catheys-Fairview "recurrences" discussed in foregoing parts of this work are convincing illustrations. And there are many other instances of confusing similarity in general aspect of long separated faunal invasions. Compare, for example, the late Stones River (Lebanon and Carters) fauna with the Lowville fauna. These two appearances of the Gulf of Mexico fauna are in general very much alike, and except for the constant presence in each of a few well marked diagnostic species their discrimination would be a matter of grave difficulty. And yet, as shown on pages 554 to 557, the Lowville is sufficiently younger than the last of the Stones River to permit deposition of the Blount group, aggregating between 3,000 and 4,000 feet of limestone, shale, and sandstone in east Tennessee.

When a decided generic change is noted between two immediately superposed faunas it means one of three things. Often it signifies that a great stratigraphic hiatus lies between them. In other cases the two faunas invaded from different oceanic realms. In the remaining cases some diastrophic change occurred within an oceanic realm, in different parts of which two or more facies may have been developing simultaneously. In the first two conditions the causes are obviously sufficient to produce the noted effects. The last cases, however, are not so simple. In these we have no sure means of determining what actually took place. It may have been an extraordinary restriction of normal habitat, as when great emergence of the continental shelf occurred. Or it may be that some land barrier, perhaps an intercontinental connection, was submerged, permitting influx of types previously excluded. Again, a passage, as, for instance, that between the Caribbean Sea and the Gulf of Mexico, may have been closed at times when the Antilles were connected with the Floridian peninsula. Finally, an intercontinental land connection may have been temporarily established, as between eastern Canada and northern Europe, which would have favored shore migration that at other times was impossible. Any of these possible conditions, if they occurred, as it is believed they did, must have had a decided effect on the composition of faunas that invaded the inland seas.

Considering the many possible conditions that are competent to cause breaks in the faunal record of local stratigraphic sequences, whose magnitude is in nowise proportionate to their respective time values, the varying import, hence the inadequacy, of fossil evidence as a primary factor in the construction of a detailed geological time scale is apparent. The only possible basis for this scale lies in the stratigraphic relations of the beds containing the fossils and in the movements which have occasioned the shifting of the seas in which the beds were laid down and in which the fossil organisms lived when conditions were favorable. The remains of the latter are endowed with taxonomic attributes only when their stratigraphic relations have been determined; and at that, except in a broad and generalized way, the application of the life sequence thus worked out is limited by provincial boundaries. A new sequence must be worked out for each of the other provinces.

The stratigraphic taxonomist, therefore, must be first and always a stratigrapher. But, so I may not be misunderstood, I shall add that a geologist can not be a self-competent stratigrapher unless he is equipped with a considerable working knowledge of fossils. It is not required of him to know that a certain guide fossil is a *Hebertella* and not a *Dinorthis,* or that the species which was formerly called *Orthothetes subplanus* is now referred to under the name *Schuchertella subplana.* Such changes are in systematic biology and have no vital effect on stratigraphic taxonomy. What he needs is the ability to discriminate closely between allied species and faunas and to recognize the individual forms and their associations when he sees them again.

Whatever the cause of the faunal change, be it great or small, local or general, it probably was occasioned by some crustal deformation, and movements of the lithosphere necessarily were accompanied by corresponding displacements of the strandline. It is therefore by the correlation of these displacements, which involves the use of all the criteria—the physical no less than the organic—that we finally establish the sequence of marine deposits and define the stratigraphic units, whose classification is the chief object of this work.

[*Editors' Note:* Material has been omitted at this point.]

Cardinal principle of the new stratigraphy.—The point that should be continually borne in mind is that the accessible stratigraphic sequence, at whatever locality and however obscure the breaks, is always incomplete. The more complete the stratigraphic record the more numerous the hiatuses, the fewer the breaks the greater their average time values; and the relative conspicuity of the stratigraphic break is never a safe indication of its importance. A sandstone may be followed without intervention of other deposits by another sandstone, or it may be two shales or two limestones that are in such close contact that continuous deposition is suggested; and yet a great time break with emergence may often be proved to have separated them. (See pages 526 to 532.)

Neither is the relative volume of deposits nor their areal extent a safe criterion. Two or three periods may be locally represented in a few feet of sediments. This is really a common occurrence on the flanks of the Nashville dome, where the Silurian and Devonian are often represented by thin overlap wedges. On the flanks of the Ozark dome, indeed, not only the Silurian and Devonian but the Ordovician and Canadian systems as well are locally absent and never strongly represented. Here, then, four periods may be represented in a varying sequence of deposits aggregating but a few hundred feet in thickness. If these fragmentary records were all that was known of these systems, it would be very difficult if not impossible to prove their true significance in the time scale. Fortunately, each is represented elsewhere by thousands of feet of sediments. More fortunately still, we can prove by satisfactory evidence that the thin wedges represent only small parts—usually the upper part—of widely spreading formations of their respective periods, which happened to extend by overlap to accessible situations on these flanks from some other area in which a more complete record was laid down.

Of all the features of the new conception, the smallness of the continental seas and the frequency and rhythmic occurrence of the oscillations, which caused withdrawal and shifting of the seas from one basin to another and thus interrupted the process of sedimentation, will perhaps seem to the average geologist the most difficult to grasp. But these together constitute the very backbone of the revision. Without them the distant correlations largely fall to the ground or remain indefinite and unsystematized as they have always been.

It is not claimed that the basic ideas are new. The revision merely applies them more generally than was heretofore contemplated. We have long recognized that some of the Paleozoic seas were small, that they occasionally shifted their boundaries and that there must be something akin to rhythm in the great deformative movements that were more or less clearly suggested in the stony record. But these phenomena were treated as exceptional and local and not as definitely connected with a general systematically developed diastrophic process. That the strandline moved in obedience to the gradual development of this process and registered its stages, and that it therefore offered a definite and easily applied basis for stratigraphic correlation and classification, was only partially acknowledged. Chamberlin and Salisbury recognized the importance of the factor in so far as the major divisions of the stratigraphic column are concerned, but the demonstration of its applicability to minor divisions no less than the major has been left to the present occasion.

Method of dividing geologic time and definiteness of time units.—In stratigraphic taxonomy the first essential is the selection of some method or principle determining when a geologic age has ended and a newer age begun. Obviously such a principle applies equally to all the divisions of the time scale, the larger divisions, up to periods and eras, being but combinations of the minor units. In other words, the base or top, as the case may be, of a terminal unit of a group, series, system, or era at the same time correspondingly delimits the division of higher rank of which it forms a part.

In practice the stratigraphic terms system, series, group, and formation are treated as respectively equivalent to the time terms period, epoch, stage, and age. In fact, however, there is a difference. The latter refer to parts that taken in serial order completely fill geologic time. The former, on the contrary, account for only so much of this time as is represented by sedimentation in places now accessible. New discoveries. however, may at any time necessitate intercalation of distinguishable and hitherto unknown stratigraphic units and corresponding revision of the time units. Moreover, some of the inaccessibly recorded parts are occasionally penetrated in deep wells, but with such relatively unimportant exceptions they are truly "lost intervals."

Under this conception a geological age or time unit of the lowest rank determinable by diastrophic movements may be defined as a definite span of time. Displacement of the strandline being accepted as the dominant criterion a geological age is regarded as having closed when the marine waters were largely or wholly withdrawn from one or more of the continental basins, the succeeding new age as having opened when the sea again began to advance in the same or in other basins. Accuracy in the delimitation of time divisions therefore depends solely on our ability to correlate exactly the stratigraphic breaks in neighboring and distant basins. First we must be able to decide that the stratigraphic hiatus in one case is, geologically speaking, small, in another great. Then we must be able to say just how much of the hiatus is elsewhere represented by deposit. Finally, we should know whether or not a stratigraphic break is discernible at the base of the most complete depositional record known of a given age. When all these factors have been determined then, and only then, will we have exhausted the possibilities in the way of exactitude in delineation of time units.

Although the subject, in so far as the Ordovician in America is concerned, is far from being exhausted, it is yet true that great progress in detailed correlation has been made in the classification of the rocks of this period. And what is possible in the Ordovician is certainly more readily achievable in the case of the more widely accessible and as a rule more highly fossiliferous younger systems. It is only the older systems, which are largely confined to disturbed areas and in which fossils are less generally distributed, that present greater difficulties.

Revision of methods as well as facts of stratigraphic classification.—It will have been observed that my revision affects the methods as well as the facts of stratigraphic correlation. By method, I mean the manner of determining what constitutes a geologic period or system, an epoch or series, or a stage or group, and, more particularly, how the boundaries of these time and stratigraphic divisions are to be drawn. Concisely stated, the method followed is to divide the stratigraphic sequence at the first plane beneath the introduction of a new fauna or beneath a marked faunal change that exhibits evidence of diastrophic movements. If the plane marks a great faunal break, providing the compared faunas invaded from the same oceanic basin, and especially when the plane corresponds also to a considerable change in the provincial boundaries that had prevailed during the greater part of the preceding period, then it seems to me it marks the beginning of a new system. It is on such grounds that the old Mississippian is divided into two systems—the

Waverlyan and the Tennessean—also, that the Eopaleozoic is divided into four systems instead of but two.

Why the Waverlyan and Tennessean are systems and not series.—A system is divided into series on similar grounds, only in these cases the breaks are commonly of a lower order and the change in provincial boundaries less extensive. We may illustrate the difference by comparing the relations of the Waverlyan to the Tennessean on the one hand and the interrelations of the series divisions of these two systems on the other. Thus, during the Waverlyan, the continental seas were developed in southeastern North America in three fairly well distinguished basins or faunal provinces. Each of these provinces, furthermore, is characterized by its own lithological sequence. At times, also, the seas shifted so that only two or perhaps but one of the basins was submerged. Named from their characteristic formations, the eastern basin, with its shales and sandstones, may be called the Cuyahoga basin; the middle area, with its shales and cherty limestones, may be called the Fort Payne basin; the western province, with shaly beds beneath and at the top and rather pure crystalline limestone in the middle, may be called the Burlington province.

It may be questioned whether all three of these basins contained marine waters at any one time during this period. I believe this occurred during the Chattanoogan and, to a considerable extent also, in the New Providence-Fern Glen-upper Cuyahoga age, and again in the Keokuk-Fort Payne-Logan-Grand Falls age. The point is not of vital importance here. It is enough to know that, as a rule, the Waverlyan seas occupied considerable parts of at least two of the basins, and that the greatest shifting during the period occurred at times marking the close of the series into which the system is divided. Thus, as shown in the Maverlyan correlation table, the Chattanoogan ends with its most widely distributed stage; the Kinderhookian is characterized by frequent oscillations and ends with its smallest (Chouteau) stage; the Osagian begins with the extensive Fern Glen stage, continues with the restricted Burlington stages, and ends with another eastward tilt, during which the Keokuk was deposited in Iowa and Illinois, the Grand Falls in Missouri and the cherty Fort Payne limestone in areas to the southeast of these States.

As indicated, the Waverlyan oscillations merely shifted the seas about within the area covered by the three basins. The movements consisted almost entirely of simple tilting of the area as a whole. The break between the Keokuk and the Warsaw stage of the succeeding Tennessean period, however, as described on page 588, is marked less by tilting than by warping of the surface. The Warsaw, and to a greater extent the

Spergen, disregarded the boundaries of the faunal and lithological provinces which prevailed during the Waverlyan and spread with surprising uniformity of expression from Missouri to Alabama and thence northeastward in the Allegheny basin to west Virginia and perhaps to Maryland. There was a change, also, in the northward extent of the seas, none of the Tennessean stages reaching as far in that direction as did certain Waverlyan formations. In short, a new set of conditions was introduced in southeastern North America with the Warsaw that thereafter prevailed—with minor oscillations and tilting—to the close of the Chesterian.

CONSISTENCY OF METHOD IN DRAWING STRATIGRAPHIC BOUNDARIES

Discussion.—The method of dividing geologic time has varied largely according to the information and individual opinions of authors. Doubtless this is as it should be, for it would be a sad condition indeed if the science of geology, with special reference to its taxonomic aspects, should ever become so crystallized and fossilized that any scheme of classification, however well it may satisfy contemporaneous knowledge, were accepted as final. Fixity in method is death to progress, and, to say the least, it is presumptuous to even hope that any one has said or ever will say the last word on a scientific subject.

But there is one feature of method that may be fixed without detriment to science, namely, consistency in application. Of course, the success of our endeavor to be consistent depends largely on the state of our information. To that extent therefore it is just as liable to failure as any other feature of method. But it fails only in details, which can be corrected from time to time. Consistency therefore is a fundamental factor that can stand whatever happens to the scheme temporarily adopted.

In the old classifications undue prominence and rank was given to the younger formations. This is because they cover wider areas, are more commonly accessible, and contain a larger number and more beautifully preserved fossils than the older formations. Their fossils also were more readily comparable with living organisms, while the relative strangeness of the Paleozoic species naturally delayed the full recognition of their geologic significance. Moreover, most of the large cities of Europe are situated on Cenozoic and Cretaceous deposits, a circumstance that obviously contributed to their early investigation. As for the Triassic and Jurassic, these like the overlying Cretaceous and Tertiary formations, are well developed and highly fossiliferous in the Alpine regions of Europe.

whose grandeur of scenery and general interest attracted many geologists. Naturally we make the most of that on which we are best informed; and obviously, too, relative accessibility and ease of determining the sequence of deposits are matters on which the state of our knowledge largely depends. Thus compared the contrast in relative stability of classification between the Cenozoic and Mesozoic rocks on the one hand and the Eopaleozoic formations on the other is sufficiently evident to make further comment unnecessary.

Geological ages—that is, the minor divisions of the time scale—being distinguished from preceding and succeeding ages by the reversal of the movement of the strandline it seems desirable, on the ground of consistency, that the same principle should govern in deciding the limits of the larger time divisions. In other words, that we should endeavor to draw the boundaries between stages, epochs, and periods at horizons indicating the greatest and most widely recognizable stratigraphic hiatus or the greatest shifting and change in pattern of seas in the comprising part of the column. In the case of periods it should be drawn with due regard to the matter of rhythm in the progress of geological processes. (See pages 398-403 and 599-607.) In the case of eras it should be at the most generally recognizable horizon of this kind occurring in the midst of beds that in composition and structure indicate activity of diastrophic processes of the highest grade.

Hitherto there has been no uniformity in method of drawing boundaries of time and stratigraphic divisions below the grade of eras. One author drew the base of stratigraphic divisions at horizons thought to represent the beginning of their typical physical and organic development; another began the new formation, group, series, or system at the first introduction of species or faunas, or at the base of beds that were thought to be introductory to conditions assumed on the ground of experience to be typical of the new age. The former method, which commonly includes that of classification by lithologic criteria and the often very differently resulting method of classification by percentage or dominance of characteristic fossils, was possibly justifiable under the old conception of large continental seas and continuous deposition through one period into the next. Under the newer conception of small continental seas and frequent interruption of sedimentation it was seen to lead to so many inconsistencies and errors in correlation and classification that it is now thought wholly indefensible. Its greatest source of error lies in the fact that it entirely ignores the diastrophic factor of geologic history when this factor is not clearly indicated by sharp changes in the character of sediments or by the structural relations of the beds. At their best the hitherto prevailing methods of drawing stratigraphic boundaries often obscure general as well as local geologic history, because the facts relating to changes in sedimentation are not brought out in their proper relationship. In many such cases, referring more particularly to formations, the so-called "transition beds" have been referred to either the top of the lower or the base of the upper formation without considering the probability of a stratigraphic hiatus beneath them. If the transition beds happen to be initial deposits of the upper formation, but agree lithologically more closely with the underlying formation, to which therefore they have been referred, the geologically important break is ignored while a relatively insignificant boundary is recognized. Many instances of this kind might be cited. In the case of larger divisions it happens occasionally that the fauna of a certain formation consists almost entirely of species whose alliance, for instance, are strongly Silurian, while elsewhere a synchronous or even a younger fauna is dominated by species recalling only Ordovician types. According to the old method one would be referred to the Silurian, the other to the Ordovician. On similar grounds contemporaneous formations have been referred to different series or groups. Oftener, however, general similarity in faunas found in locally superposed formations, or in geographically separated formations, which on the ground of such similarity may have been correlated, has caused their reference to the same group or series when in fact they belong to distinct groups or series. Examples of all of these conditions have been discussed and the difficulties explained in preceding parts of this work. And many more are indicated on the correlation tables appearing farther on. At the risk of repetition a few may be cited here.

[*Editors' Note:* Material has been omitted at this point.]

The principle of introductory facies of faunas and deposits—Discussion and illustrations.—The relations of the Sainte Genevieve to the succeeding beds of the Chester series are essentially the same as those of the Warsaw to the Spergen. They are also like those of the Richmondian to the Niagaran and of the Helderbergian to the middle Devonian. More exactly, though, they are comparable with the relations of the Lowville to the middle and upper Black River Mohawkian formations and of the Keyser limestone to the succeeding members of the Helderbergian. In all of these cases the first named stratigraphic unit—at least its lower part—commonly resembles the next underlying deposits more than it does the overlying beds. Yet, if our classification is to be chronoligic and based on diastrophic principles in which neither the purely lithologic nor the paleontologic criteria are of paramount importance, these units should be referred to the younger and not the older group, series, or system, as the case may be. In each of the mentioned cases I have endeavored to associate and classify the formations in accord with the principle that, quoting from Weller (op. cit., page 290), "the time boundary between two geological ages (periods) should be marked by the time of maximum withdrawal of the sea or the subsequent readvance during which new sets of conditions were introduced." From this quotation, which most geologists will indorse, it is seen that we agree in principle and theory. Differences arise chiefly in practice, when principles are frequently forgotten and mere likeness of deposits and absence of conspicuous evidence of diastrophic breaks in sedimentation is likely to dominate conclusions. In the case of the Lowville-Stones River contact its importance and really great significance has been shown in a preceding part (see pages 553 to 557). The Cayugan-Helderbergian contact also is now conceded to be unconformable in all known exposures and the usual hiatus between the two is admitted to be of greater time value than had been supposed. The same, again, is to be said of the Maysville-Richmond contact. The structural evidence of the stratigraphic hiatus in all of these instances is surely quite as clear as in the late Niagaran-Onondaga contact at Louisville, Kentucky, and at Newsom, Tennessee, and in the similar contacts of Silurian and Devonian beds at Buffalo and elsewhere in New York—illustrations cited by Schuchert and generally accepted by geologists as satisfactorily established cases of long but inconspicuously marked time breaks. Under the circumstances, especially since stratigraphers are now practically a unit in placing the Helderbergian at the base of the Devonian, it seems to me we should insist on consistent treatment of all such cases. In other words, if the Helderbergian is Devonian and the Lowville is the lower part of the Black River group, then the Richmondian is Silurian, the Keyser is Helderbergian, the Warsaw is Tennessean, and the Sainte Genevieve is early Chesterian.

In applying the principle of introductory deposits to parts of formations, particularly when the underlying formation is of clastic matter and the overlying of limestone, general practice has been far from consistent. As a rule, however, these inconsistencies have resulted through lack of data and information. With plenty of determinative fossils, a mistake of this kind is unlikely. Without good fossil evidence, positive determinations are often difficult, but it is seldom indeed that close study of the structural relations of the "passage" beds fails to solve the taxonomic problem.

It may be set down as a general proposition that when a break in sedimentation is indicated at the base of a clastic bed whose top grades, without evidence of interrupted submergence, slowly or rapidly into less clastic deposits, then the new formation or group should include the former as an initial deposit, even though it is lithically much more like the underlying formation than it is like the overlying sediments. The Kiefer sandstone in Maryland affords an excellent illustration. This sandstone has hitherto been referred to as the top member of the Clinton. On investigation it was established that the Kiefer is a locally developed deposit, that it passes without break into the overlying sandy shales of the McKenzie formation, and that it is bounded below by an unconformity which in places cuts out the Rochester member or formation of the Clinton group. The intervening hiatus is regarded as representing the upper or Chicago group of the Niagaran series. On these grounds, which are supplemented by good faunal evidence, the Kiefer sandstone is now classified as a local basal member or facies of the McKenzie formation.

The Hardin sandstone in west central Tennessee and the Sylamore sandstone in north Arkansas are similarly referred to the Chattanooga shale. These, however, are true basal deposits of a widely overlapping formation, differing from the preceding case in that the age of the bed varies from place to place according to the stage of the transgression. The relation of the Tellico sandstone to the Ottosee shale is more like that of the Kiefer to the McKenzie, only in this case the lower introductory member is of sufficient thickness and areal extent to rank as a separate formation.

In applying the principle great caution is always to be observed in determining the presence or absence of a stratigraphic break at the top of the clastic bed. Thus the contact of the Saint Peter sandstone and the succeeding limestone in the upper Mississippi Valley often suggests gradation from the sandstone through shaly sandstone and calcareous shale to the limestone. In fact, however, there is an important hiatus at the top of the Saint Peter, and the apparent transition suggested by the overlying mixed material is occasioned solely by the sandy character of the redeposited regolith of preceding Saint Peter land areas. In this case, then, only the variable and mixed intermediate bed, and not the Saint Peter itself, is to be regarded as the introductory deposit of the overlapping limestone formation. The Oriskany-middle Devonian sequence in the Appalachian Valley and in New York exhibits precisely analogous conditions.

The first appearance of fossil species and faunas.—As a general conception the first introduction of organic types is regarded as marking the beginning and not the closing term of geologic ages. It is deemed a valid interpretation even when the general composition of a fauna which contains a considerable number of such types is greatly like some older facies. This opinion and consequent practice is based (1) on experience, which teaches that reappearances (survivals) of faunal facies are more common in stratigraphic history than are preexistences, and (2) on the fact that the sedimentary record was frequently broken by inaccessibly or poorly recorded intervals in which new things were evolved without complete extinction of the old. The essential truth of the conception is clearly proved by prevailing views respecting reappearances of faunas, whose discrimination depends on a relatively small number of diagnostic new forms by which each is distinguished from all the others and not on comparisons involving their more numerous associates.

Time was required to evolve these new forms, so that their first appearance in continental seas may as a rule be supposed to have followed an inaccessibly recorded period during which they came into being. Or it may be that they are migrants from some other province whose previous exclusion was due to some physical barrier; or, if they happen to be pelagic forms, like the graptolites and *Styliolina,* their invasion of continental basins may have been prevented by the absence of current thoroughfares. In either of the latter cases we may justly infer not only that ill-recorded time intervened, but also that diastrophic movements occurred which opened the way for freer intermigration. The introduction of new types therefore is always of possibly high significance and generally worth searching inquiry. And from the standpoint of stratigraphic taxonomy it is always of greater significance than is the presence of surviving older species. The new things show that something has happened, and that the happening was unusual is clearly indicated also by the partial extinction of the preceding fauna.

[*Editors' Note:* Material has been omitted at this point.]

CLASSIFICATION BY DIASTROPHIC MOVEMENTS

Maximum, major, and minor movements.—All diastrophic movements and processes have ever been characterized by periods of activity alternating with periods of relative quiescence. Periodicity, then, is a fundamental factor of geologic history. Further, all diastrophic processes are rhythmic in operation and recurrence, because they are occasioned by the necessarily rhythmic action of terrestrial forces. In some the meter is long, in others relatively short; in some the effects are pronounced and relatively impulsive, in others gentle and gradual in development. Diastrophic movements thus offer a basis for classification.

There were *maximum movements* or "revolutions." These were marked by strong crustal deformation with decided horizontal movement, especially toward the equator, by deepening of oceanic basins, and the formation of new geosynclines or the resubmergence of basins long unused; also by orogenic activity in marginal tracts of continents. Widespread emergence prevailed and the orogenic movements affected areas farther inland than did any preceding them. Erosive processes were especially active in the succeeding period or periods.

These maximum movements did not take place all at once, even in any particular area, but they were distributed through long periods in which diastrophic processes were more than usually active. Though long, the first, second, and fourth of these periods of activity were of much shorter duration than the periods of relative quiescence with which they alternated. The third seems to be more important than the others, the criteria on which extraordinary activity is postulated having begun earlier and continued longer. Stratigraphically, they are distinguished by extraordinary accumulations of detrital deposits in continental marine basins, and more particularly by the fact that apart from minor variations these deposits were laid down in areas farther inland from the borders of the continents than before, and perhaps since. The visible development of these criteria is modified by such theoretic considerations as the "inland migration of the belt of folding." (See discussion, pages 467 and 477.)

Four periods of maximum diastrophic activity are indicated. The first began in the closing stages of the Proterozoic and continued almost through the Cambrian. The second began toward the close of the Ordovician and continued through the early part of the Silurian. The third began late in the Devonian, continued to the close of the Jura-Trias, and attained its maximum at the close of the Tennessean. The fourth marks the transition from the Cretaceous to the Tertiary. The great time intervals separating the culminating stages of these movements correspond essentially to the eras of the geologic time scale.

Then there were *major movements* which seem to have been chiefly of the nature of broad adjustments to strains engendered during the periods of maximum activity. They were marked by deepening of submarginal geosynclines, accentuation of already existing primary axes, and great sea withdrawals—all phenomena occasioned by landward suboceanic spreading. Commonly, these conditions were followed by a stage in which seaward continental creep prevailed, with profuse and extensive crustal warping and consequent great changes in pattern of continental seas. The major movements define systems, series, and to a large extent, also, the groups into which the stratified column is divided. It is perhaps unnecessary to state that at certain times they are included in the operation of the maximum movements.

Finally, there were many *minor movements,* under which term we may refer to tilting of land masses, whether these involved the greater part of a continent or only some subordinate "positive" part of same, and to all movements that are relatively local in their effects on the strandline. Theoretically, these minor movements define geological formations or ages; also groups of formations which are distinguished on lithological grounds, but seem to have been deposited without break. They are commonly marked by retreat and subsequent advance of the sea without decided change in provincial boundaries, by deepening or emergence, as the case may be, of the subsidiary troughs of a geosyncline, and by local changes in the lithological facies of formations.

Obviously there is no sharp distinction between these three grades of movements. The minor movements vary to the point where they may justly be called major, and these again grade into those of the highest or maximum rank. Their taxonomic value, therefore, is correspondingly indefinite and peculiarly liable to personal differences in their estimation. The most generally admitted, if not the most convincing, part of the evidence from which diastrophic movements are inferred, namely, the clastic deposits, varies so greatly from place to place that wide difference of opinion respecting the taxonomic significance of the several occurrences is to be expected. The theory of inland migration of belts of active folding, also the proved small relief of the interior areas of the continents, helps greatly in explaining the paucity of detrital deposits at localities and times when much more might be expected. Fortunately, we have two other classes of evidence—(1) that showing stratigraphic overlaps, and (2) that showing the geographic derivation and distribution of fossil faunas and floras—that give no less convincing testimony in proving the frequency and varied kinds of movements. These two classes of evidence bear directly on the displacements of the strandline, and the variations in direction and extent of its transgressions give the surest proof of the instability of the lithosphere. Further, since the rhythmic periodicity of diastrophic processes is almost universally accepted as a fundamental fact in geologic science, these displacements logically become not only the ultimate but also the immediate natural basis of exact stratigraphic taxonomy.

[*Editors' Note:* Material has been omitted at this point.]

32

Reprinted from pages 29-36 of *The Jurassic System in Great Britain*
Clarendon Press, Oxford, 1933, 681p.

THE CLASSIFICATION OF THE ENGLISH JURASSIC ROCKS, AND THE PARTITION OF JURASSIC TIME

W. J. Arkell WED

[*Editors' Note:* In the original, material precedes this excerpt.]

The whole crux of the matter upon which the two schools of opinion differ is the factor of *time*. Ecological studies at the present day take no account of time as the geologist knows it. They presuppose a stability of animal distribution which the palaeontologist cannot tolerate. Thus it is interesting to notice that those who invoke partial distribution to explain faunal dissimilarity in the rocks picture the past ages in which the geologist is accustomed to soar without restriction as a series of kaleidoscoped presents, rather like a reel of motion-picture films, each hemera small and rigid. Prof. Stamp complains that an ammonite or a brachiopod would not have time to attain wide distribution in 'such a short space of time as a hemera';[1] later he again refers to a hemera as 'a brief space of time'.[2]

The views of the other school are strongly contrasted with this. Prof. Trueman believes that 'a hemera must represent some thousands of years',[3] or again 'must represent a very long period of time, certainly many thousands of years'.[4] Buckman goes farther and says 'What is there to prevent giving to a hemera a length of time like a million years'.[5]

Given sufficient time, a floating and free-swimming species like an ammonite would disseminate its shells, borne along by the wind and the tides after its death, over vast areas of the earth's surface. Moreover, the geographical barriers, which are the principal factor in the separation of local faunas, undergo, in the course of time, removal by erosion and earth-movements, thus giving free play to migration.

Every serious student of palaeontology and geology comes to take this extreme slowness of the accumulation of the sedimentary series for granted. Buckman put the case characteristically in 1922, three years before Stamp's criticisms were made:

'A hemera, though taken as the chronological unit, must be regarded as a very lengthy stretch of time. Migration of Ammonites would be a slow process; but . . . the rate of Ammonite migration to that of deposition was like the flight of an aeroplane to the progress of brick-laying.

'Present-day phenomena of deposition or of faunal dispersal are very unsafe guides. Geological strata are made by the net result of a constant battle of addition versus subtraction, in which are seen, locally, the small, slow victories of addition, after many vicissitudes. The same applies to modern faunal irregularities—they cannot be true criteria of what the ultimate geological record in the rocks will be: they are only records of temporary local phenomena, observed during a length of time quite negligible in comparison with the length of a hemera.' [6]

[1] L. D. Stamp, 1925, loc. cit., p. 19.
[2] Ibid., p. 23.
[3] A. E. Trueman, 1923, *P.G.A.*, vol. xxxiv, p. 196.
[4] Ibid., p. 206.
[5] S. S. Buckman, 1925, *T.A.*, vol. v, p. 70.
[6] Ibid., vol. iv, p. 24.

[*Editors' Note:* Material has been omitted at this point.]

The gist of the argument in favour of the utility of dissimilar faunas for the detection of stratal failure may be summed up as follows: A certain species is found in remotely separated localities, such as in the Lower Inferior Oolite of Dorset and the Hebrides, and it is thus proved to have been of widespread distribution and not merely a local form. It is contended that the period of its dominance was so long that the species would have had time to penetrate to all parts of the sea-bed between the places where it is found, and over a wide area beyond, during at least some part of its hemera. If it is absent from any area where the lithological facies is the same and there are no obvious impediments to its migration (such as coral reefs), then it is contended that the strata which would have contained the missing species have been removed by erosion.

The most serious objections to correlating by the polyhemeral system arise, not out of the theoretical considerations advanced by what might be termed the ecology school, but from the practical results obtained by the field-palaeontologists themselves, in the course of their collecting in distant regions. As, with the aid of the newly acquired refinement in discrimination of species, detailed work is extended beyond the classic areas, an increasing body of evidence is being collected tending to show that the succession of ammonite species is not everywhere repeated in the same order.

[*Editors' Note:* Material has been omitted at this point.]

If, as Dr. Spath believes, these results are soundly established, they prove (what might have been expected) that the ascertained acme of a given species was a local event depending on favourable local conditions. Three species may have attained their acmes in one basin of deposition in the order A–B–C, in another in the order C–B–A, and another B–A–C, in another C–A–B, in another B–C–A, in another A–C–B. Such possibilities cut at the roots of polyhemeral chronology, for if the hemera, founded on the acme of a species, is likely to have occurred at different times in different places, it is clearly useless as a time-unit. Epiboles therefore lose their value in correlation over all but short distances.

The results obtained at Champfromier are not surprising when we come to examine the idea of the acme of a species, upon which the hemera is based. There is no evidence that the *visible* acme in any district corresponds even approximately with the complete acme of the species, in the total area over which it migrated. Still less does it bear any constant relation to the total range, which is the only concrete and sometimes determinable quantity. Almost invariably the species perforce chosen as hemeral indices are unrelated to those either above or below. They appear upon the scene without antecedents and disappear as suddenly, leaving no immediate descendants behind them in the succeeding strata. Now, unless all these species were specially created and extinguished (if evolution has been continuous), they must have had antecedents and descendants somewhere. It is unlikely that every species entered our district immediately after it came into existence or left it at the moment of its modification to form a new species elsewhere: it is improbable, in other words, that every lineage remained in one particular district during exactly the time of existence of a single species, no more and no less. Since they did not remain longer, we must assume they did not remain so long. Therefore all that we can see is a part, and an unknown proportion, of any biozone.

When the epiboles are thus analysed, it is not surprising to find that many of the species, which appear to have a restricted range in certain localities, prove to have a much longer range—become 'zone-breakers'—when the area of observation is extended.[1] Rather, it seems inevitable that they should. It follows from the fact that the local acme (epibole) of such a species is only a fraction of the complete biozone, that any fraction may be represented in any place—the epibole in different places may be on different horizons within the much larger biozone. Put in different words, the hemera during which the local epibole was formed may represent different parts of the biochron in different places.

Now, in view of this, the fundamental fallacy in the polyhemeral system of correlation seems to be, that it takes no account of the probability of several, or even any number, of species belonging to unrelated stocks having thrived and migrated simultaneously in the same basin. Yet if the Jurassic seas were anything like those of the present day, this postulate should be at the forefront of any speculations. The paths of the migrating species and stocks would intersect in criss-cross fashion as shown in Fig. 1, and the order of their appearance in any given district would be entirely fortuitous.

Behind this fallacy there lies confusion of thought in the definition of a hemera. The original definitions make no distinction between 'local hemeræ' and 'ideal hemeræ'. From what has just been said it will be evident that the two are by no means the same thing. Yet Buckman, on recognizing a thin epibole, assumed that the 'hemera' during which it was formed coincided with the 'hemera' or time of acme of the species over the whole area of its occurrence. As we have seen, such an assumption is quite unjustifiable. The two conceptions are so entirely different that there should be two distinct time-terms to express them.

The Germans overcome this difficulty, for they have two different words for these distinct concepts. They do not speak of hemeræ and epiboles, however, but prefer to think in terms of total range, using biozones (of which the equivalent time-units, as we have seen, are biochrons). In practice, since neither acme nor total range can generally be accurately defined, as has been said, the result is the same. The ideal biochron is as elusive as the ideal hemera. The visible range at any given places bears no constant relation to the complete or absolute total range of the species, considered over the whole area of its occurrence. Consequently the Germans do not use the word biozone

[1] 'Such species of ammonites, that have shown themselves to indicate a distinct zone in certain districts, but transgress the boundaries of that zone in other districts of the same zoological province, I call zone-breaking species (zonenbrechende Spezies).'—C. Deiner, 1918, loc. cit, p. 113.

for the visible fragments or parts of biozones with which they usually have to deal. For these they have another, an excellent, word—Teilzones.[1]

A Teilzone is a part, and an unknown proportion, of the theoretically complete biozone. All so-called 'zones' founded on species unrelated to those above or below (that is, on species whose antecedents and descendants are unknown) are Teilzones. The term is therefore precisely what we want in order to clear up the existing confusion in the use of 'hemeræ'. The corresponding time-unit would be most logically expressed by some such term as 'Teilchron'.

At present almost the whole of our zonal table is built up of these fragmentary units or Teilzones, and it is all-important not to lose sight of their

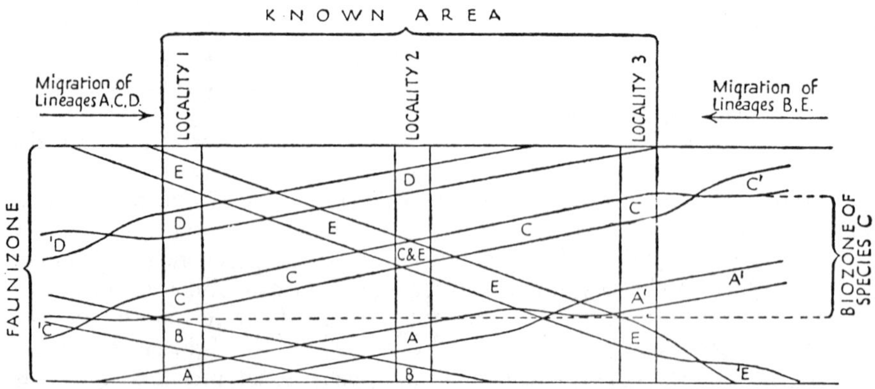

FIG. 1. Diagram illustrating the relations between faunizones, biozones and epiboles. At locality 1, within a faunizone is a sequence of 5 local epiboles, marking the local acmes of five unrelated species, A, B, C, D, E. At localities 2 and 3 the sequence is found to be different. The diagram illustrates the suggested explanation of this, the species being represented by long pencils.

The lineage to which A belongs is the only one that gives rise to a new mutation or transient (A′) within the area covered by the three localities. All the other species both came into existence and were modified to form descendant species or transients outside the known area. Within the known area all the species are therefore represented only by teilzones.

(The evolution of the several species is incorrectly represented as occurring in jerks for the sake of diagrammatic clearness.)

limitations. Only very rarely can we dispense with them—only when we are able to deal with lineages.

When we can detect lineages running up through a stratified series and so can get to know the antecedents and descendants of a species, then only can we define its range and be satisfied that the visible range is also the absolute range. For obviously the biozone of a species cannot embrace either antecedent or descendant species of the same lineage, and so bounds are set to the biozone. Unfortunately, opportunities for making use of lineages in zonal work are extremely rare.

[*Editors' Note:* Material has been omitted at this point.]

[1] See especially H. Frebold, 1924, *Centralblatt für Min.*, &c., pp. 313–20. I am unable to suggest an adequate and unugly English translation.

There can be no doubt that in practical stratigraphy the old zones or faunizones, broad enough to contain a number of epiboles or teilzones, have many advantages. In the first place, they accord with the subdivisions universally employed in the other systems; in the second place they are frequently all that it is possible to recognize or map, and they usually suffice for elucidating the structure of a district. For this reason they should be known by all geologists, who need not trouble about (and could scarcely memorize) some hundreds of local epiboles. Lastly, they do give a tolerably true picture of the palaeontological sequence. They will therefore be adhered to wherever possible in the stratigraphical part of this book.

A faunizone, however, as Lange remarks, is nothing more than the sum of the biozones of a number of species, and its use is therefore subject to all the same reservations and limitations as a biozone. Most of the faunizones with which the stratigrapher has to deal are fragmentary—they are only 'faunizone-teilzones', made up of bundles of teilzones (see fig. 1).

It is noteworthy that the distinction between faunizones and the smaller local units or epiboles, founded on single species, although obvious in a comparatively uniform clay-formation such as the Lower Lias, breaks down when we come to deal with strata of more varied lithology. For instance, in the Inferior Oolite Series, a varied limestone formation, the smallest subdivision that can be made by means of the ammonites (the epiboles) are only 11 or 12 in number (see pp. 189 and 230). But these subdivisions can be, and have been, followed all along the outcrop in the South of England, often without finding any specimen of the index-fossil, simply by the general assemblage of fossils, with the aid, locally, of the lithology. Such stratal divisions answer in every way to the definition of true zones or faunizones. Here, therefore, the faunizones and the epiboles are in practice the same thing.

[*Editors' Note:* Material has been omitted at this point.]

From this a general law may be formulated, that in a slowly-deposited clay formation the faunizones are much larger than the epiboles, while in varied and presumably more rapidly-formed deposits, the epiboles may equal or exceed the faunizones in thickness.

[*Editors' Note:* Material has been omitted at this point.]

It is therefore probably true to say of ammonites that in a very general way the frequency in the change of dominant species is some measure of the passage of time, even though it can only be gauged by local teilzones. Their migrations seem to have been determined by forces largely independent of local conditions, and therefore, when they are available, they provide by far the best zonal indices.

VI. OTHER CRITICISMS

The tables of ammonite sequences which we have been considering, based as they are on careful field-work, represent definite and valuable additions to knowledge; for even if the restricted vertical distribution of ammonites observed in one locality be found to have changed in another, the mere discovery of this fact is knowledge gained. All such detailed field-investigations carry out the sound precept of Quenstedt, that we must get to know our successions minutely, as the very first step in stratigraphical geology. It is absolutely essential that this work should be undertaken, for until the detailed knowledge of the rocks and faunas it provides is available for large areas, generalizations are impossible. The task of establishing the palaeontological succession in monotonous clay-formations is a thankless one. Those who delight in proclaiming that all the results are worthless would therefore do well to keep their triumphant tones for more deserving objects.

Tributes to the value of the results accruing from this detailed work come from all parts of the world, and it is clear that without Buckman's chronology even general correlation between distant continents would often be impossible.

[*Editors' Note:* Material has been omitted at this point.]

Hitherto, however, we have confined our attention to hemeral tables based on actual field-work. Unfortunately other hemeral tables have been propounded without the necessary basis in fact, and for such work scorn need not be spared. It is one of the greatest misfortunes for Jurassic geology that when increasing age and frailty prevented Buckman from continuing active field-work, he lost sight of the distinction between results obtained with hammer, collecting bag and field notebook, and those arrived at by speculation and deduction from matrices at home. The extent of the calamity is magnified by the fact that this loss of grasp of the importance of keeping surmise distinct from fact came when his reputation was at its highest, and the two kinds of results are almost inextricably interwoven in his later published works. Only those with intimate local knowledge of the English Upper Jurassic rocks can hope to distinguish the two. Foreign geologists have been all too ready to adopt Buckman's conclusions indiscriminately and an appalling amount of error has already been propagated in the world's literature; but they cannot be blamed for this. If the following chapters help in some small degree to guide future workers to distinguish between fact and fable, this book will not have been written in vain.

That Buckman, who had tramped the Cotswolds and the Sherborne country from end to end and knew every quarry intimately, whose earlier work was built up solely on sound field-work, could also be the author of his last paper on 'Some Faunal Horizons in Cornbrash'[2] and of some of the later parts of the fifth and sixth volumes of *Type Ammonites*, is difficult to believe.

[*Editors' Note:* Material has been omitted at this point.]

[2] S. S. Buckman, 1927, *Q.J.G.S.*, vol. lxxxiii.

AUTHOR CITATION INDEX

Adams, F. A., 14
Arduino, G., 14

Barrell, J., 208
Bergman, T., 14
Berry, W. B. N., 41
Brocchi, G., 41
Brongniart, A., 14, 41, 84
Buckman, S. S., 253, 351, 356

Carrozi, A. V., 14
Challinor, J., 14, 41, 84
Chamberlin, T. C., 297, 307, 330
Clark, J. W., 14, 84
Conkin, B. M., 5, 41, 208, 268
Conkin, J. E., 5, 41, 208, 268
Conybeare, W. D., 41, 84
Cox, L. R., 14
Cuvier, G., 14, 41

d'Halloy, J. B. d'O., 41, 84
d'Orbigny, A., 84
Davis, A. G., 14
Deiner, C., 353
Deshayes, G. P., 41, 84
Desmarest, N., 14
Dunbar, C. O., 84, 268

Eyles, V. A., 14

Farey, J., 14, 48
Forbes, E., 218
Frebold, H., 354
Füchsel, G. C., 14

Gaudry, A., 248
Geikie, Sir A., 14
Gignoux, M., 208
Grabau, A. W., 5, 6, 14, 208, 268
Guettard, J., 14

Hayes, C. W., 321
Hedberg, H. D., 14, 84, 268
Hooke, R., 6, 14
Hughes, T. M'K., 14, 84
Hutton, J., 14

Jameson, R., 15

Lamarck, J. B., 15
Lapworth, C., 208
Lavoisier, A., 15
Lehmann, J. G., 15
Lister, M., 6, 15

Monty, C. L. V., 84
Moore, C., 208
Murchison, R. I., 41, 84, 208

Oppel, A., 208, 268

Phillips, J., 15
Phillips, W., 41, 84
Playfair, J., 15
Porter, R., 15, 41

Richardson, W., 41, 48
Rodgers, J. , 84, 268

Salisbury, R. D., 307
Schlotheim, E. F., 41, 208
Schneer, C. J., 15
Schuchert, C., 306, 308
Sedgwick, A., 15
Sheppard, R., 15
Sherborn, D. C., 84
Smith, W., 15, 41
Stamp, L. D., 351
Steno, N., 6, 15
Strachey, J., 6, 15

Tomkieff, S. I., 15
Townsend, J., 15
Trueman, A. E., 208

Ulrich, E. O., 321

Weller, J. M. , 84
Werner, A. G., 15
Willis, B., 298, 299, 302, 323, 330
Wilson, E., 253
Woodward, H. B., 41
Woodward, J., 15

Zittel, K. von, 6, 15

SUBJECT INDEX

Age(s), determination of, 75, 80, 144-152, 253-255
Ammonites
  best zonal indices, 355
  in Lias formation, 122-123, 125
  in Marlstone formation, 124-125
  usefulness of, 222, 251, 254
Assise, definition of, 245

Baseleveling, of continents, 274-275, 285-288
Biozone
  bounds set if lineage known, 354
  as proposed and discussed by Buckman, 207, 258-259
Boundary, boundaries
  agreement of criteria, 222
  based on littoral deposits, 206
  based on regional successions, 206
  determination of, 222
    paleontologic, lithologic, and diastrophic criteria, 265
    fossils and diastrophic evidence, 295
  frequently overlooked, 331
  lithologic, caused by crustal movements, 294
  systemic
    nature of, 235
    physical evidence of, 305
  time, criteria for, 335

Cambrian-Silurian controversy, 229-237
Catastrophism, 38-39, 48, 52-53, 82
Cenozoic era, 247
Chronology, units of, 253-255
Chronostratigraphy
  column, basis for, 5
  determinations, methods for, 266-267
  unit, concept of, 78, 106, 215
    natural basis for, 262
Classification
  of British rocks, greatest change in, 178-179
  geochronology, based on most complete sections, 265
Clastic beds, overlying breaks, 266
Contemporaneous, meaning in stratigraphy, 224
Correlation
  in four stages, 289-290
  inexactness in distant ones, 289
  intercontinental, 309
  interfingering in, 315
  limits of, 263-264, 295
  minutely detailed, value of, 250
  by negative displacements of strandline, 264-265, 312-313
  overlaps in, 315
  principles of, 75, 76, 79, 80
  process, 331
  by use of organic, lithologic, and diastrophic evidence, 331

Deformations, inherited and successive, 285-286
Diastrophism
  boundary determination, basis for, 206
  criteria, summary of, 332-333
  manifestation in different regions, 302-303
  migration of strandline, 296
  movements
    active and impulsive, 264, 309-310, 348
    Chamberlin's and Willis's ideas discussed by Ulrich, 330-331
    correlation by, 330-331
    cycles of, 301
    gentle and gradual, 264, 309-310, 348
    maximum, major, and minor, 348-350
  paleontology and stratigraphy, basis for, 206
  periodic, not perpetual, 287-288
  processes
    evidenced by moving strandline, 339
    exact stratigraphic taxonomy, basis of, 350
    general systematic development, 339

rejuvenation of continents, restoration of ocean basin capacity, 285-287
ultimate basis of correlation, 283-291, 301
Disconformity
concept of, 190, 215
description of, 12, 37
evidence of, 190-191
relationship, concept of, 76-77, 150-151
term proposed by Grabau, 263
undetected, 204

Earth, history of
Cuvier's basis for, 65
explained by present, 12
interpreted by facts found, not present conditions, 312
Eocene Epoch, 109-110, 115
Epibole, 207
local part of biozone, 353
Epoch, 245, 248n
Era, 246-247

Facies
of assemblage of stages, 193
concept of, 11, 79, 80, 192
fossils, concept of, 9, 186-187, 204, 220
introductory, 265-266, 345-348
littoral and pelagic in Jurassic, 140-141
term proposed by Gressly, 75, 79, 137-139
Fauna
annihilation of, 192
change
by crustal deformation, 265-294, 337
due to physical change, 277-278
generic meaning of, 336-337
comparison of, 301
distinctness, 235
lists, 165
migration, time involved, 334-335
provinces, 218-219
synchronism of, 243-244
Faunizones
advantages, 355
limitations of, 355
term proposed and discussed by Buckman, 207, 258-259
Field-investigation, detailed, 356
Formations
age, 99, 112, 161
aspects of, 163. *See also* Facies
change in, 40
definition of, 37, 40, 45
Lias
stratigraphical arrangement of fossils in, 121-123
marlstone, fossils in, 123-124
overlapping, initial deposits of, 320-325
frequently lacking, 323-325
littoral conditions recorded by, 322-323
passages between, 154
recent, synoptical table of, 115
Fossiliferous grauwacké, 117-118, 158
Fossilization, nature of, 95-97
Fossils
carboniferous, in old red sandstone, 182
care in collecting, and identifying, 203, 252
comparison with recent shells, 28-29, 46, 51, 78, 86, 112-113
described and figured erroneously, 165
"extraneous," 11, 58
as indices to strata, 13, 42, 168
littoral, 276-277
for drawing boundaries, 246, 248
as media or criteria of chronology, 9
pelagic and littoral, 12
record, incompleteness of, 217
reworked, 38
species, 169
role in stratigraphy, 204, 296
in strata, 22, 47-48, 54-55, 88-90, 182, 198-199
stratigraphical limits of, 166
uses
for correlation, 11, 30, 75-76, 241, 296
to determine relative age, 87-88, 90-91, 100-102, 216, 242
to evaluate gaps, 242
to interpret environments of deposition, 241
to locate boundaries, 331-332

Genera, distribution of, 187
Geological Society of London, founding and purpose, 36
Geologic maps, 13, 40, 153, 188
Geologic record, gaps in, 266
Geologic time
criteria for classification of, 263
divided by stratigraphic breaks, 339-343
measured by stratigraphic units, 294
Geologic time scale, 224, 294
based on faunal, diastrophic and lithologic criteria, 294
based on fossils, 224
natural bases for division of, 267, 269-282
Groups, 246-247
Mesozoic, 247
Paleozoic, 247

Hemera, 206, 207, 208, 250-252, 254, 256-259
length of, 351

time of, compared to migration of Ammonites, 351
Hiatuses, time value of, determined by fossils, 325

Jurassic terrains, 194, 196-197
divided into stages, 195-197
facies in, 141

Lithologic change
in rocks of same geologic position, 186
cause of, 186
Lower Paleozoic Era, 233-240
alternative division of, 232-233
rocks and fossils, tripartite division of, 228-229
Lyell, Buckman's comment on work of, 207-208

Marine deposits, overlap of, 315-318, 320
Measured sections, detailed, 203
Miocene Epoch, 108-109, 115
Murchison, Lapworth's comments on work of, 226-227

Ocean basins
formation of, 272
increased by crustal readjustments, 293
"Oceanic thrusts," 264, 300
Old red sandstone, changes in, 187-188
Oscillation, of seas and lands, 305

Paleogeographic maps, 245
Paleozoic rocks, an uninterrupted sequence of systems, 189
Palezoic seas, nature of, 265
Paraconformity, 263
concept of, 204
Paracontinuities, of Conkin and Conkin, 4-5, 40, 83, 206, 263
basal sands overlying, 264
example of, 264, 315-316, 317
frequently overlooked, 265
resulting from gentle diastrophic movements, 264
separate "transition sequences," 265
Pelagic deposits and fossils, used for grouping related strata, 206, 245-246, 248
Period(s)
definition, 246
distinctness of, causes of, 110-111
eustatic sea level change, 304
evidence for, 223-224
intermediate gradations between, 111
Physical change, effect on life, 242, 245

Pliocene, Newer and Older, 107-108, 114-115
Polyhemeral chronology of Buckman, criticized by Arkell, 352-353
Pre-Cambrian, 233
Protozoic rocks, 163-164
Pulsation theory, 268

Quaternary, definition of, 247
Quenstedt, work of, reply to criticism by d'Orbigny, 211-214

Radiometric dates, value in stratigraphy, 266-267
Recent, definition of, 78, 106-107, 114
Regressions, of sea
at boundaries, 312
causes of, 262
evidence of, 314, 327
frequent, 326-327

Sea level
changes
evidence of diastrophic events, 264
tectonic, greater at close of periods, 304
eustatic, change, 304
movement of, 303-304
Sedgwick, Lapworth's comments on work of, 226-227
Series, definition of, 246
Species
acme of, 255, 352
hemera based on, 352
causes of change in, 62, 77, 94-98, 168
contemporaneity of
evidence for, 252
determination of, in fossils, 103, 211
discrimination of, 337
dispersal time, 267
distribution of, 187
extinction of, 219
gradual, 66, 81, 94-98, 110, 113, 168-169, 171
gradual change in, 171
migration of, 243
origin of, 219
paracme of, 255
representative, 205, 225, 241
successive creation of, 193
unequal distribution of, 222
zone-breaking, 353
zone of maximum development of, 220-221
Stage(s)
age recognized by fossils, 192
Bajocian, 197

Bathonian, 197
Callovian, 197
Corallian, 197
definitions of, 44, 82-83, 198-199, 245
divided into zones by Oppel, 209
Jurassic, 194-197
separation, causes of, 195-196
Kimmeridgian, 197
Liasian, 196
mixtures of species, in 198
of d'Orbigny, criticized by Quenstedt, 213
Oxfordian, 197
Portlandian, 197
separation of, evidence for, 82, 190-191
Sinemurian, 196
stratigraphical equivalent of an age, 257
Toarcian, 196-197
Strandline
affected by active and gentle diastrophic movements, 310-311
displacements
by sea level change, 304
connected with body deformation, 303, 306
negative, used for correlation, 312-313, 332
evidence for shift of, 313
Strata
comparison of ancient and more recent, 50
description, 20-23, 24-30, 31-34
identified by fossils, 165-166
traced by fossils, 10-11
Stratification, 68-71
Stratigraphical, first use of term, 13
Stratigraphy, stratigraphic, 39, 82, 199
breaks
causes of, 235
correlation of, 340
inconspicuous, 338, 345
classification
conceptual development of, 2
diastrophic and structural criteria for, 264
fossil foundation of, 293
hiatus, nature of, 340
historical development of, 1-5, 8-13, 36-40, 75-83, 202-208, 262-268
methodology, 204
paleontology, 192
succession, earliest, 10
taxonomy, based on diastrophism, 348-350
terms and time terms, difference in, 339
units, defined by strandline displacements, 337
"Suboceanic spreading," 264, 303
Substage, 245
System(s)
definitions, 78, 158-159, 246, 248n
Devonian, 81
rocks and fossils, 175-189
zoological evidence, 178
distinguished by fossils peculiar to them, 168
in England and Wales, eight proposed by Murchison, 153-154
Ordovician, 205, 238-239
need for, 226-240
Silurian system
fossils, similarity of 169-171
Lower, 118-119
Murchison's study of, 155-162
named by Murchison, 118
rocks, lithologic variation in, 160-161
Upper, 118-119

Teilzones, 353-354
Terrains, facies of, 137-139
Tertiary
chronology, 78, 106-112, 115
fossils, compared with living species, 103-106
Time
cause of change in species, 171
divisions
natural bases for, 269, 271-272
not absolute, 270
missing, 205, 215, 243
Transgressions
causes of, 262, 305-306
in Paleozoic, 306
evidence of, 314
suggested by Cuvier, 62-64
nature of, 263
rapid and extensive, 300
variation in areas submerged by, 307-309
Transgressions and regressions, "universal" and rhythmic, 267-268, 278-279
Transgressive deposits, 288-289
Transition beds, breaks in or under, 265, 344
Transition rocks, 117-118, 158
Type localities, concept of, 79, 116-117, 196
Type section, concept of, 167

Unconformity, unconformities
angular
causes of, 235
earliest illustrations of, 10
as evidence of cycles in earth history found by Hutton, 10
resulting from diastrophism, 264

use in mapping, 235
use of, 215
evidence for obscure type of, 328-329

Wealden group
description, 126-128, 133-134, 136
marine deposits coeval with, 129-130

Zone(s)
concept of, 3, 76, 78, 91-92
definition, 245
everywhere essentially same age, 267
of Moore, 206-207, 250, 257
of Oppel, 203, 209
method of determining, 209-210
and stages, of deLapparent, 205
term criticized by Buckman, 251, 258-259
various meanings of term, 258

Zoological change, relative magnitude of, as basis for chronological scale, 228
various meanings of term, 258

About the Editors

BARBARA M. CONKIN is associate professor and coordinator of the Geology/Geography Program at Jefferson Community College, Louisville, Kentucky, where she has taught since 1968, and is a Research Associate at the University of Louisville. She received the B.A. in geology from Smith College in 1950 and the M.S. in geology under Professors Cecil G. Lalicker and Raymond C. Moore at the University of Kansas in 1954. She has co-authored over seventy papers on Paleozoic paleontology and stratigraphy. She illustrated part of the *Treatise on Invertebrate Paleontology* in 1952 and presently illustrates and writes for the University of Louisville Studies in Paleontology and Stratigraphy. In addition, she has been a consulting geologist.

JAMES E. CONKIN is professor of geology at the University of Louisville where he has taught since 1957. He founded the Department of Geology in 1964 and served as its chairman from 1964 to 1968. His undergraduate degree in geology is from the University of Kentucky, 1950; the M.S. is from the University of Kansas, 1953; and the Ph.D., 1960, is from the University of Cincinnati where he worked under Professor K. E. Caster. He was employed in 1952 by the U.S. Geological Survey in the western United States and in 1957 by the Geological Survey of Canada in New Brunswick, and was a micropaleontologist on the Gulf Coast of the United States for United Gas Corporation from 1953 to 1956. During 1964 and 1965, he held a senior Fulbright Fellowship at the University of Tasmania to study the Permian of Australia. In 1980, he was awarded a research fellowship at the Nanjing Institute of Geology and Palaeontology by the Committee on Scholarly Communication with the People's Republic of China, sponsored by the National Academy of Sciences and the Academia Sinica, to study Devonian charophytes and Paleozoic foraminiferans in Guangxi and Jiangsu, South China. His current researches, with Barbara M. Conkin and others, include Paleozoic arenaceous foraminiferans in general, Silurian foraminiferans of Gotland (with Project Ecostratigraphy), orientation of fossil charophytes, unconformities, bone beds, Ordovician and Devonian metabentonites and boundary determination, and correlations within the Ordovician, Silurian, Devonian, and Lower Mississippian.

The Conkins are editors of the University of Louisville Studies in Paleontology and Stratigraphy, which they originated in 1973.